Corrosion Resistance

Corrosion Resistance

Edited by **Guy Lennon**

New York

Published by NY Research Press,
23 West, 55th Street, Suite 816,
New York, NY 10019, USA
www.nyresearchpress.com

Corrosion Resistance
Edited by Guy Lennon

International Standard Book Number: 978-1-63238-102-6 (Hardback)

Printed in the United States of America.

Contents

Preface

The book presents the state-of-the-art techniques, advancements and research progress of corrosion studies in a vast range of research and application areas. Topics contributed by the authors are corrosion features and its resistant nature. Apart from the conventional corrosion study, the book also discusses energy, fuel cell, reinforcing fibers and anticorrosive coatings. The discussed operations of corrosion resistance substances will also add value to the reader's knowledge.

This book unites the global concepts and researches in an organized manner for a comprehensive understanding of the subject. It is a ripe text for all researchers, students, scientists or anyone else who is interested in acquiring a better knowledge of this dynamic field.

I extend my sincere thanks to the contributors for such eloquent research chapters. Finally, I thank my family for being a source of support and help.

Editor

1

A Systematic Study and Characterization of Advanced Corrosion Resistance Materials and Their Applications for Plasma Etching Processes in Semiconductor Silicon Wafer Fabrication

Hong Shih
Etch Products Group, Lam Research Corporation, Fremont, California, USA

1. Introduction

Corrosion resistance is a quantitative measure of materials under study in a special corrosion environment. With a continuous development in semiconductor IC industry on silicon wafer fabrication and the rapid shrinkage of silicon wafer feature size as of to 32nm, 25nm and even smaller, the requirement on **corrosion resistance chamber materials** under high density plasma becomes extremely critical and difficult. Therefore, the study, characterization and new development of **corrosion resistance chamber materials** have been a critical task for technologists in semiconductor IC industry. Without the correct selection of **corrosion resistance chamber materials**, it is impossible for semiconductor IC industry to achieve current technology levels. Among steps of semiconductor wafer fabrication, plasma dry etching is the most difficult and comprehensive step which has a very high standard for the selection of **corrosion resistance chamber materials**.

Different from the traditional corrosion study, materials under high density plasma during dry etching processes should meet a comprehensive requirement. First of all, chamber materials must demonstrate a very **high corrosion/erosion resistance** under high density plasma during etching processes as well as in the defined wet chemicals. Since different etching processes use different reactive gases and chamber conditions, **chamber materials selected** have to vary in order to meet the variations of etch processes and chamber conditions. Secondly, chamber materials should have low particles and defects during etching processes because the particles and defects generated from chamber materials will fall on the silicon wafer, serve as the killer defects, and cause the loss of wafer production yield. Thirdly, chamber materials should avoid metal contamination issues on silicon wafer. The high metal contamination generated from chamber materials such as Na, K, Fe, Ni, Cr, Cu et al will electrically shorten the dies on a silicon wafer and directly impact wafer production yield. In addition to the above requirements for **advance corrosion resistance chamber materials**, chamber etching process stability and transparence, chamber impedance matching and stability, thermal and dielectric properties, capable of surface texturing, microstructure, wet cleaning compatibility, resistance to in-situ waferless plasma dry cleaning (WAC), RF coupling/grounding efficiency, adhesion of etch by-products and polymer, bonding strength of surface coatings, fundamental mechanical properties, manufacture ability and reproducibility, and cost of the materials have to be considered as a whole. After reviewing the overall requirements of chamber materials, one can see that it is not an easy task to find a suitable chamber material for semiconductor IC wafer fabrication which can meet all the above requirements. A comprehensive study has to be performed in order to find and to determine **the best chamber materials** among the existing materials in the world for a special etching application. Due to the complexity, the qualification processes of a new advanced corrosion resistance material for plasma etching processes are not only very time-consuming, but also very expensive. The fundamentals and applications of plasma dry etching and the applications on equipment of semiconductor silicon wafer fabrication have been described and studied extensively [1-20].

Let's take some examples. During metal etch processes (etching aluminum line), Cl_2 and BCl_3 are the main reactive gases to etch aluminum. Ar, N_2, CF_4, CHF_3, C_2H_4, or O_2 are also used during etching and WAC processes. Therefore, **the selected chamber materials** have to demonstrate **high corrosion (and erosion) resistance** to these gases under the high density plasma. For silicon etch processes, SF_6, NF_3, HBr and HCl are the main reactive gases used to etch silicon. Other gases may also be used in the etching and WAC processes. **The selected chamber materials** should have a **high corrosion resistance** to both F-based gases and HBr corrosion. In particular, the corrosion of HBr mixed with a very tiny amount of water on the heat effected zone of stainless steel has been an issue for a long time. For dielectric etching processes, C_xF_x based reactive gases are usually used with a high applied power in order to etch oxide. Chamber materials selected have to show high corrosion and erosion resistance at a relatively high temperature and high power. For special etch processes such as metal hard mask etch, MRAM etch, high K etch and Bevel etch, special process gases and chamber conditions are applied. Therefore, the requirements to **corrosion resistance chamber materials** may be different. Since some plasma etching processes even etch noble metals such as Pt, Ru and Ir, one has to find chamber materials which can survive in these aggressive plasma etching conditions. Therefore, chamber materials which are submitted to sputtering, chemical etching, ion-enhanced etching, as well as ion-enhanced inhibitor etching have to be studied and characterized thoroughly for each special etching applications. There is no any material which can meet all plasma etching applications. In summary, some of the key requiements of chamber materials is listed below [21-39]:

- Low erosion rate under vigorous plasma bombardment.
- Low chemical reaction rate under many chemistries such as
- Cl_2/BCl_3-containing plasma,
- Fluorine-containing plasma,
- $HBr/HCl/Cl_2$-containing plasma,
- Oxygen-containing plasma.
- Low transition metal transport to the workpiece.
- Low or zero particle contamination from surfaces.
- Strong interface bonding of surface coatings for long part lifetime.
- Excellent and repeatable dielectric properties for RF energy coupling.
- Pore-free ceramic materials and low porosity surface coating to avoid undercut corrosion and to eliminate substrate attack.
- Excellent adhesion of etch by-products and polymers.
- Excellent corrosion resistance in wet chemistry cleaning.

- Cost effective in manufacturing.
- Excellent repeatability from part to part and wafer to wafer.

The relationship among chamber materials used in semiconductor etching equipment, etching, wet cleaning, sputtering, and etch by-products is shown in Fig. 1[21, 22, 35, 36].

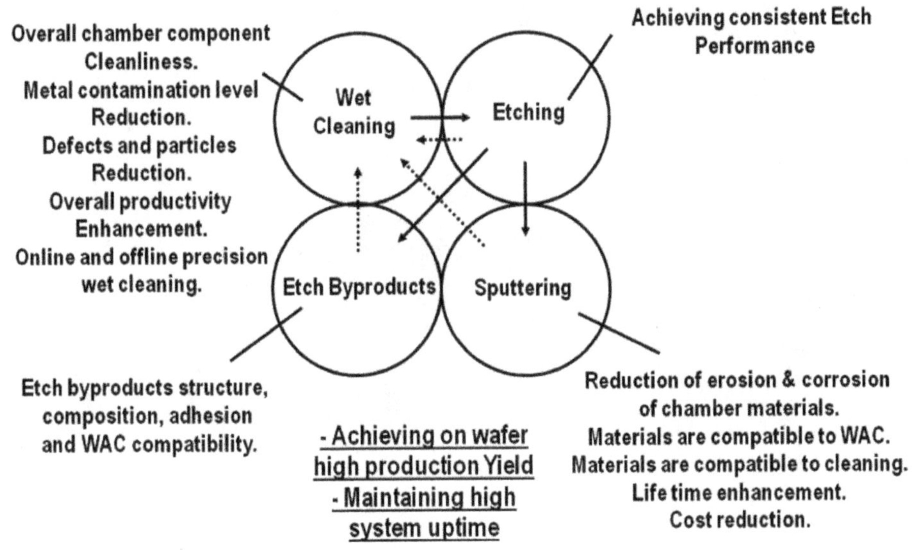

Fig. 1. The relationship of chamber materials, etching processes, precision wet cleaning and etch byproducts in a plasma etching chamber [21, 22, 35, 36].

For etching process requirement, a metal etch film stack and common issues are shown in Fig. 2

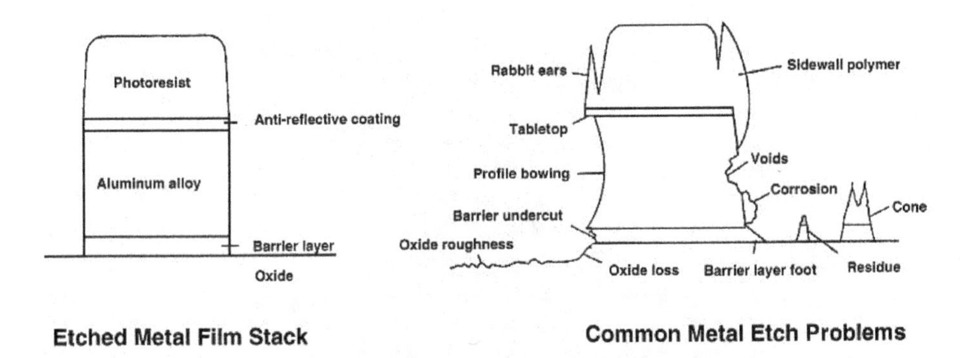

Fig. 2. Aluminum metal film stack and common issues in etching processes [21, 22].

The killer defects which are generated during metal etching processes fall on metal lines and cause the loss of production yield in wafer fabrication. The killer defects may either come from chamber materials or etch by-products [21, 22, 23, 25, 27].

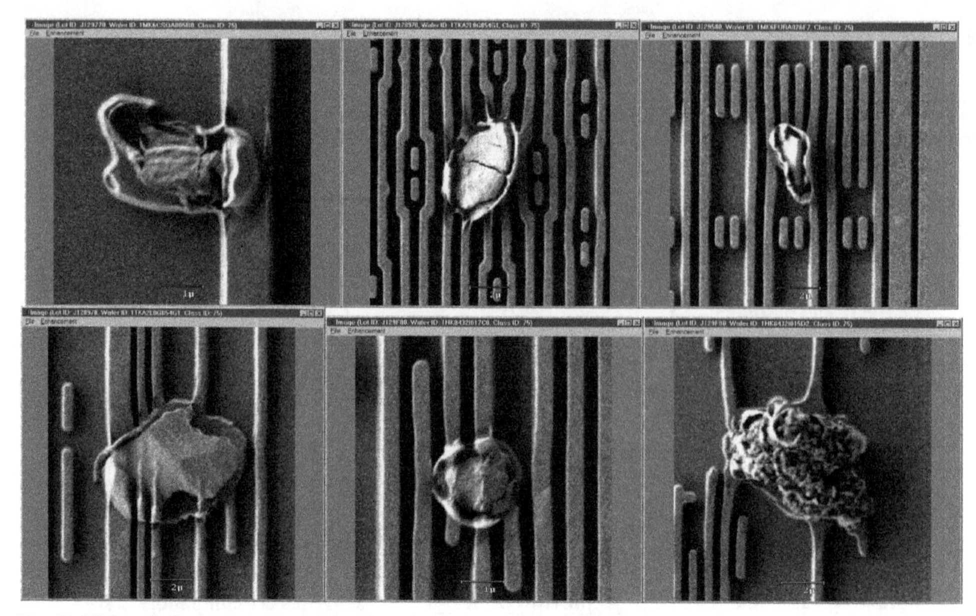

Fig. 3. Killer defects generated in aluminum metal etch processes.

The corrosion/erosion patterns of chamber materials showed three different patterns under plasma. Fig. 4 shows the three different patterns [21, 22, 28, 35, 36].

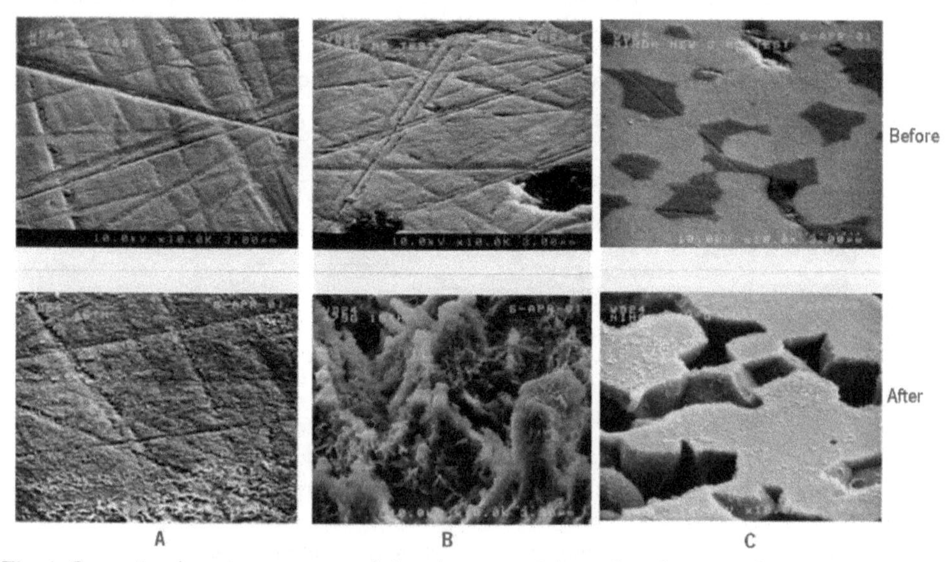

Fig. 4. Corrosion/erosion patterns of chamber materials under plasma etching (pictures are at 10,000x magnification). Model A indicates a uniform corrosion/erosion which can either be higher or low; Model B shows the attack at grains of materials; and Model C shows the attack at grain boundaries of materials.

In pattern A, chamber materials can be etched/sputtered by plasma uniformly. The etch rate can be very low or very high. The etch rate depends on the plasma chemistry, process recipe, and materials. For example, high purity Y_2O_3 has showed very high plasma resistance in both Metal and Silicon etch processes. A uniform corrosion/erosion pattern is observed [21, 22, 25, 30]. For anodized aluminum, a very high corrosion/erosion rate is observed under BCl_3-containing plasma during metal etch processes. In fact, an anodized aluminum film with a 75 μm in thickness (hot deionized water sealed) can only hold up to 1,800 wafers in some etch process recipes in production. This became a severe problem on the lifetime of anodized aluminum in aluminum etch processes. For Silicon etch processes, the lifetime of anodized aluminum has no issue because there is no obvious attack of reactive gases to anodized aluminum in Silicon etch processes. The only concern is the formation of AlOF on anodized aluminum surface when SF_6 and NF_3 are used in the etching processes. The formed AlOF can either have chamber particle issue or cause etch process shift due to the surface impedance change on anodized aluminum surface. The wet cleaning to fully remove AlOF film on anodized aluminum surface is very critical to achieve a consistent and reliable etching performance on wafer fabrication. Fig. 5 shows an anodized aluminum metal etch chamber after 1,800 wafer fabrication in production. The anodized aluminum is fully removed under Cl_2/BCl_3 high density plasma [21, 22, 25, 30]. The major attack of anodized aluminum is due to the chemical reaction between BCl_3 and Al_2O_3 under the high density plasma. The reaction rate of the attack to anodized aluminum highly depends on the gas concentration of BCl_3 and the plasma density. Chamber erosion test indicates that Cl_2 has little attack to anodized aluminum [21, 22, 25, 30].

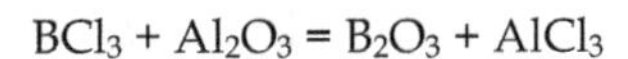

$$BCl_3 + Al_2O_3 = B_2O_3 + AlCl_3$$

Fig. 5. Anodized aluminum is fully removed under Cl_2/BCl_3 plasma after only 1,800 wafers in production (about 60 RF hours). The special attacking pattern depends on the local plasma density and gas concentration.

The high density plasma reaction rate of BCl_3 with anodized aluminum or high purity alumina at different flow is shown in Fig. 6. The high reaction rate occurs on chamber top window due to both high density plasma and gas flow. On the chamber wall, the reaction rate of BCl_3 with Al_2O_3 is almost a liner relationship, but the reaction rate is much lower than that on the chamber top window. It also indicates that without BCl_3 flow, the reaction rate of Cl_2 plasma has almost no attack to anodized aluminum or to high purity alumina. In the plasma reaction rate study, the total flow is fixed as of 205 sccm. The Argon gas flow is fixed at 40 sccm. The test starts at 165 sccm Cl_2 flow and zero flow of BCl_3, then 155 sccm Cl_2 flow and 10 sccm BCl_3 flow, until the final flow of Cl_2 is 85 sccm and BCl_3 flow is 80 sccm. The test coupons are either on chamber top window or on the chamber wall. Nine different types of anodized aluminum and high purity alumina are used in the test [21, 22, 25, 30]. The reaction rate is in the unit of mils per RF hour.

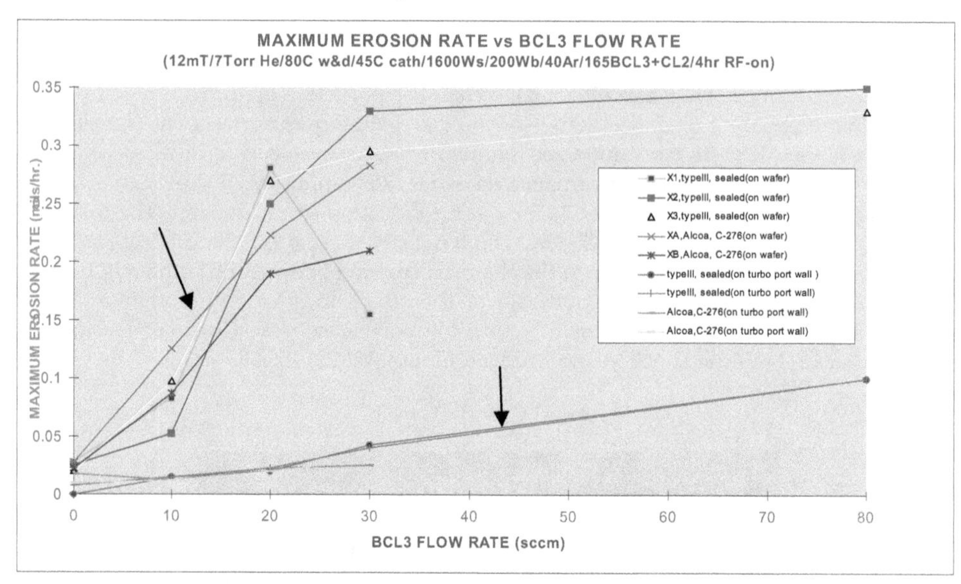

Fig. 6. The maximum reaction rate of Al_2O_3 at different BCl_3 gas flow under high density plasma [21, 22, 25, 30].

In pattern B, chamber materials suffered the attack of grains under plasma. CVD SiC grains can be attacked by Cl_2-containing plasma and SiC material cannot be used in aluminum etch processes as a chamber material. Grains of high purity ceramic (99.5% or higher alumina) can also be attached by BCl_3 in metal etch processes and the glass phases such as SiO_2, CaO, and MgO remain. It is obvious that BCl_3 can attack anodized aluminum and alumina under high density plasma. For high purity AlN, AlN grains are attacked by fluorine-containing plasma such as SF_6 and NF_3, the grain boundaries remain.

In pattern C, chamber materials are attacked at grain boundaries only. A typical example is high purity alumina (99.5% or higher in Al_2O_3), glass phases such as SiO_2, MgO, and CaO can react with fluorine-containing gases. In this case, grains of alumina remain. The formation of AlOF may occur on alumina surface. Fig. 7 shows a ceramic ESC surface which is covered by a layer of AlOF after exposure to plasma in silicon etch processes [35, 36].

A 33% atomic% of F is detected on electrostatic chuck ceramic surface (high purity alumina) indicating the formation of AlOF on high purity alumina surface under fluorine-containing plasma. The chemical treatment to remove AlOF using TMAH (tetramethylammonia hydroxide) is also demonstrated in Fig. 8 [35, 36, 40].

Fig. 7. A uniform AlOF film (rainbow color) covers the ceramic surface of a used electrostatic chuck after silicon etch processes.

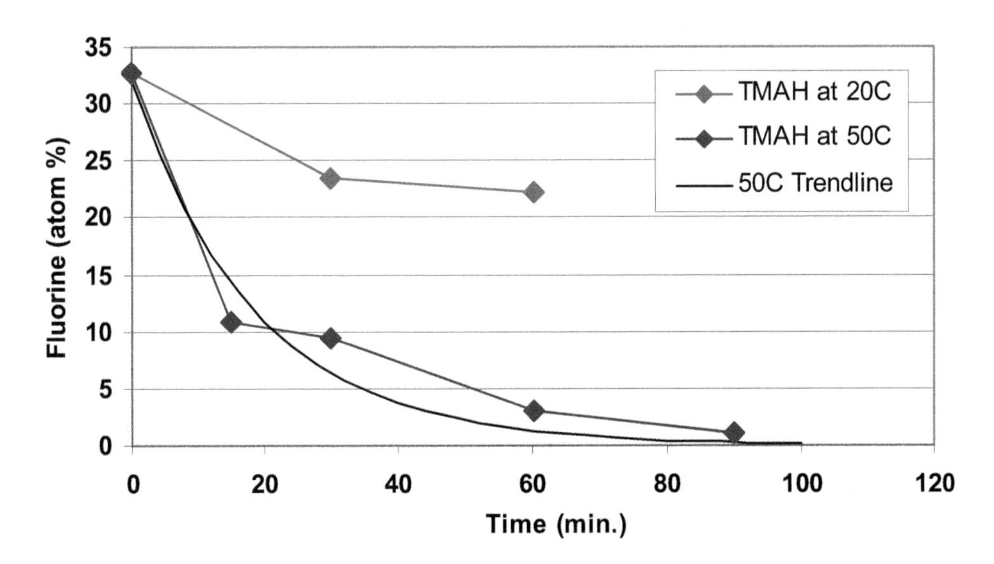

Fig. 8. The ceramic surface of a used electrostatic chuck contains about 33 atom% fluorine on the surface film with a rainbow color.

Since the limitation of the space of this chapter, anodized aluminum, boron carbide, and Y_2O_3 as chamber materials will be demonstrated.

2. Experimental and discussion

Since the limitation of the space of this chapter, anodized aluminum and boron carbide coating as chamber materials will be demonstrated and discussed in details.

All the ceramic and CVD test coupons (except anodized aluminum coupons) are polished to mirror surface finish with the average surface roughness less than 1.0 μ-in in Ra. The anodized aluminum coupons are anodized with the surface roughness less than 32 μ-in (as-received). The thermal spray coatings keep the as-coated surface. Ceramic and CVD coated coupons weigh the pre-test weight. Anodized aluminum and thermal spray coating coupons were measured to obtain the average coating layer thickness before test. All the test coupons were soaked in IPA for 5 minutes, wiped by both IPA and acetone, rinsed by deionized water (DIW) for 1 minute and baked at 110°C for 30 minutes. A special thermal conductive tape was used to mount the test coupons on locations in the etching chamber. Three locations were selected to mount the test coupons. Test coupons are mounted on chamber top window, chamber wall, and electrostatic chuck surface, respectively. Etching systems used in this study include Applied Materials 200mm and 300mm etching tools and Lam 2300 etching tools. During materials characterization on chamber wall, on chamber top window and on electrostatic chuck, a dummy aluminum wafer was used to cover the electrostatic ceramic surface. The etching process recipe keeps running for three minutes, followed by a cooling down process for about two minutes, than repeat the etching recipe. The minimum process time (RF hours) cycled is 120 RF hours and the longest process time cycled is 200 RF hours.

A typical process recipe under a 200mm etch tool is shown below [21, 22, 25, 30]:

Step 1. Plasma Etching & Dechuck Steps

- 12mT/85Cl_2/80BCl_3/40Ar/1600W_s/200W_b/45°C cathode/80°C wall & top window/7 Torr He flow/180 seconds.
- 100Ar/TFO/500W_s/100W_b/5sec.

Step 2. Cooling Down Step

- 12mT/200Ar/45°C cathode/80°C wall & top window/120 seconds.

Repeat process recipe (step 1 and step 2) until the accumulated RF hours achieve 120 RF hours or 200 RF hours, respectively.

After plasma etching processes, all the test coupons were removed from the chamber. A post wet cleaning was carried out to remove polymer, etch by-products, and other contaminants. All the coupons were then DIW rinsed and baked at 110°C for 30 minutes. Post weight measurements were carried out to obtain the average thickness loss per RF hour. For anodized aluminum and spray coating coupons, post thickness measurements were carried out in order to obtain the coating thickness loss per RF hour.

All the coupons are studied by SEM before and after plasma etching process. The corrosion/erosion rates of different test coupons are recorded and compared as mils per RF hour. Test coupons on etch chamber top window, chamber wall and on electrostatic chuck surfaces are shown in Fig. 9.

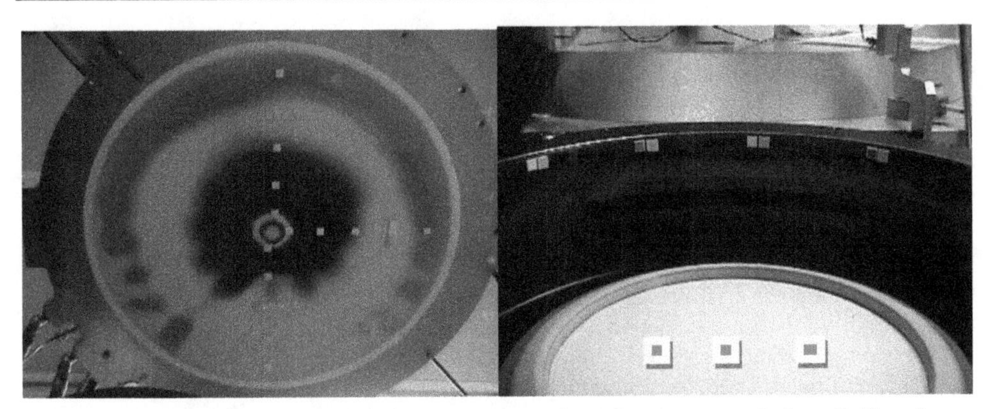

Fig. 9. Test coupons in etching chamber are mounted on chamber top window (left) and on chamber wall (right) and on the dummy aluminum wafer on an electrostatic chuck surface (right, white surface).

Fig. 10 shows the test results of various materials obtained from worldwide suppliers. The letters of A, B, C, D et al represent the suppliers and their materials. Agreements were signed for not allowing to release the names of the worldwide suppliers and their materials. The plasma etching rate is in the unit of mils (1 mil = 25.4 μm). It is obvious that either YAG (solid solution of Al_2O_3 and Y_2O_3) and solid Y_2O_3 can reduce the plasma etching rate at the order of 40-50 times in comparison with the previously used chamber materials such as high purity alumina. That is the reason why Y_2O_3 has been as one of the leading chamber materials in plasma etching tools in the past 10 years for the leading semiconductor etching equipment companies.

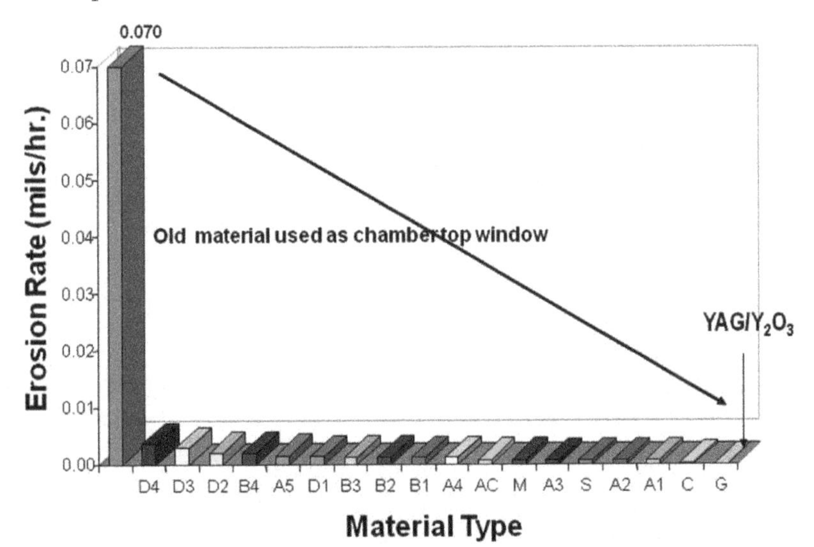

Fig. 10. Test results of new and old chamber materials in plasma etching on chamber top window. The etch rate reduction of new chamber materials can reduce the etching rate by 40 to 50 times.

For test results on chamber wall, the etching rate of anodized aluminum from various suppliers w/wo hot DI water seal is between 0.050 to 0.070 mils / RF hour. For boron carbide coating through a thermal spray method, the etching rate is below 0.001 mils/RF hour. For sintered or hot pressed boron carbide, the etching rate is between 0.0001 to 0.0007 mils/RF hours. It is also obvious that the plasma etching resistance of boron carbide can improve the plasma resistance by 50 times or higher. In fact, boron carbide coated chamber has been using at worldwide wafer fabrication customer sites since 1998. 50 to 100 times chamber life improvement has been demonstrated since 1998 up to today [21, 22, 25, 30, 41].

In order to select the best configuration of surface coatings such as B_4C (boron carbide), three configurations are considered. Configuration 1 is the coating of B_4C on bare aluminum surface. Configuration 2 is the B_4C coating on anodized aluminum surface. Configuration 3 is the B_4C coating on anodized aluminum surface and then HL126 sealant is used to seal the pores in the spray coating layer. HL126 contains methacrylate esters and it can fill very tiny pores. The metal contamination levels of HL126 is pretty low. All metal levels are below 1 ppm except the sodium level at 57 ppm. Permabond HL126 is a high strength and low viscosity anaerobic threadlocker. Its properties are listed in the attached table below:

PHYSICAL PROPERTIES OF THE UNCURED ADHESIVE *	
Properties	
Base Resin,	Methacrylate Esters
Solid, %	100
Color	Green
Viscosity, cP, 25°C (77°F)	20
Consistency	Liquid
Gap Filling, in	0.005
Specific Gravity	1.09
Flash Point, °C (°F)	>110(230)
Shelf Life stored at or below 27°C (80°F), months	12
Electric Properties	
Dielectric strength, MV/m	11
Electric Resistance, ohm-cm	10^{15}
Performance properties of the cured sealant	
Operating temperature °C (°F)	150 (300)

Table 1. Properties of HL126 sealant

The corrosion resistance of boron carbide coated coupons after plasma etching is tested by HCl bubble test method which was first proposed by Shih in 1992 and was used as a standard technique in anodization study for IC industry in 1994 [42]. The fundamental concept of the defined HCl bubble test method can be explained as follows. The dilute HCl solution can go through the pores and micro-cracks on coating and anodized aluminum layer to react with bare aluminum under the coating or under the anodized aluminum. When HCl reacts with aluminum alloy, hydrogen bubbles will generate. Streams of hydrogen bubbles can be observed and the time to start the continuous hydrogen bubbles can be recorded and compared for different coating configuration and different types of anodized aluminum before and after plasma etching processes. Shih [43, 44] has set up the method at two major semiconductor equipment companies since 1994 and the method has been widely accepted by worldwide anodization suppliers. The method is simple, low cost and fast in comparison with ASTM standard salt spray test method [45, 46]. The test results show that boron carbide coating on anodized aluminum and sealed with HL126 provide the best corrosion resistance among the four configurations as shown in Fig. 11 [25].

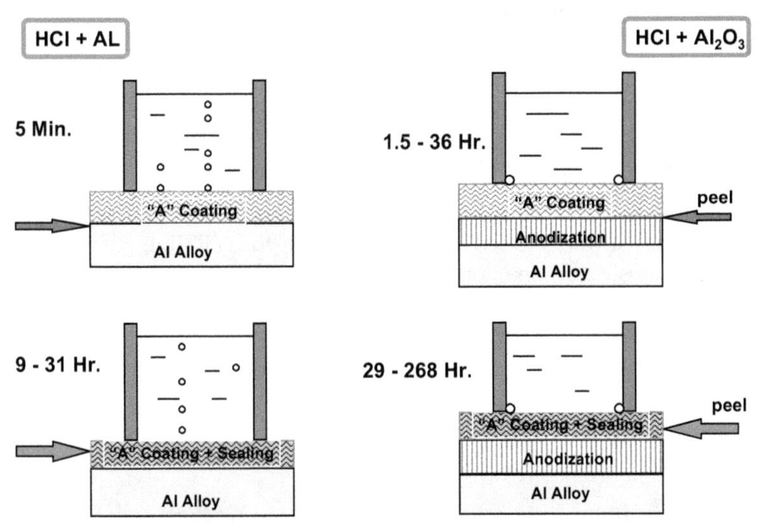

Fig. 11. After plasma etching for 200 RF hours, Boron carbide coated anodized aluminum sealed with HL126 sealant provides the best corrosion in all configuration.

The wet cleaning compatibility of four configurations is also tested by soaking the large size B_4C coated rings in saturated $AlCl_3$ solution at pH=0 for 90 minutes, then put the rings in an environmental chamber to monitor the time when boron carbide coating starts to peel off. The test sequence is shown in Fig. 12 [25].

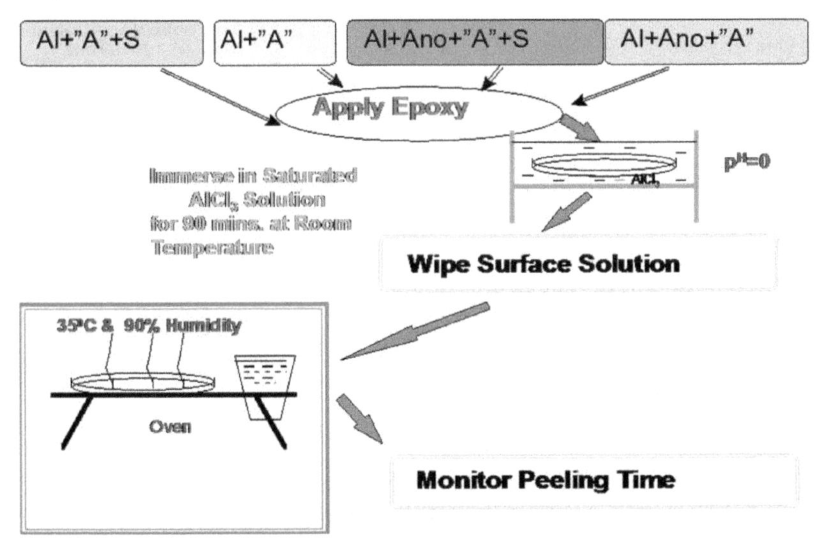

Fig. 12. Wet cleaning compatibility test of four configurations of boron carbide coated rings.

For coating on bare aluminum alloy, the entire coating layer peeled off during immersion in the saturated $AlCl_3$ solution at pH=0.0. The coating on anodized aluminum can hold 45 hours in the environmental chamber and the coating layer peeled off completely at 47 hours. Both coating on bare aluminum alloy and on anodized aluminum with the use of HL126 sealant can hold up to 114 hours in the environmental chamber without any failure. At 114 hours, the environmental chamber test was stopped. From the test results of HCl bubble test and wet cleaning compatibility test, coating on anodized aluminum with the use of HL 126 sealant can provide the best corrosion resistance. This configuration is selected as the final configuration as the new chamber wall material.

In order to qualify boron carbide coating as a new chamber material, many aspects have to be considered. One of the concerns is the impact to ICF (ion current flux). Three configurations are considered and compared in the etching chamber. The ICF of anodized aluminum chamber is used as the baseline. Boron carbide coatings on bare aluminum or on anodized aluminum are studied through ICF measurements. The results showed that the three configurations have the compatible ICF. The results of ICF measurements are shown in Fig. 13 [21, 22, 25, 30].

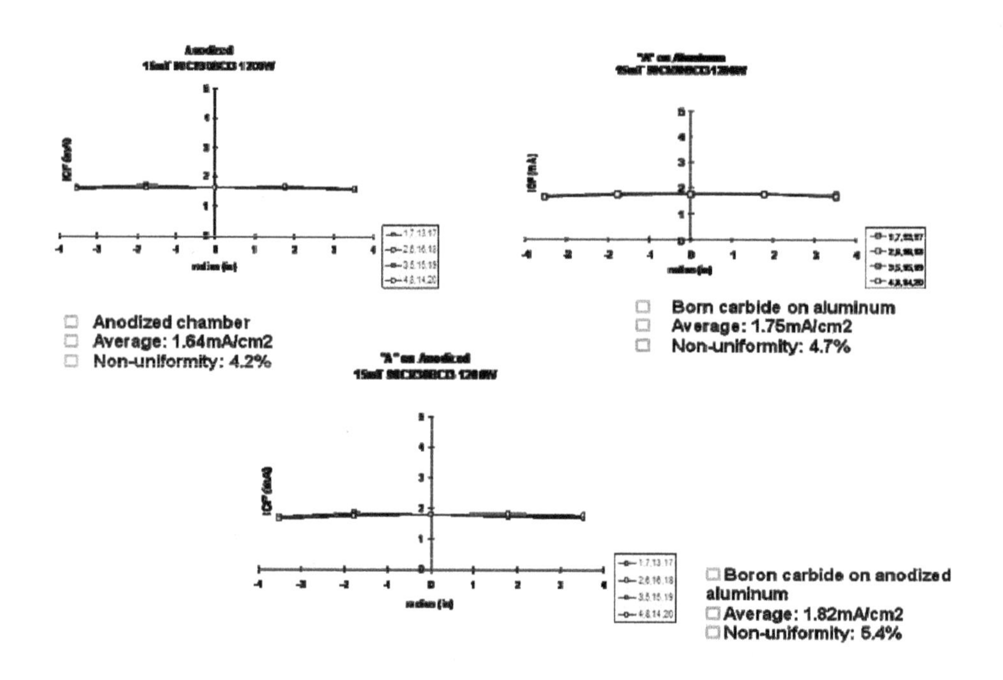

Fig. 13. ICF measurements on the wafer during the use of three configuration chambers.

Another concern is the potential damage to gate oxide. The leakage current measurements on the gate oxide show that born carbide coating does not introduce damage to gate oxide.

The measurements of leakage current of gate oxide are shown in Fig. 14 [21, 22, 25, 30].

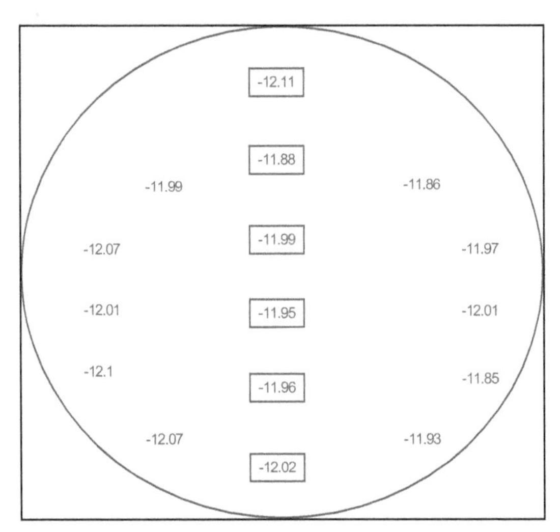

Fig. 14. The leakage current of gate oxide in log scale indicates that there is no damage to gate oxide when a born carbide coated chamber is used.

The metal contamination using a boron carbide coated chamber has shown meeting the specification of metal contaminations such as Ca, Co, Cr, Cu, Fe, K, Mg, Mn, Mo, Na, Ni, Ti, and Zn in 1,000, 2,000, and 3,000 wafer marathons, respectively.

There is no metal contamination introduced when a high purity boron carbide coating is introduced as the new chamber wall coating.

The monitoring data of on-wafer aluminum etch rate, etch rate non-uniformity, defect and particle performance, and thickness measurements of boron carbide coating before and after plasma etching processes are shown in the following figures and tables. In Fig. 15, the particle data during a 3,000 wafer marathon are provided and compared with the specification requested by customers. It is obvious that new B_4C coated chamber wall can meet the requirement of particles. In this study, particles at and larger than 0.2 μm are recorded. The B_4C coated chamber wall can also provide excellent aluminum etch rate and etch rate non-uniformity through the entire 3,000 wafer marathon as shown in Fig.16 [47].

The boron carbide coated chamber is also qualified through a 2,000 wafer marathon for etching of 0.15 μm feature size. Excellent aluminum etching performance is demonstrated as shown in Fig. 17 [48].

On a 300mm etch tool, boron carbide coated chamber was also used in a 1,000 wafer marathon. The boron carbide coated chamber meets all the requirements including aluminum etch rate and etch rate non-uniformity, etch profiles, defects and particles, metal contamination [49]. The particle performance at 0.12 μm or larger is the critical requirement. It is obvious that the boron carbide coated chamber can meet the requirement. The up limit of particle allowance at 0.12 μm or larger is defined as 50 adders/per wafer.

After plasma etching O_2/Cl_2 for 120 RF hours, the thickness of pre and post boron carbide coating on anodized aluminum is measured and the data are listed in Table 2

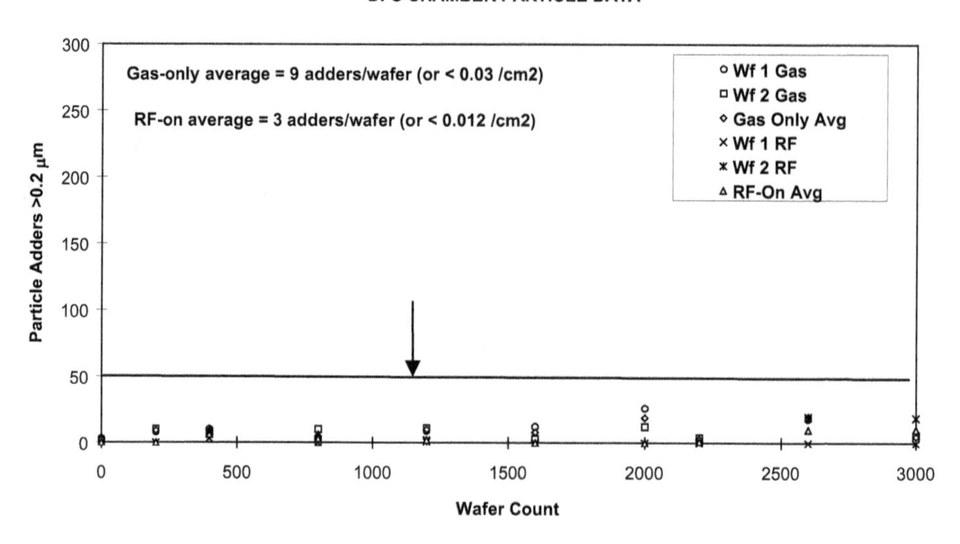

Fig. 15. Gas-only and RF-on particles during a 3,000 wafer marathon. The up limit of allowance of defect and particles is defined as 50 adders/ per wafer.

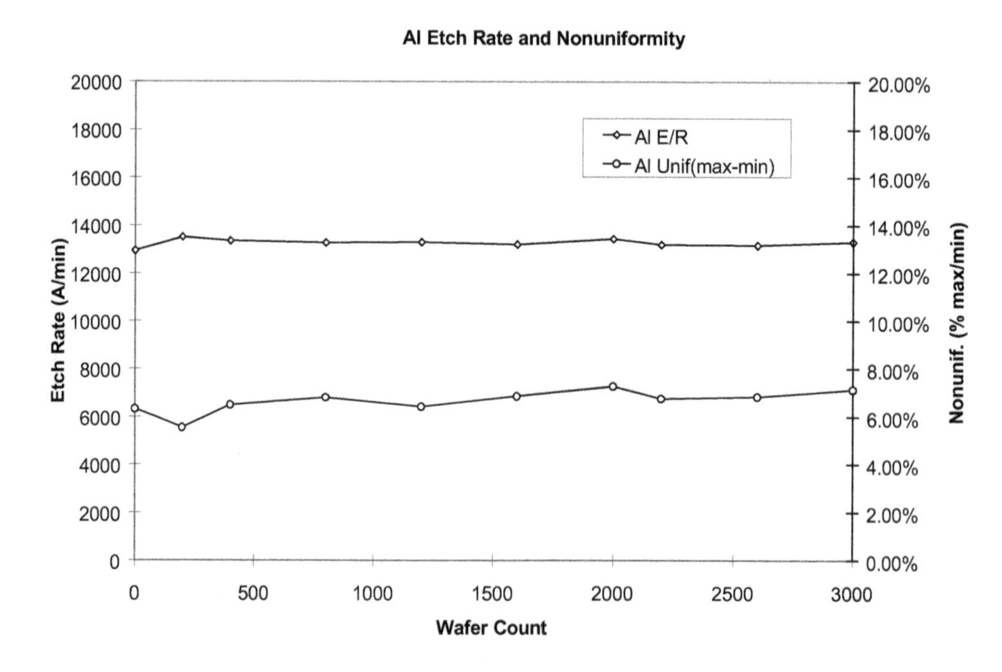

Fig. 16. Al etch rate and etch rate non-uniformity during a 3,000 wafer marathon.

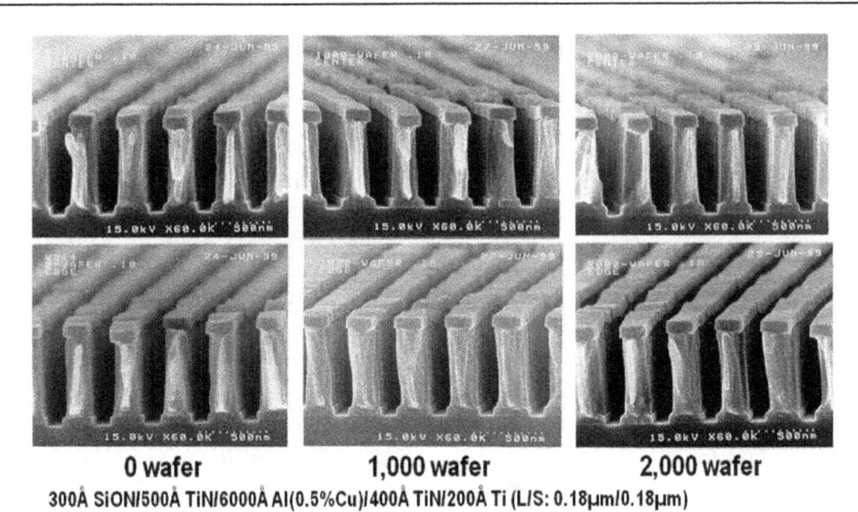

Fig. 17. The boron carbide coated chamber shows an excellent aluminum etch performance on a feature size as of 0.15 µm through a 2,000 wafer marathon.

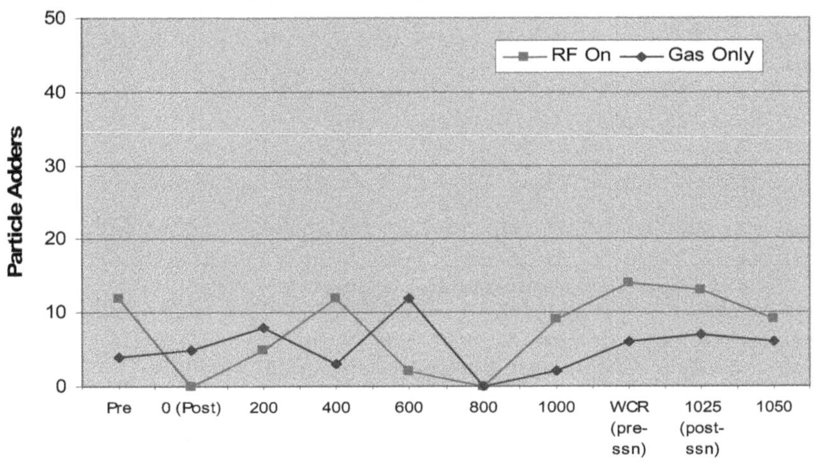

Fig. 18. Particle adders as of ≥ 0.12 µm size in a 1,000 wafer marathon on a 300mm etch tool. The boron carbide coating on anodized aluminum with HL126 sealant is used as the chamber coating to replace anodized aluminum.

Location	Pre (mils)	STD(mils)	Post (mils)	STD(mils)	Delta
A (10 data)	17.60	0.413	17.98	0.236	+0.38
B (10 data)	17.55	0.363	17.52	0.225	- 0.03
C (10 data)	18.76	0.287	18.56	0.201	- 0.20
30 data	**17.97 (average)**		**18.02 (average)**		**+0.05**

Table 2. The overall coating thickness before and after plasma etching under O_2/Cl_2 plasma for 120 RF hours.

It is obvious that there is little coating thickness loss after 120 RF hours under O_2/Cl_2 plasma. The main purpose of O_2/Cl_2 plasma is to test the performance of HL126 sealant under O_2/Cl_2 plasma condition.

After the detail study through a thorough process qualification, the new boron carbide coated chamber wall is used to replace the previously anodized aluminum surface. The new ceramic material such as YAG or Y_2O_3 is used to replace original high purity alumina. This configuration was introduced to semiconductor wafer fabrication for evaluation. Excellent etch performance, enhanced defect and particle reduction, and 50 to 100 times chamber lifetime improvement are reported. The production yield of the wafer fabrication also improved about 7% in production at the customer site (see Fig.19) [41]. The following data provide some of the information. The sequence of the data collection is as follows:

Baseline configuration using the old chamber hard ware submitted to gas-only and RF-on particle measurements without seasoning. After 1st RF-on particle measurement, five oxide wafers were used for seasoning the chamber, then RF-on particles were measured again. Two PR wafers were used to seasoning the chamber before final RF-on particle measurement. The test data are shown in Table 3 [41].

Condition	Gas-only w/o seasoning	RF-on(1) w/o seasoning	RF-on(2) 5 ox seasoning	RF-on(3) 2 PR seasoning
Old chamber configuration	10	34	53	48
New chamber Configuration	2	5	6	5

*: Unit in particle counts/wafer and particle adders at 0.2 μm or larger are recorded.

Table 3. Gas-only and RF-on particles of old and new chamber configurations

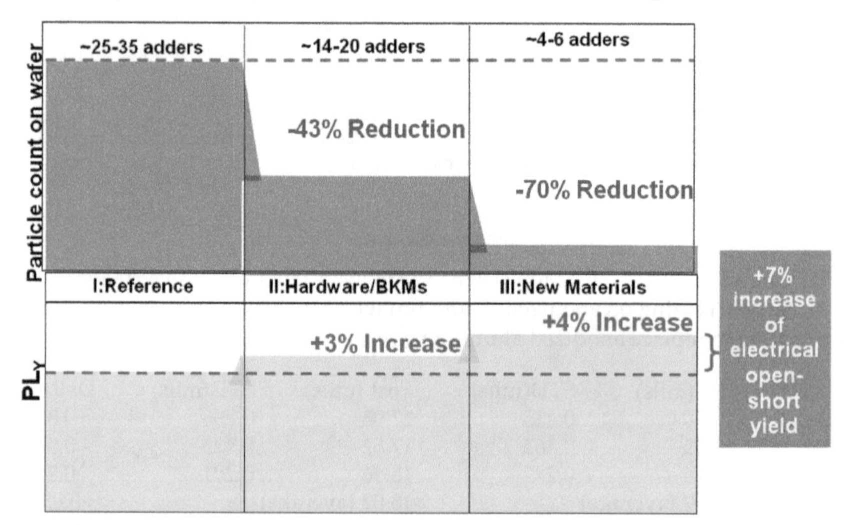

Fig. 19. Production yield improvement at wafer fabrication when new chamber material, hardware and best-known method are implemented.

About 7% production yield is reported in comparison with the old chamber configuration. The lifetime of chamber of chamber wall and chamber top window can improve about 50 times [41].

The new boron carbide coating has been introducing to worldwide wafer fabrication for over 10 years with over 1,000 chambers introduced to wafer fabrication in IC industry. The chamber lifetime has demonstrated to improve from the worse case as of 60 RF hours (1,800 wafers) under BCl_3/Cl_2 etching plasma to over 4,000 RF hours or longer in semiconductor wafer fabrication in the world. It also demonstrates that the chamber materials play a critical role in semiconductor etching equipment, particularly, for the cost reduction. A short comparison of anodized aluminum and born carbide coating is highlighted in Table 4 [21, 22, 25, 30].

Items	Anodized aluminum	Boron Carbide Coating
Maximum etch rate	0.07 mils/RF hour	0.001 mils/RF hour
Mininum lifetime	1,800 wafers	120,000 wafers
Particle performance	normal*	better
Metal contamination	normal	better
Micro-cracks	yes	no, but with coating pores
Coating bonding	very high	less than anodized Al
Surface roughness	normal	higher
Polymer adhesion	normal	better
Wet cleaning recovery	normal	normal
Process performance	normal	normal
Production yield	normal	better
Gate oxide damage	no	no
Water adsorption	normal	normal
Micro-hardness (100g)	360-450	3,000
Localized attack	yes, through cracks	no, with HL126 sealant.
Etch process window	normal	normal
Risk of undercut corrosion	no	no, with HL126 sealant
HCl bubble time (5wt% HCl solution)	≤ 10 minutes (non hot DIW seal) 30 minutes to 24 hours (after hot DIW seal)	> 50 hours, with HL126 sealant
Effect of base aluminum alloys to coating quality	yes, large impact	no

Table 4. Comparison of Anodized Aluminum and Boron Carbide Coating

Anodized aluminum has been using as the major etching tools surface coatings since 1980. It still received a lot of applications in plasma etching tools because of its low cost, easy to manufacture, easy to make large or small sizes of the parts, wide applications, easy to refurbish, and achieving good quality control at different suppliers in the world. Therefore, the study of anodized aluminum has always been a major task for the major semiconductor etching tool manufacturers. For high purity Y_2O_3 thermal spray coating, it has been qualified and applied as one of the major chamber components in plasma etching tools in the past 10 years. It is still one of the major materials as coating or as a solid sintered material which is used in plasma etching tools. At Lam Research Corporation, great attentions have been paid in the improvements and the new development of anodized aluminum and Y_2O_3 coatings.

These studied have been highlighted in John and Hong's presentation [28, 31-33, 35-39] and the publications of the new anodized aluminum study with Mansfeld et al [38, 39].

The study and characterization of anodized aluminum and the methodology are shown below. But the techniques are not limited to the techniques listed below:

- Admittance measurements on anodized aluminum to check sealing quality.
- Micro-hardness on surface and through the anodized aluminum layer.
- SEM cross section and Eddy current meter for anodized layer thickness.
- X-ray diffraction for the phase analysis of anodized aluminum.
- SEM cross section of micro-structure and secondary phases observation.
- TEM analysis to estimate the barrier layer thickness.
- EDX analysis for element analysis on the surface or through the layer.
- ICPMS analysis to obtain the surface cleanliness before and after cleaning.
- GD-OES analysis of the depth profile of elements in anodized layer.
- HCl bubble test to obtain the acidic corrosion resistance of anodized layer.
- Dielectric voltage breakdown of the anodic layer.
- Color and color uniformity of the anodic film.
- Electrochemical impedance spectroscopy to obtain the overall impedance.
- Surface roughness and coating thickness and their variations.
- Taber abrasion test to obtain wear resistance of the anodic film.
- Microhardness on surface and through coating cross section.
- Coating weight.
- Erosion/corrosion rate under high density plasma with different chemistries.
- Raw aluminum alloys analysis through different manufacturing processes.
- Intermetallic inclusions and their chemical composition analysis.
- Thermal property of anodic film after thermal cycling at different temperatures.

Although there are so many techniques used in the anodized aluminum study, there are only key techniques which are selected as a routine quality monitoring of worldwide anodization suppliers. The basic techniques are surface roughness, thickness of anodic film, color and color uniformity, dielectric voltage breakdown, acidic corrosion resistance through HCl bubble test, electrochemical impedance in 3.5wt% NaCl solution, surface micro-hardness, SEM cross section to observe the anodic layer micro-cracks, and admittance under 3.5wt% K_2SO_4 solution at 1000 Hz. For the surface cleanliness of anodization, ICPMS analysis of post precision wet cleaning has been used as a standard technique for metal contamination control. Since the requirements to anodized aluminum quality, corrosion resistance, and surface cleanliness for plasma etching tools are much strict and higher than the traditional industry applications, improvements of corrosion resistance and surface cleanliness are always the tasks. Lam Research has defined the surface cleanliness and the corrosion resistance of anodized aluminum specification for a standard type III and advanced anodized aluminum [28, 31-33, 35-39, 44].

The reaction mechanism of aluminum oxidation is summarized by Macdonald [50] as a reasonable model. The oxides grow as bilayer structures with an inner layer due to movement of oxygen vacancies from metal/film interface and an outer layer due to the movement of cations outward from the film/environmental interface. The vacancy concentrations vary exponentially with distance. The cathode consumes electrons by evoloving hydrogen and reducing oxygen. Barrier layer grows into metal phase via reaction.

Outer film grows via precipitation of Al^{3+} due to hydrolysis. The fundamental reactions for anodized aluminum systems are shown as follows:

Metal	film	Environment
(1) $m + V_M^{x'} = M_M + V_m + Xe^-$		(3) $M_M = M^{x+} + V_m^{x'}$
(2) $m = M_M + (x/2)\, V_o'' + Xe^-$		(4) $V_o'' + H_2O = O_o + 2H$

The principal crystallographic defects are (1) vacancies: $V_o''$and $V_M^{x'}$ for $MO_{x/2}$; (2) interstitials: O_i^{2-} and M_i^{x+}. In fact, oxide films can be described as exponentially-doped semiconductor junctions. The fundamentals and process optimization of anodized aluminum have been studied thoroughly by Brace, Thompson, Wood, Mansfeld, and recent years by Shih through the comprehensive studies of anodization of different aluminum alloys, different anodization processes, and different manufacturing processes [51 – 60]. The interface model of anodized aluminum with hot DIW seal has been described by Mansfeld, Kendig, Shih and others [61- 72] as shown in Fig.20.

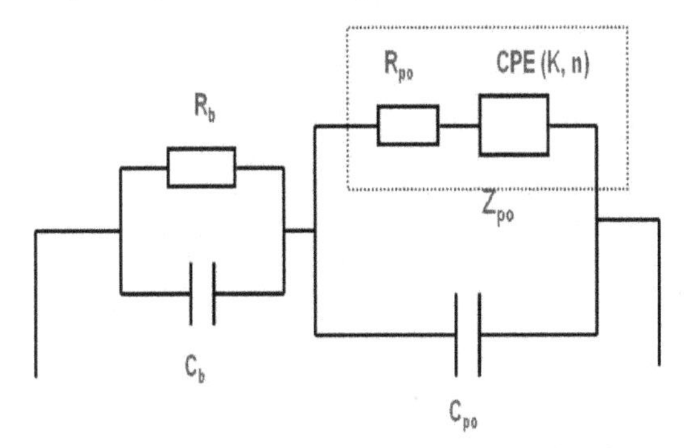

Fig. 20. The typical interface model of anodized aluminum with a hot DI water seal.

$$Z(\omega) = R_s + R_b/\{1+(j\omega C_b R_b)^{\alpha_2}\}+(R_{po}+CPE)/\{1+(j\omega C_{po}(R_{po}+CPE))^{\alpha_1}\}$$

where $C_b = \varepsilon^o \varepsilon_b A/D_b$; $C_{po} = \varepsilon^o \varepsilon_{po} A/D_{po}$ and $CPE = k(j\omega)^n$

$\varepsilon^o = 8.854 \times 10^{-14}$ F/cm and is the permittivity of free space.

In Fig.20, C_b and R_b are barrier layer capacitance and resistance, respectively. R_{po} and CPE are the total impedance of the porous layer defined as Z_{po} which equals to R_{po} + CPE. C_{po} is the capacitance of the porous layer. CPE represents the constant phase element (CPE). A two-time constant interface model and suitable values of R_b and Z_{po} indicate a good quality of anodized aluminum. Z_{po} values highly depend on the quality control of hot DI water sealing process and it is very important for the improvement of the corrosion resistance of anodized aluminum [73-82]. R_b values depend on the voltage applied during anodization as well as the overall process control during anodization. A uniform and thick barrier layer helps to improve the dielectric voltage breakdown of the anodized aluminum. Mansfeld and Shih [63-69] developed a software package specially for the analysis of electrochemical impedance spectroscopy (EIS) data of anodized aluminum and the software has been widely applied for EIS data analysis. The EIS data of the new anodized aluminum developed and qualified at Lam Research Corporation show that the anodized aluminum has no corrosion in 3.5wt% NaCl (similar to seawater) for 365 days as shown in Fig. 21 [28, 38].

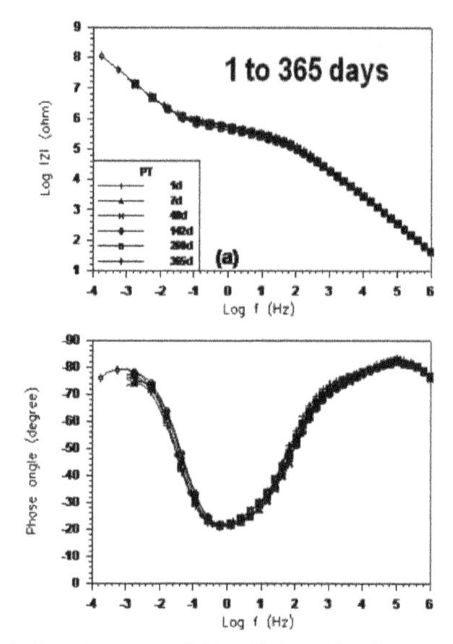

Fig. 21. EIS data of anodized aluminum in 3.5wt% NaCl solution for 365 days.

The overall impedance and HCl bubble test results are shown in Table 5. EIS data of three test coupons after immersion in 365 days in 3.5wt% NaCl solution are analyzed using the software written by Shih and Mansfeld called "ANODAL" [63-69].

Coupon ID	Z_{po} (ohm-cm^2)	R_b (ohm-cm^2)	HCl bubble time in hours
365D005	1.462x10^7	1.257x10^{10}	> 24
365D107	1.568x10^7	1.641x10^{10}	> 24
365D073	1.396x10^7	1.363x10^{10}	> 24

Table 5. The overall impedance and HCl Bubble Time of Test Coupons After Immersion in 3.5wt% NaCl solution for 365 days (coupons were prepared in three different batches of anodization processes) [78].

The Bode-plots of the three EIS data after 365 day's immersion in 3.5wt% NaCl solution is shown in Fig. 22. The anodized aluminum shows an excellent corrosion resistance and high quality of process control.

The complete EIS data analysis of the three test anodized aluminum samples is listed in Table 6 below. It is obvious that a consistent and an excellent corrosion resistance on both porous layer and barrier layer have been demonstrated. It is very important to improve the overall corrosion resistance of anodized aluminum through a well-controlled hot DI water sealing process. The parameters of hot DI water sealing tank water purity level, temperature range, sealing time, hot DI water pH value, and the pre-cleaning of the anodized aluminum before loading to the hot DI water tank will impact the quality of the sealing quality. The anodized anodization as a chamber coating for semiconductor IC industry moved from previously used non-sealed type III anodization or other types of non-sealed anodization to a well-controlled hot DI water sealed anodization for over 15 years because the hot DI water sealed anodized aluminum has demonstrated much better overall corrosion resistance in plasma etching chamber.

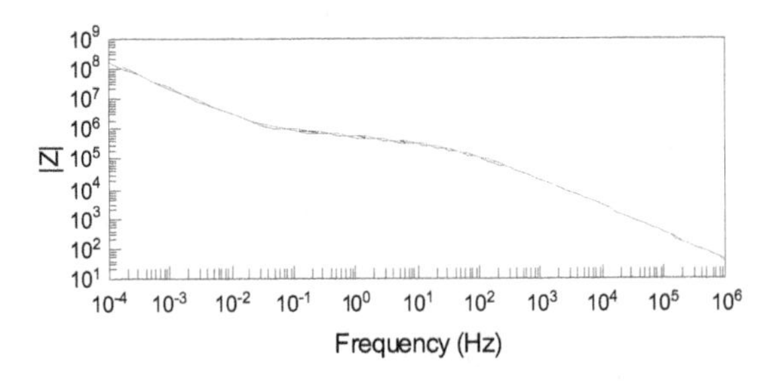

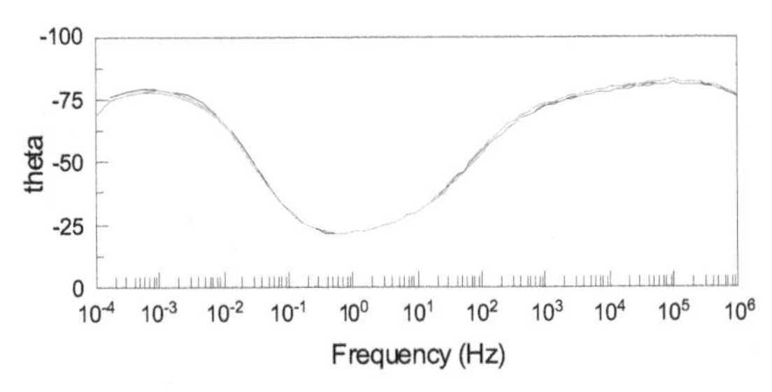

Fig. 22. EIS Bode-plots of advanced anodized aluminum coupons in 3.5wt% NaCl solution for 365 days. Black – 365D005; Red – 365D107; Blue – 365D073.

Parameters	D005	D107	D073
C_{po} (nF)	9.15	9.20	9.33
C_b (μF)	8.45	9.19	8.97
Z_{po} (Mohm)	0.731	0.784	0.698
R_b (Mohm)	628.4	820.5	681.5
n	0.1814	0.1733	0.1733
α_1	0.887	0.900	0.887
α_2	0.958	0.930	0.940
A (area in cm^2)	20.0	20.0	20.0
Chi-sq	2.785×10^{-3}	3.761×10^{-3}	2.775×10^{-3}
Z_{po} (ohm-cm^2)	1.462×10^7	1.568×10^7	1.396×10^7
R_b (ohm-cm^2)	1.257×10^{10}	1.641×10^{10}	1.363×10^{10}

Table 6. Detailed of EIS Data Analysis of test Samples D005, D107 and D073 after 365 Day's Immersion in 3.5wt% NaCl Solution

In Table 6, α is called frequency dispersion which is related to surface inhomogeneties with different dimensions [83]. Chi-sq is the fitting error between the experimental data and fitted data at each frequency. The detailed calculation is shown as below [84] and is the sum of the fitting error at each frequency multiplying 100 and dividing the total data points.

$$\text{Chi-sq} = (100/N)\ \Sigma\ \{[\ |Z_{exp}(f_i) - Z_{fit}(f_i)|]/Z_{exp}(f_i)\}$$

TEM is also used to obtain the barrier layer thickness of different types of anodized aluminum [80]. In Fig. 23, a standard type III anodization achieves about 50nm thickness of the barrier layer. The thickness of a new anodization can be as thick as 100nm due to the higher voltage applied in the anodization process. The thicker barrier layer can provide a higher barrier layer resistance during the EIS study as shown in Table 5 and Table 6. By combining both an excellent hot DI water seal processing to obtain an excellent corrosion resistance of the porous layer and a thicker barrier layer of the anodic film, the anodic film can hold 365 days in seawater without corrosion.

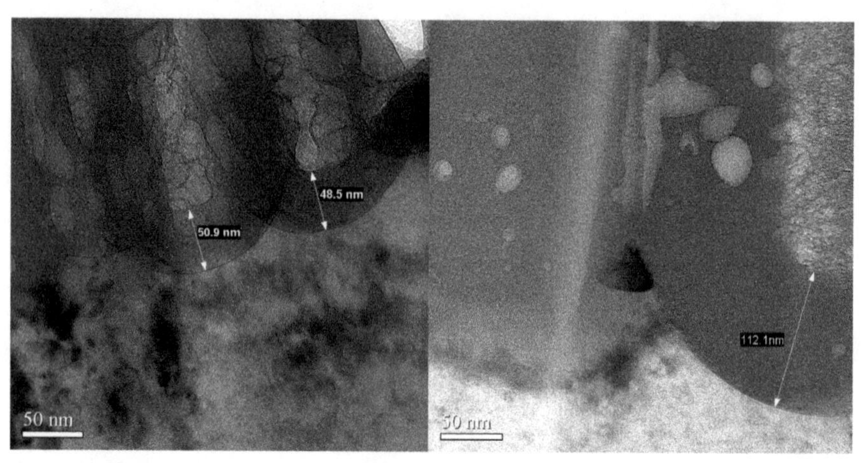

Fig. 23. TEM pictures of a standard type III hard anodization (left) and a mixed acid anodization (right) [80]

For acidic corrosion resistance of anodized aluminum, HCl bubble test is an easy and very effective method to obtain the corrosion resistance. From Fig. 24 below, one can see the good and poor anodized aluminum under the solution of 5wt% HCl solution (28, 31-33, 44, 78]. On the left of Fig.24, there is no any hydrogen bubble generated under the attack of a strong acid within 2 hours immersion. It indicates a high quality of anodized aluminum. On the right of Fig. 24, anodized aluminum generates a lot of hydrogen bubbles in 5wt% HCl solution only after 10 minutes immersion in the acid. It indicates a poor anodized aluminum.

Fig. 24. Acidic corrosion resistance of two test coupons of anodized aluminum. On the left, anodized aluminum does not show any acidic corrosion in two hours and on the right, anodized aluminum shows severe acidic corrosion after only 10 minutes immersion in the acid.

The HCl bubble test can be processed at any position of etching chamber before and after etching process. In Fig. 25, one process chamber is studied on its corrosion resistance in 5.0wt% HCl solution [76]. This method has received a wide application for the corrosion resistance study of anodized aluminum.

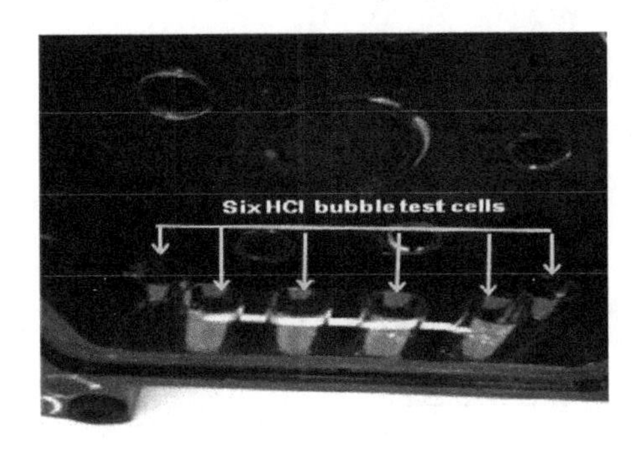

Fig. 25. HCl bubble test on a used process chamber after 12,000 wafers processing. Six locations are selected to run the HCl bubble test [76].

A systematic study of anodized aluminum made of Al6061-T6 11″ thick block was carried out. It is obvious that HCl bubble test can be studied at different thickness positions to compare the differences of corrosion resistance [77]. The detail positions of different tests are shown in Fig. 26. Eleven different test methods are applied to the study.

HCl bubble test can be carried at different thickness to evaluate the corrosion resistance at different thickness in a thick aluminum block [77].

The thermal properties of anodized aluminum have also been studied. One of the typical studies was published in the work with Mansfeld [39]. A lot of studied have been carried out at Lam through the years [39, 73-82]. All these studies show that anodized aluminum film can be degraded through the use at a relatively high temperature. Both porous layer and barrier layer can be impacted depending on the operation temperature. Radii before anodizing on aluminum parts have to be controlled and the micro-cracks at corners and edges depend on the type of anodized aluminum and final thickness of anodic layer.

High purity Y_2O_3 has advantages in comparison with anodized aluminum and ceramic such as high purity alumina in many aspects. First of all, it can reduce the plasma etching rate for both metal etch and silicon etch by a factor of 40 times. It can bring cost saving in etch tools for semiconductor wafer fabrication. It can also reduce metal contamination too. The comparison of anodized aluminum and thermal spray coating of high purity Y_2O_3 are summarized by John and Shih [28, 31]. The advantages of Y_2O_3 coated anodized aluminum are as follows:

- Particle and defect reduction due to the elimination of aluminum fluoride.
- Metal contamination reduction due to its lower transition metal content.
- Better resistance to dielectric breakdown due to a thicker coating.
- Chamber material lifetime improvement due to much lower etch rate under plasma.
- Cost reduction due to extensive chamber materials lifetime.

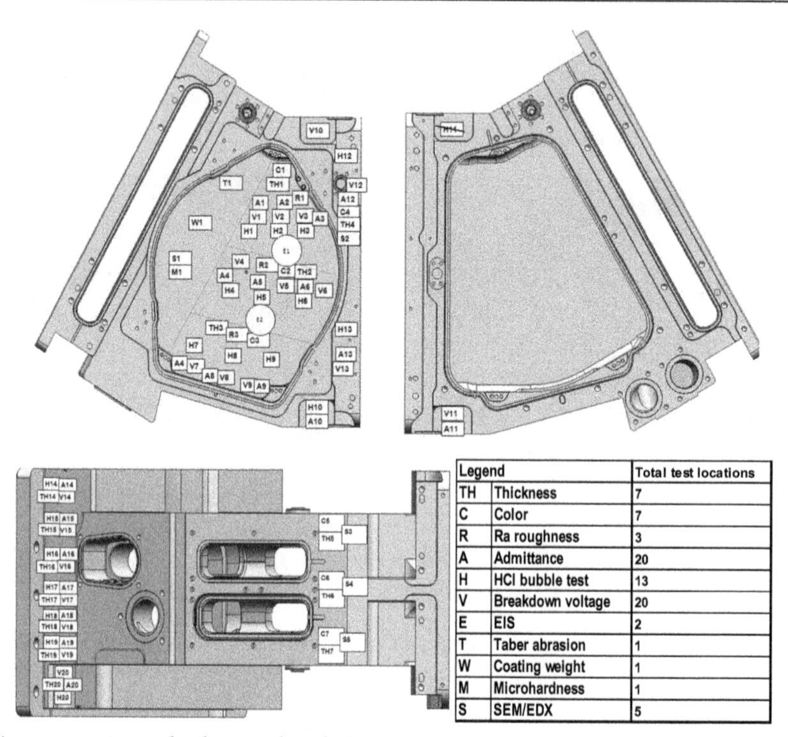

Legend		Total test locations
TH	Thickness	7
C	Color	7
R	Ra roughness	3
A	Admittance	20
H	HCl bubble test	13
V	Breakdown voltage	20
E	EIS	2
T	Taber abrasion	1
W	Coating weight	1
M	Microhardness	1
S	SEM/EDX	5

Fig. 26. A systematic study the anodized aluminum of Al6061-T6 11" thick block.

Although thermal spray and sintered Y_2O_3 has been widely using as one of the chamber materials in wafer fabrication, the study of this material as well as its coating has never been stopped because of the challenges. These studies for semiconductor IC wafer fabrication contain the following studies, but not limit to these techniques [28, 78].

- Impurity levels of raw powder.
- Particle size distribution of the raw powder.
- Impurity levels of the raw powder.
- Environmental control of coating processes.
- Optimization of coating process to reduce porosity and to eliminate non-melted particles.
- Coating thickness, roughness, color and the uniformities.
- Bonding strength and bending strength through pulling and bending tests.
- Dielectric voltage breakdown.
- Acidic corrosion resistance under 5wt% HCl solution.
- Micro-hardness.
- Overall admittance (coating/anodize/aluminum).
- ICPMS analysis for surface cleanliness.
- Surface particles after precision wet cleaning.
- Wet chemicals compatibility study.
- Porosity estimation through SEM cross section analysis.
- Overall impedance and interface model of coating through EIS study.
- Plasma resistance under both BCl_3/Cl_2 and SF_6/HBr gases.

In order to study Y_2O_3 coating on anodized aluminum, the following electrochemical cell configuration is used to study the overall impedance and interface model of the coated samples or parts as shown in Fig. 27 [28, 78].

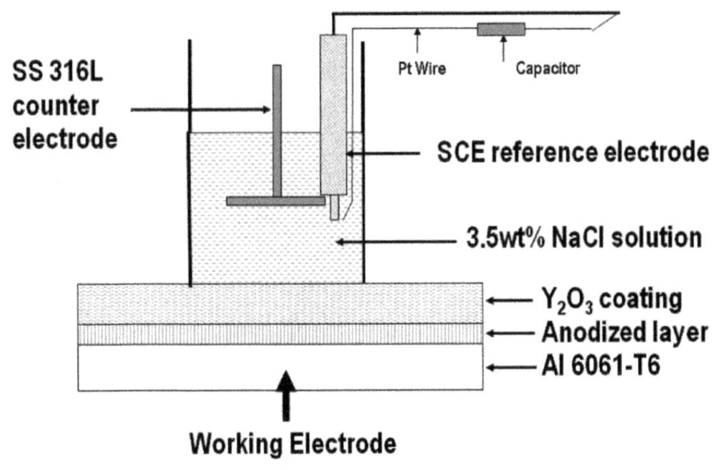

Fig. 27. Electrochemical cell configuration during EIS study of Y_2O_3 coated anodized aluminum in 3.5wt% NaCl solution [78].

An interface model of Y_2O_3 coated anodize aluminum shows a three-time constant interface model indicating a Y_2O_3 coated layer, the porous layer of anodized aluminum, and the barrier layer of anodized aluminum as shown in Fig. 28 [28, 78]

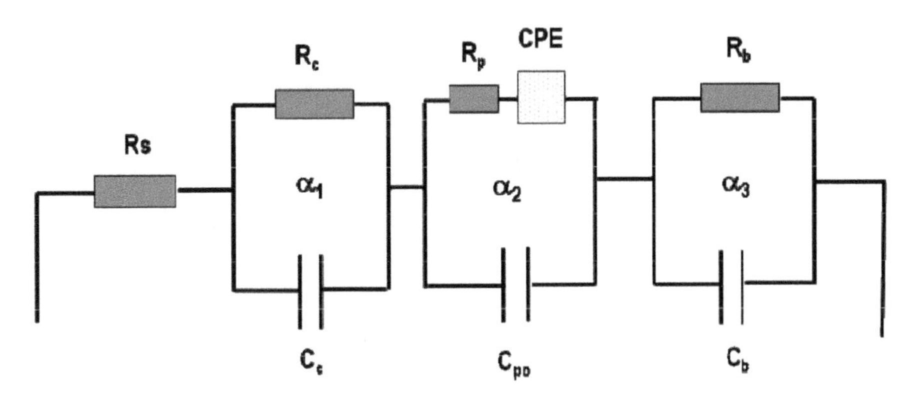

Fig. 28. The proposed interface model of Y_2O_3 coated anodized aluminum in EIS study

Assuming that C_c, C_{po} and C_b are capacitances which represent the capacitances of Y_2O_3 coating layer, porous layer of anodized aluminum and barrier layer of anodized aluminum, respectively. The interface parameters can be obtained and the coating quality can be monitored. The interface model can be described as the following equation [28, 78].

$$Z(\omega) = R_b/\{1+(j\omega C_b R_b)^{\alpha 3}\} + (R_p + CPE)/\{1+(j\omega C_{po}(R_p+CPE))^{\alpha 2}\} + R_c/\{1+(j\omega C_c R_c)^{\alpha 1}\} + R_s$$

Where Z is the total impedance, R_b is the barrier layer resistance of anodization, R_p is the porous layer resistance of porous layer of anodized aluminum, CPE is the constant phase element of the porous layer, R_c is the coating resistance, and R_s is the solution resistance.

Soaking three spraycoated Y_2O_3 on anodized aluminum in 3.5wt% NaCl solution for 7 days, the EIS data are shown in Fig.29. The EIS data indicate that samples coated at different time have the similar overall impedance and the quality control of coating process is consistent. The complete analysis of the EIS data using a three-time constant interface model is shown below (Table 7).

Parameters	Sample 1	Sample 2	Sample 3
R_s (ohm)	2.0	2.0	2.0
R_c (ohm)	40	40	40
R_p (ohm)	$1.2x10^5$	$1.1x10^5$	$1.2x10^5$
R_b (ohm)	$5.0x10^7$	$5.0x10^7$	$5.1x10^7$
K (ohm)	$8.0x10^4$	$6.6x10^4$	$7.5x10^4$
A (area in cm^2)	20.0	20.0	20.0
Z_c (ohm-cm^2)	800	800	800
Z_{po} (ohm-cm^2)	$4.0x10^6$	$3.5x10^6$	$3.9x10^6$
Z_b (ohm-cm^2)	$1.0x10^9$	$1.0x10^9$	$1.0x10^9$

Table 7. Interface parameters of three spraycoated Y_2O_3 on anodized aluminum

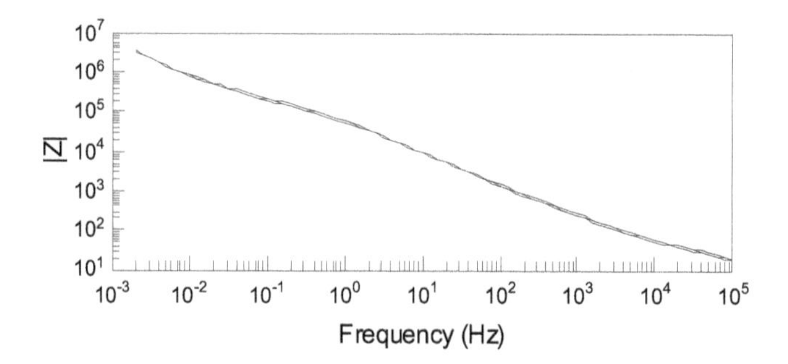

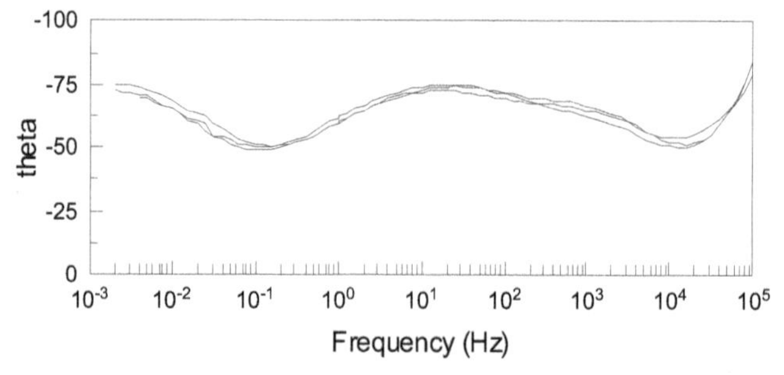

Fig. 29. EIS data of three spraycoated Y_2O_3 on anodized aluminum in 3.5wt% NaCl solution for 7 days. Excellent coating quality control is observed through the EIS study [28, 78].

3. Conclusions

- The revolutionary chamber materials study under high density plasma has opened a new scientific field in the characterization of materials. Meeting the comprehensive requirements of plasma etching tools in semiconductor wafer fabrication with the technology node shrinkage is not an easy task. The efforts and methodology developed through these studies have built up the foundation in the advanced materials characterization, development and application.
- A systematic study of chamber materials for plasma etching tools has been demonstrated. The materials studied include anodized aluminum, boron carbide coating, Y_2O_3 coating and sintered Y_2O_3 as well as the solid solution of aluminum oxide and yttrium oxide (YAG).
- In addition to the discussion of systematic study of anodized aluminum, an advanced anodized aluminum developed has shown no corrosion in seawater environment for 365 days.
- Thermal spray coating of boron carbide has been studied thoroughly. It has been introducing to worldwide semiconductor wafer fabrication since 1998. The coating has demonstrated over 50 times even 100 times lifetime improvement in production of semiconductor wafer fabrication and has been one of major chamber materials up to today.
- Y_2O_3 coating has also been studied thoroughly through the years since 2002. It has demonstrated longer chamber lifetime under plasma for both metal etch and silicon etch applications. It has been introduced to worldwide wafer fabrication since 2002 with a great cost saving and improved etch performance. As of today, it is still the major chamber coating of the current plasma etching tools.
- Lam Research Corporation has been putting great efforts and support for the new chamber materials development. It becomes more critical for semiconductor plasma etching equipment companies to develop new and advanced materials for current and next generation plasma etching feature size applications.

4. Acknowledgement

The author would like to express great thanks to Professor H. W. Pickering and Professor D. D. Macdonald for their guidance during author's Ph.D. study at the Pennsylvania State University between 1981 and 1986. Great thanks to Professor F. B. Mansfeld for author's post doctor and research work at University of Southern California between 1986 and 1990. Thanks to Dr. M. W. Kendig and Professor W. J. Lorenz. Many thanks to Dr. Richard Gottscho, Dr. John Daugherty, and Dr. Vahid Vahedi at Lam Research Corporation. Both Lam Research Corporation and Applied Materials provided the author the opportunity to carry on the systematic study of advanced materials under high density plasma on plasma etching tools. The author would like to express thanks to many individuals during the study of chamber materials through the past 18 years. There have been hundreds of individuals who gave support and encouragement to the author. Due to the limitation of space, the author cannot list all the individuals. Some of the individuals are Dr. Duane Outka, Dr. Tuochuan Huang, Chris Chang, Patrick Barber, Declan Hayes, Dr. Shun Wu, Dr. Harmeet Singh, Dr. Yan Fang, Dr. Armen Avoyan, Dr. Siwen Li, Shenjian Liu, Dr. Qian Fu, Dr. Steve Lin, Nianci Han, Dr. Peter Loewenhardt, Dr. Diana Ma, Mike Morita, Tom Stevenson, Sivakami Ramanathan, John Mike Kerns, Dean Larson, Alan Ronne, Hilary Haruff, David

Song, Dr. You Wang, Dr. Yan Chun, Dr. Qi Li, Dr. Allan Zhao, Joe Sommers, Dr. Griff O'Neil, Dr. Raphael Casaes, Dr. T. W. Kim, Dr. Lin Xu, Dr. Catherine Zhou, Dr. Yijun Du, Dr. Daxing Ren, Ho Fang, May He, Josh Cormier, Fangli Hao, Dr. Steve Mak, Dr. Gerald Yin, Dr. Xikun Wang, Dr. Danny Lu, Yoshi Tanase, Dr. Samantha Tan, Dr. Fuhe Li, Shang-I Chou, Tim. Su, John Holland, Peter Holland, Jason Augustine, Dr. Y. L. Huang, Lucy Chen, Jie Yuan, Hui Chen, Li Xu, Charlie Botti, Han Sellege, Dr. Arthur W. Brace, and many others.

5. List of related patents

[1] Hong Shih and M. S. Mekhjian, "Method and Apparatus for Determining the Quality of A Coating", USA Patent No. 5,373,734, December 20, 1994.

[2] Hong Shih and M. S. Mekhjian, "Apparatus for Measuring A Coating on A Plate", USA Patent No. 5,493,904, February 27, 1996.

[3] Hong Shih, N. Han, S. Mak, and G. Yin, "Boron Carbide Parts and Coatings in A Plasma Reactor", USA Patent No. 6,120,640, September 19, 2000.

[4] Hong Shih, Nianci Han, Jie Yuan, Danny Lu, and Diana Ma, "Ceramic Composition for An Apparatus and Method for Processing A Substrate", USA Patent No. 6,352,611, March 5th, 2002.

[5] N. Han, Hong Shih, J. Yuan, D. Lu, and D. Ma, "Ceramic Composition for An Apparatus and Method for Processing A Substrate", USA Patent No. 6,123,791, September 26, 2000.

[6] Hong Shih, Joe Sommers, and Diana Ma, "Method for Monitoring the Quality of A Protective Coating In A Reactor Chamber", USA Patent No. 6,466,881, October 15, 2002.

[7] N. Han, Hong Shih, J. Y. Sun, Li Xu "Diamond Coated Parts in A Plasma Reactor", USA Patent No. 6,508,911, January 21, 2003.

[8] Hong Shih, Nianci Han, Jie Yuan, Joe Sommers, Diana Ma, Paul Vollmer, "Corrosion-Resistant Protective Coating for An Apparatus And Method for Processing A Substrate", USA Patent No. 6,592,707, July 15, 2003.

[9] Nianci Han, Hong Shih, Jie Yuan, Danny Lu, and Diana Ma, "Substrate Processing Using a Member Comprising An Oxide of A Group IIIB Metal", USA Patent No. 6,641,697, November 4th, 2003.

[10] Hong Shih and Nianci Han, "Coating Boron Carbide on Aluminum", USA Patent No. 6,808,747, October 26, 2004.

[11] N. Han, Hong Shih, and Li Xu, "Process Chamber Having Component with Yttrium-Aluminum Coating", USA Patent No. 6,942,929, September 13, 2005.

[12] Hong Shih, Duane Outka, Shenjian Liu, and John Daugherty, "Extending Lifetime of Yttrium Oxide As A Plasma Chamber Material", Lam Research Patent Application, Publication Number: US20080169588, Publication Date : 07/17/2008: Database : US Patents.

[13] Harmeet Singh, John Daugherty, Vahid Vahedi, and Hong Shih, "Components for Use in A Plasma Chamber Having Reduced Particle Generation and Method of Making", Lam Research Patent Application, Publication Number: US20090261065, Publication Date : 10/22/2009: Database : US Patents.

[14] Hong Shih, "Method for Refurbishing Bipolar Electrostatic Chuck", Assignee: Lam Research Corporation, Publication Number US 2010/0088872 A1, April 15, 2010.

[15] Hong Shih, Saurabh Ullal, Tuochuan Huang, Yan Fang, and Jon McChesney, "System and Method for Testing An Electrostatic Chuck", Assignee: Lam Research Corporation, Publication Number US 20100090711 A1, April 15, 2010.

[16] Hong Shih, Qian Fu, Tuochuan Huang, Raphael Casaes, and Duane Outka, "Extending Storage Time of Used Y2O3 Coated Ceramic Window before Wet Cleaning", US Patent No 7,976,641, July 12, 2011.

[17] Josh Cormier, Fangli Hao, Hong Shih, Allan Ronne, John Daugherty, and Tuochuan Huang, "Automatic Hydrogen Bubble Detection and Recording System for the Study of Acidic Corrosion Resistance for Anodized Aluminum and Surface Coatings", Lam Research Patent Invention Disclosure P2232, Filed for US Patent on January 4, 2011.

[18] John M. Kerns, Yan Fang, Hong Shih, and Allan Ronne, "Silicon Gas Line Coating for Semiconductor Applications", Lam US Patent Application, Docket No: 2287, May 5, 2011.

[19] Hong Shih, Duane Outka, Shenjian Liu, and John Daugherty, "Extending Lifetime of Yttrium Oxide As A Plasma Chamber Material", Lam Research Patent Application, Publication Number : (WO 2008/088670), Publication Date: 07/17/2008

[20] Hong Shih, Saurabh Ullal, Tuochuan Huang, Yan Fang, and Jon McChesney, "System and Method for Testing An Electrostatic Chuck", Assignee: Lam Research Corporation, Publication Number: WO 2010/042908, April 15, 2010.

[21] Hong Shih, Nianci Han, Jennifer Y. Sun, and Li Xu, "Diamond Coated Parts in A Plasma Reactor", Assignee: Applied Materials, Publication Number: WO2001/013404, February 2001.

[22] N. Han, Hong Shih, Danny Lu, and Diana Ma, "A Ceramic Composition of An Apparatus and Method for Processing A Substrate", Applied Materials, Publication Number: WO2000/007216, February, 2000.

[23] Hong Shih, N. Han, S. Mak, and G. Yin, "Boron Carbide Parts and Coatings in A Plasma Reactor", European Patent Number. EP0849767, Publication Date: June 24, 1998.

[24] Hong Shih, N. Han, S. Mak, and G. Yin, "Boron Carbide Parts and Coatings in A Plasma Reactor", European Patent Number. EP0849767, Publication Date: March 21, 2001.

[25] Tuochuan Huang, Daxing Ren, Hong Shih, Catherine Zhou, Chun Yan, Enrico Magni, Bi Ming Yen, Jerome Hubacek, Danny Lim and Dougyong Sung, "Methods for Silicon Electrode Assembly Etch Rate and Etch Uniformity Recovery", European Patent Publication No. 1848597, October 31, 2007.

[26] Hong Shih, Nianci Han, Jie Yuan, Danny Lu, and Diana Ma, "Ceramic Composition for An Apparatus and Method for Processing A Substrate", Applied Materials, July 2001: Korea 1020017001225.

[27] Tuochuan Huang, Daxing Ren, Hong Shih, Catherine Zhou, Chun Yan, Enrico Magni, Bi Ming Yen, Jerome Hubacek, Danny Lim and Dougyong Sung, "Methods for Silicon Electrode Assembly Etch Rate and Etch Uniformity Recovery", Lam Research Corporation, China Patent Application No: CN200580044362, March 2008.

[28] Hong Shih, Duane Outka, Shenjian Liu, and John Daugherty, "Extending Lifetime of Yttrium Oxide As A Plasma Chamber Material", Lam Research Patent Application, China Patent Number CN200880002182, November 2009.

[29] Hong Shih, N. Han, S. Mak, and G. Yin, "Boron Carbide Parts and Coatings in A Plasma Reactor", Assignee: Applied Materials, Singapore Patent Application Number: 9704073-7, November 17, 1997.

[30] Hong Shih, Duane Outka, Shenjian Liu, and John Daugherty, "Extending Lifetime of Yttrium Oxide As A Plasma Chamber Material", Lam Research Patent Application, Taiwan Patent Number TW097100617, December 2008.

[31] Hong Shih, N. Han, S. Mak, and G. Yin, "Boron Carbide Parts and Coatings in A Plasma Reactor", Assignee: Applied Materials, Taiwan Patent Application Number: TW086114934, November, 1999.

[32] Hong Shih, Joe Sommers, and Diana Ma, "Method for Monitoring the Quality of A Protective Coating In A Reactor Chamber", Applied Materials, Taiwan Patent Application No: TW089107053, May 2002..

[33] Hong Shih, Nianci Han, Jie Yuan, Joe Sommers, Diana Ma, Paul Vollmer, "A Corrosion-Resistant Protective Coating for An Apparatus and Method for Processing a Substrate", Applied Materials, Taiwan Patent Application No: TW089106928, November, 2011.

[34] Hong Shih, Nianci Han, Jie Yuan, Danny Lu, and Diana Ma, "Ceramic Composition for An Apparatus and Method for Processing A Substrate", Assignee: Applied Materials, Taiwan Patent Publication Number: TW088112796, August 2005.

[35] Hong Shih, Duane Outka, Shenjian Liu, and John Daugherty, "Extending Lifetime of Yttrium Oxide As A Plasma Chamber Material", Lam Research Patent Application, Taiwan Patent Number TW097100617, December 2008.

[36] Hong Shih, Saurabh Ullal, Tuochuan Huang, Yan Fang, and Jon McChesney, "System and Method for Testing An Electrostatic Chuck", Assignee: Lam Research Corporation, Taiwan Patent Publication Number: TW096115736, June 2008.

[37] Hong Shih, Saurabh Ullal, Tuochuan Huang, Yan Fang, and Jon McChesney, "System and Method for Testing An Electrostatic Chuck", Assignee: Lam Research Corporation, Taiwan Patent Publication Number: TW098142359, March 2010.

6. References

[1] M. A. Lieberman and A. J. Lichtenberg, "Principles of Plasma Discharges and Materials Processing", John Wiley & Sons, Inc., Second Edition, April 14, 2005.

[2] J. W. Coburn, "Plasma Etching and Reactive Ion Etching: The Fundamentals and Applications", A Short Course on Plasma Etching, 2008.

[3] M. A. Lieberman, "A Short Course on the Principle of Plasma Discharges and Materials Processing", A Short Course on Plasma Etching, 2008.

[4] M. A. Lieberman and R. A. Gottscho, in Physics of Thin Films, edited. By M. Francombe and J. Vossen, Academic Press, 1993.

[5] R. A. Gottscho, C. W. Jurgensen, and D. J. Vitkavage, "Microscopic Uniformity in Plasma Etching", J. Vac. Sci. Technol. B10(5), 2133, 1992.

[6] J. W. Coburn, J. Vac. Sci. Technol. B12, 1384, 1994.

[7] C. Lee, D. B. Graves, and M. A. Lieberman, and D. W. Hess, J. Electrochem. Soc. 141, 1994.

[8] C. Lee, D. B. Graves, and M. A. Lieberman, Plasma Chemistry and Plasma Processing, Vol. 16, No. 1, 1996.

[9] B. N. Chapman, "Glow Discharge Processes", John Wiley & Sons, Inc., 1980.

[10] O. A. Popov, "High Density Plasma Sources", Noyes Publications, 2000.

[11] R. J. Shul and S. J. Pearton, editors, "Handbook of Advanced Plasma Processing Techniques", Springer-Verlag, 2000.

[12] H. Singh, J. W. Coburn, and D. B. Graves, J. Vac. Sci. Technol. A17(5), 2447, 1999.

[13] M. Sugawara, "Plasma Etching: Fundamental and Applications (Semiconductor Science and Technology), Oxford Science Publications, 1998.

[14] J. E. Daugherty and D. B. Graves, J. Vac. Sci. Technol. A11, 1126, 1993.

[15] Hong Xiao, "Introduction to Semiconductor Manufacturing Technology", Pearson Education International, Prentice Hall, New Jersey, 2001.

[16] C. Y. Chang, "Semiconductor Manufacturing Equipment", Wunan Publishing Co., Ltd., Taiwan, November 2000.
[17] C. Y. Chang, "Deep Submicron Silicon Processing Technology", Wunan Publishing Co., Ltd., Taiwan, May 2002.
[18] Masanori Kikuchi, "Graphics of Semiconductor", Nippon Jitsagyo Publishing, Co., Ltd., 2000.
[19] "Zukai Handotai Guide", edited by Toshiba Semiconductorsha, Toshiba Corporation, 2001, Seibundo Shinko-sha Publishing Co., Ltd., Toyko, 2001, translated by Zhou Yong Xu and published by Princeton International Publishing Co., Ltd., Taiwan, 2004.
[20] "Arute 21 Handoutai Device", edited by Hiromu Haruki, Published by Ohmsham Ltd., 1999.
[21] H. Shih, "Materials Characterization under High Density Plasma", Keynote presentation on American Ceramic Society, Coconut Beach, Florida, February 26, 2004.
[22] H. Shih, "A Materials Study and Characterization of Semiconductor Wafer Fabrication", Presentation at the Pennsylvania State University for the 2004 McFarland Award, April 24, 2004.
[23] H. Shih, "Technology Development of Materials Characterization in Wafer Fabrication Equipment", Technical Presentation on 2001 IC Equipment Supply Chain Symposium and Tainan Manufacturing Center Opening", Hosted by Applied Materials, Taiwan, May 4th, 2001.
[24] H. Shih, N. Han, S. Mak, and G. Yin, "Development and Characterization of Materials for Sub-Micron Semiconductor Etch Application under High Density Plasma", Presentation on the 13th International Symposium on Plasma Chemistry", June 22, Beijing, China, 1997.
[25] H. Shih and D. Ma, "Revolutionary Chamber Materials for Metal Etch", SEMICON West Oral Presentation, Moscone Convention Center, San Francisco, California, June 1998.
[26] H. Shih, "Defect Density Reduction of 0.18μm and Beyond", Presentation on SEMICON Korea, Seoul, Korea, February 15-17, 2000.
[27] H. Shih, "Defect Density Reduction and Productivity Enhancement for 0.18μm and Beyond", Presentation on AMSEA 7th Annual Technical Seminar at Singapore, May, 2000.
[28] H. Shih and J. Daugherty, "Systematic Study of Yttrium Oxide Coating on Anodized Aluminum Surfaces", Keynote Presentation on International Thermal Spray Coating (ITSC) Conference & Exposition, May 4-7, 2009, Las Vegas, Navada, USA.
[29] H. Shih, "Can Semiconductor IC Equipment Survive without Thermal Spray Coatings", Guest Editorial in Advanced Materials & Processes, Vol 168, No 5, May 2010.
[30] H. Shih, N. Han, S. Mak, E. Polar, G. Yin, "Significant Lifetime Enhancement of Dielectric Materials for Next Generation Etch Chamber Materials", Applied Materials ET Conference Paper No. 455, Oral Presentation at San Jose State University, San Jose, California, 1998.
[31] H. Shih and Patrick Barber, "Materials Laboratory Development for Anodized Aluminum Study and Supplier Qualification - Part one and Part Two", Lam Research Confidential Technical Report, May 31, 2003.
[32] H. Shih, T. C. Huang, S. Wu, and J. Daugherty, "Summary of Corrosion Study During 180 days Immersion in 3.5wt% NaCl Solution", Lam Research Confidential Technical Report, Lam Research Corporation, September 28, 2006.
[33] H. Shih and J. Daugherty, "EIS Data Explanation of Anodized Aluminum 6061-T6 After Thermal Cycling", Lam Research Confidential Technical Report, Lam Research Corporation, January 26, 2009.

[34] Mike Kerns, Yan Fang, and Hong Shih, "Summary of Corrosion Tests on Silicolly Coated SS316L Gas Lines", Lam Research Confidential Technical Report, Lam Research Corporation, July 29, 2010.

[35] H. Shih, "Precision Wet Cleaning at Lam Research Corporation", Technical Presentation at SMIC, Beijing and Shanghai, August, 16 to 19, 2011.

[36] H. Shih, "Precision Wet Cleaning and Wet Cleaning Technical Support for TSMC", Technical Presentation at TSMC, Taiwan, July 5th, 2011, Hsinchu, Taiwan.

[37] H. Shih, "Advanced Chamber Materials for Plasma Etching Applications", Lam Research Confidential Technical Report, February 6, 2011.

[38] Y. L. Huang, H. Shih, T. C. Huang, J. Daugherty, S. Wu, S. Ramanathan, C. Chang, F. Mansfeld, "Evaluation of the Properties of Anodized Aluminum 6061-T6 Using Electrochemical Impedance Spectroscopy (EIS)", Journal of Corrosion Science, Vo. 50, Issue 12, p. 3569-3575, 2008.

[39] Y. L. Huang, H. Shih, J. Daugherty, F. Mansfeld, " Evaluation of the Properties of Anodized Aluminum 6061 Subjected to Thermal Cycling Treatment Using Electrochemical Impedance Spectroscopy", Journal of Corrosion Science, Vo. 51, Issue 10, p. 2493-2501, October, 2009.

[40] D. Outka and H. Shih, "Surface Analysis of Electrostatic Chuck Surface and Wet Cleaning Development", Lam Confidential Technical Report, August 20, 2008.

[41] Y. Takakura, T. Miyauchi, T. Ono, J. Hachiya, S. Kitamura, A. Endo, S. Park, N. Han and H. Shih, "Wafer Defect Reduction with DPS Metal Etch", Applied Materials ET Paper 471, Canada, 2000.

[42] H. Shih, "Intel Anodized Aluminum GDP Failure Analysis through Eddy Current, Surface Resistivity, Admittance, Hardness, HCl Bubble and Electrochemical Impedance Spectroscopy", Applied Materials Confidential Technical Report, January 4, 1994.

[43] H. Shih, J. Sommers, S. Lin, and D. Ma, "Analysis and Improved Specification for Critical Anodization on Metal Etch Process Parts", Applied Materials ET Conference Paper No. 453/484, San Jose State University, 1998.

[44] H. Shih, D. Outka, and J. Daugherty, "Specification for Hard Anodized Aluminum Coatings Using Mixed Acid for Critical Chamber Components", Lam Research specification 202-047671-001, January 17, 2006.

[45] MIL-A-8625, "Military Specification : Anodic Coatings for Aluminum and Aluminum Alloys.

[46] ASTM B117 – Method of Salt Spray (Fog) Testing.

[47] H. Shih, N. Han, S. Mak, and G. Yin, "Overview of A-Coat Development for DPS Metal Etch", Applied Materials ET Conference Oral Presentation at San Jose State University, 1997.

[48] H. Shih, N. Han, J. Yuan, and Q. Li, "DPS Chamber Lifetime Enhancement – A Coating Characterization and Burn-in", Applied Materials ET Conference Paper No. 588, San Jose State University, 1997.

[49] H. Shih, R. Xie, R. Koch, X. K. Wang, H. Chen, G. W. Ding, C. Sun, L. Chen, S. Arias, E. Chiang, K. Kawaguchi, M. Jain, A. Jiang, J. Papanu, R. Hagborg, B. Hatcher, B. Aeaia, B. Ching, R. Hartlage, V. Todorov, P. Leahey, N. Arboiuz, B. Dodson, S. Mak, R. Kerns, C. Lane, J. Holland and M. Barnes, "Summary of Process and Productivity Results of 1,000 Wafer Alpha Release Burn-in of Metal Etch New 300mm System", Applied Materials ET Conference Paper No. 599, San Jose State University, 2001.

[50] D. D. Macdonald, Private communication, 1991.
[51] A. W. Brace, "Anodic Coating Defects - Their Causes and Cure", Published by Technicopy Books, England, 1992.
[52] "Aluminum and Magnesium Alloys", Vol. 02.02, Annual Book of ASTM Standards, 1993.
[53] S. Wernick, R. Pinner, and P. G. Sheasby, "The Surface Treatment and Finishing of Aluminum and It's Alloys", 5th edition, ASM International, 1987.
[54] G. E. Thompson and G. C. Wood, "Treatise on Materials Science and Technology", 23, 205-329, 1983.
[55] Hong Shih and John Daugherty, "Systematic Study of Yttrium Oxide Coating on Anodized Aluminum Surfaces", Keynote Presentation on International Thermal Spray Coating (ITSC) Conference & Exposition, May 4-7, 2009, Las Vegas, Nevada, USA.
[56] Hong Shih and John Daugherty, "Chamber Materials for Current and Future Etchers", Lam Technical Report, December 10, 2011.
[57] Hong Shih, Shun Wu, and David Song, "Coordination of HCl Bubble Test and Electrochemical Impedance Spectroscopy of Lam Worldwide Anodizers", Lam Technical Report, December 20, 2010.
[58] Lin Xu and Hong Shih, "Secondary Phase Distribution of Al6061-T6 Alloys", Lam Technical Report, September 7, 2011.
[59] Hong Shih, "Electrochemical Impedance Spectroscopy Study of AI and Altefco D-Chamber Using Alcan Ravenswood 11" Thick Block", Lam Technical Report, February 14, 2011.
[60] Hong Shih and John Daugherty, "EIS Data Explanation of Anodized Aluminum 6061-T6 After Thermal Cycling", Lam Technical Report, January 26, 2009.
[61] Hong Shih, "Aluminum Alloy Anodic Oxidation - Theory, Common Failure, and Techniques for Anodization Study in Semiconductor Equipment Applications", Lam Technical Report, December 16, 2007.
[62] F. Mansfeld and M. W. Kendig, J. Electrochem. Soc. 135, 828, 1988.
[63] H. Shih, T. C. Huang, and J. Daugherty, "Lam Research Confidential Technical Report, October 10, 2006.
[64] F. Mansfeld, H. Shih, H. Greene, and C. H. Tsai, " Analysis of EIS Data for Common Corrosion Processes", in "Electrochemical Impedance : Analysis and Interpretation", ASTM STP 1188, J. R. Scully, D. D. Silverman, and M. W. Kendig, Eds., ASTM, p23, 1993.
[65] H. Shih and F. Mansfeld, in "New Methods for Corrosion Testing of Aluminum Alloys", ASTM 1134, edited by V. S. Agatwala and G. M. Ugianksy, ASTM, 180-195, 1992.
[66] F. Mansfeld, "Analysis and Interpretation of EIS Data for Metals and Alloys", Schlumberger Technical Report 26, 1993.
[67] F. Mansfeld, H. Shih, and C. H. Tsai, "Software for Simulation and Analysis of Electrochemical Impedance Spectroscopy (EIS) Data", in Common Modeling in Corrosion", ASTM STP 1154, R. S. Munn, ed, ASTM, p.186, 1992.
[68] H. Shih, "Electrochemical Impedance Spectroscopy and Its Application for the Characterization of Anodic Layers of Aluminum Alloys on Semiconductor Manufacturing Industry", Presentation on Corrosion Asia, 94, September 26-30, Marina Mandarin, Singapore, 1994.

[69] H. Shih and H. W. Pickering, "Some Aspects of Potential and Current Distributions During AC Polarization in Electrochemical Systems", Presentation on Corrosion Asia, 94, September 26-30, Marina Mandarin, Singapore, 1994.

[70] Y. L. Huang, H. Shih, and F. Mansfeld, "Concerning the Use of Constant Phase Element (CPE) in the Analysis of Impedance Data", Materials and Corrosion, 2009, 60, No. 9999.

[71] H. Shih, H. J. Chen, and F. Mansfeld, "Data Analysis of Electrochemical Impedance Spectroscopy (EIS) in Corrosion Monitoring and Detection", Presentation on Corrosion Asia, 94, September 26-30, Marina Mandarin, Singapore, 1994.

[72] F. Mansfeld and H. Shih, "ANALEIS A/S 1.0 Module 4: ANODAL - A Software Library for the Simulation and Analysis of Electrochemical Impedance Data", February, 1992.

[73] H. Shih, T. C. Huang, S. Wu, S. Ramanathan, and J. Daugherty, "The Development of Next Generation Anodized Aluminum - Summary of Corrosion Study during Immersion in 3.5wt% NaCl Solution for 180 Days", Lam Research Confidential Technical Report, September 28, 2006.

[74] H. Shih, S. Wu, T. C. Huang, S. Ramanathan, D. Outka, D. Larson, Y. Fang, C. Zhou, and J. Daugherty, "Summary of DOE of ACME Coated Anodized Aluminum", Lam Research Confidential Technical Report, January 21, 2008.

[75] H. Shih, S. Wu, Y. J. Du, Y. Fang, and J. Daugherty, "Breakdown Voltage Measurements on A 8K Thermal Oxide Si Wafer Using Ball Tip and Needle Tip", Lam Research Confidential Technical Report, February 8, 2008.

[76] H. Shih, S. I. Chou, and R. Casaes, "D-Chamber Evaluation after Running 12,000 Wafers", Lam Research Confidential Technical Report, November 11, 2010.

[77] D. Song, H. Shih, S. Wu, and H. Haruff, "Systematic Study of D-Chamber Anodization", Lam Research Confidential Technical Report, February 14, 2011.

[78] H. Shih, "Electrochemical Impedance Spectroscopy (EIS) Technology and Applications", Lecturer for A 8 Hour Short Course Sponsored by the San Francisco Section of the Electrochemical Society, Crown Plaza Hotel, Milpitas, California, June 15, 2006.

[79] H. Shih, "Aluminum Alloy Anodic Oxidation - Theory, Common Failure and Techniques for Anodization Study in Semiconductor Equipment Application", Lam Research Confidential Technical Report, December 16, 2007.

[80] Yan Fang, Duane Outka, and Hong Shih, "TEM Analysis of Al6061-T6 Type III Anodization and the Mixed Acid Anodization", Lam Research Confidential Technical Report, January 8, 2008.

[81] H. Shih, A. Avoyan, T. C. Huang, D. Outka, and J. Daugherty, "Anodized Aluminum Cleaned with Additional Chemical Solutions to Achieve a Better Surface Cleanliness on Both Anodized Aluminum and Bare Aluminum Surfaces", Lam Research Confidential Technical Report, November 18, 2008.

[82] Siwen Li, Duane Outka, and Hong Shih, "Cracks in Anodized Aluminum Film", Lam Research Confidential Technical Reports, August 26, 2009, September 8, 2009, November 11, 2009, February 5, 2010, September 23, 2010, November 12, 2010, November 30 2010.

2

Corrosion Resistance of High Nitrogen Steels

Roman Ritzenhoff and André Hahn
Energietechnik-Essen GmbH
Germany

1. Introduction

1.1 Some basics about High Nitrogen Steels (HNS)

1.1.1 Nitrogen in steel

Nitrogen as an alloying element has been known and used in technical applications since the 1940s, initially under the premise for nickel substitution in stainless grades.

Nitrogen in low alloy steels is undesirable due to the formation of brittle nitrides. However, the use of nitrogen in high alloy steels has an array of advantages that makes it appear interesting as an alloying element. In references one find this sufficiently researched, so that in this situation only the most important points need to be summarized [Dailly & Hendry, 1998], [Energietechnik-Essen [ETE], 2011], [Allianz Industrie Forschung [AIF], 2003]:

- significant increase of strength without restricting ductility
- Improvement of corrosion resistance
- Increasing the high temperature tensile strength
- Extended / stabilized austenite form
- no formation of tension induced martensite with high cold working rates
- Inhibits the discharge of inter-metallic phases

These as HNS-Alloy (**H**igh **N**itrogen **S**teels) specific material group are characterised through an interesting material profile, i.e. a combination of strength and corrosion resistance.

A state-of-the-art production routine is P-ESR melting (pressurised electro slag remelting) which will be covered within the following section. Austenitic steels as well as martensitic steels can be manufactured and are suitable for forging and hot rolling. Some basic knowledge about the material characteristics is mandatory to avoid any potential issues at plastic deformation, heat treatment and to maintain the excellent corrosion resistance.

1.2 About this book chapter

This chapter will not cover the fundamentals of corrosion – we assume that the reader will have a basic knowledge about the principles of corrosion. This paper will provide an overview about the role of nitrogen related to corrosion of stainless steels. Since a lot of results have been published to date, common knowledge will summarized and topped up with own data and experimental results.

Finally, some typical HNS grades will be discussed with regards to their industrial application.

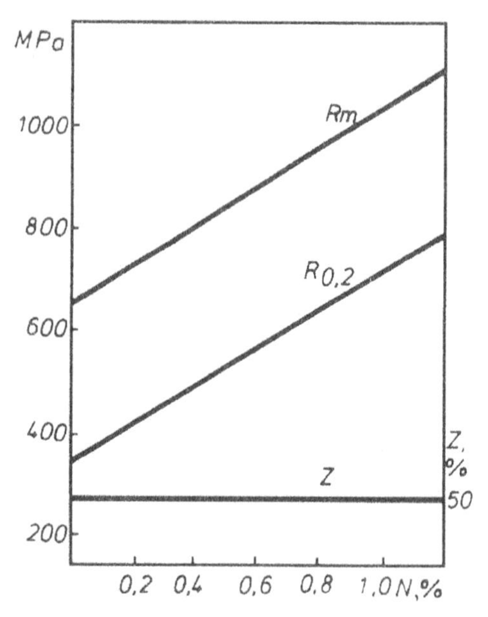

Fig. 1. Mechanical properties in dependency of different nitrogen concentrations after quenching at 1150°C, 2h in water. [Rashev et al.,2003]

2. Melting of HNS alloys

Melting of HNS alloys requires special process techniques as the nitrogen content is above the solubility limit at atmospheric pressure. The equipment development has started in the 1960´s with pressurized induction furnaces (lab scale) and has finally led to the first PESR unit in 1980. Today, PESR is state of the art due to its high process capability, good productivity, large ingot sizes and a save H&S environment [Holzgruber, 1988].

2.1 The PESR-process

Today's biggest PESR unit is located at Energietechnik Essen GmbH, Germany. Its operating pressure is max. 40 bars and can achieve ingot weights up to 20 tons and 1030 mm diameter. The functional principle is shown schematically in fig.2.

Basically, the PESR process is a conventional remelting facility that works in a pressure tank. The process is designed to meet both, an ESR refining and nitrogen pick up. The metallurgical approach is similar to a standard ESR-process, i.e. refining, low segregation, no porosity or shrinkage, defined microstructure and solidification.

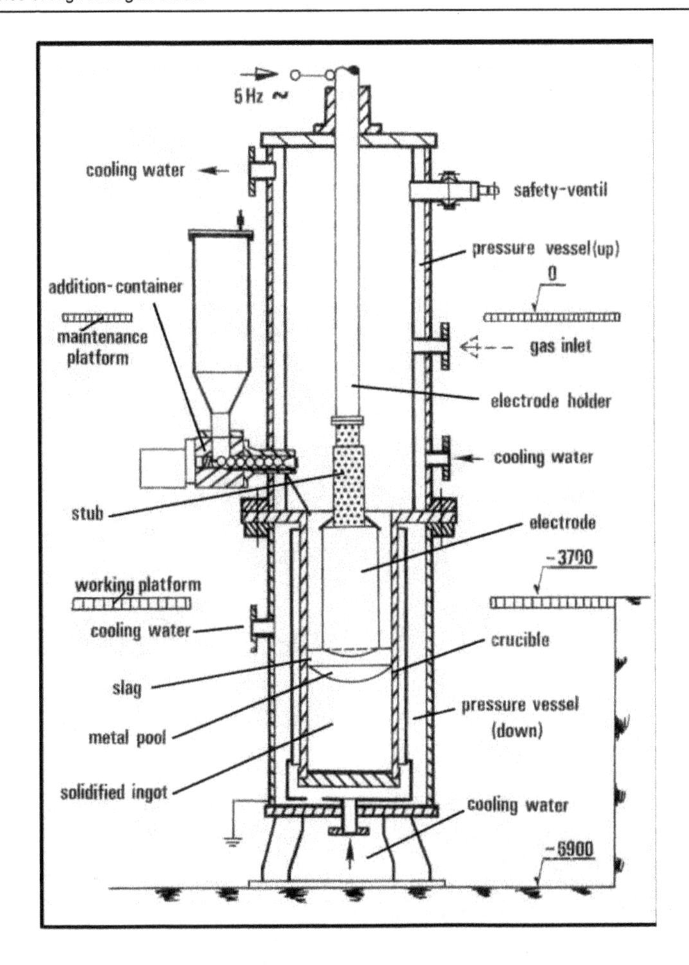

Fig. 2. Schematic design of a pressure electro slag remelting furnace (PESR).

The physical fundamentals of nitrogen pick up are specified over the Sievert square root law, accordingly that the nitrogen solubility is a function of pressure and temperature:

$$[\%N] = k \cdot \sqrt{P_{N2}} \tag{1}$$

With p_{N2}: Nitrogen partial pressure over melt in bar, k: Material constant (temperature and alloy dependent)

In real systems, the actual solubility is additionally codetermined through the alloy composition. Thermodynamic activities are used to describe the effect of the individual elements.

$$[\%N]_{Fe-X} = \frac{[\%N]_{Fe}}{f_N^X} \cdot \sqrt{p_{N2}} \tag{2}$$

With $[\%N]_{Fe-X}$: Nitrogen solubility in multi-component systems, $[\%N]_{Fe}$= 0,044% (equilibrium constant in pure Fe at 1600 °C and 1 bar)

The activity coefficient f is thereby defined as

$$\log f_N^X = e_N^X[\%X] \tag{3}$$

With e_N^X : interaction coefficient, [%X]: Concentration of the elements X in %

Fig. 3. View of the industrialized PESR -process at Energietechnik Essen GmbH for ingot weight up to 20 t und ∅1030 mm.

It is obvious as per table 1 that specific elements will increase the nitrogen solubility (e.g. manganese), while others will reduce the solubility (e.g. silicon). This has not only an impact on the nitrogen pick up at remelting but also on the precipitation of inter-metallic phases in the solid state.

Element	Coefficient e_N	
C	+ 0.125	
Si	+ 0.065	Reduction of N-Solubitly
Ni	+ 0.01	
W	- 0.0015	
Mo	- 0.01	
Mn	- 0.02	
Cr	- 0.045	Increase of N-Solubility
V	- 0.11	
Nb	- 0.06	
Ti	- 0.053	

Table 1. Activity coefficients of several elements with effect on the nitrogen solubility in steel at 1 bar.

Due to the general alloy composition, the nitrogen solubility is accordingly larger in austenitic as in ferrite or martensitic steels.

The nitrogen pick up can occur through the gas phase and also as well from a solid nitrogen carrier. The choice of a solid body nitrogen pick up medium is down to the following boundary conditions:

- Nitrogen partial pressure: high enough to allow a dissociation at ~ 40 bar
- Characteristics of the slag or flux are not allowed to change (e.g. electrical conductivity, metallurgical properties, etc)

In practice the standard Si_3N_4 is used, in exceptions CrN as well. A transfer of silicium respectively chromium in this case must be taken into consideration. The following table 2 provides a comparison about advantages and disadvantages of Si_3N_4 and gaseous nitrogen.

	Advantage	Disadvantage
Si_3N_4	• Nontoxic • Ease of operation and storage • Ease of dissociation	• Very abrasive (joints and gaskets, valves) • Kinetic of dissociation of N^{3-} -ion must be regarded. • Non-continiuous allowance on slag • Silicium transfer in melt
N_2 - Gas	• Continuous allowance possible • Simple regulation over the pressure • High equal distribution in ingot • Appropriate for Si-critical steel grades	• Slag composition very important • Sievert´sches law at high pressure not ideally achieved. • Diffusions conditions in system slag-metall must be known.

Table 2. Advantages and disadvantages different nidriding mediums.

The selection of the slag takes place after metallurgical consideration and depends on the alloy. Above all, the slag composition has importance for the nitrogen pick up of the steel.

3. Microstructrual characteristics of HNS

Nitrogen stabilizes the γ-area in a very clear way. It is undesired in low alloyed steels due to the formation of brittle phases; however it is very beneficial in terms of strengthening and corrosion resistance for high alloyed steels. It is considered to be the most efficient solid solution strengthening element. [Pickering, 1988] has reported that nitrogen is approximately twice as efficient as carbon.

[Bernauer & Speidel, 2003] et al. has published results, whereas nitrogen can improve the γ-stability by adjusting a carbon/nitrogen ratio. The high interstitial steels, i.e. containing carbon and nitrogen show a higher thermodynamic stability compared to common Cr-Mn-N-steels.

[Pickering, 1988] has investigated the influence of nitrogen within various steelbased alloys with regards to their microstructure. Therefore, nitrogen has a higher solubility in the lattice than carbon. Its presence is related with the formation of nitrides (or carbon nitrides). These nitrides tend to precipitate as small particles and - this is particular of interest for any hot forming applications - grow significantly slower than carbides. It is obvious that this will have an impact on recovery, grain growth and heat resistance. The nitrides are thermodynamically more stable than the corresponding carbides, i.e. have a lower solubility.

At first sight, the considered HNS- alloys are not significantly distinguishable than the nitrogen free varieties. The following fig. 4 & 5 show exemplary micrographs of a Mn-austenite with approx 0.65% nitrogen as well martensite with approx. 0.4% nitrogen.

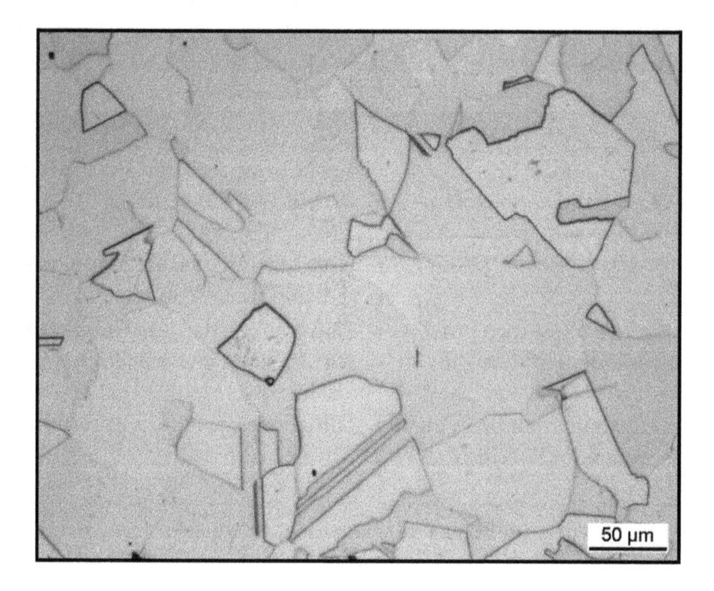

Fig. 4. Microstructure of a unformed and solutions annealed austenite with ~ 0,6 % N.

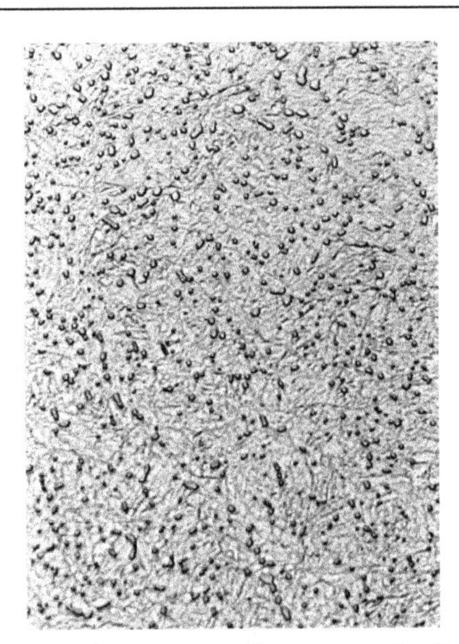

Fig. 5. Tempered microstructure of a nitrogen alloy martensite (1.4108). (M 1000:1)

However - and this is a difference to the conventional nitrogen free alloy variations - one should consider that HNS-alloys have a specific precipitation behavior. This must be kept in mind so that potential difficulties at hot forming at heat treatment can be avoided. Additionally, any precipitation will affect the corrosion resistance so a good understanding of the alloy is mandatory to maintain the alloy characteristic.

3.1 Atomic structure of nitrogen alloyed steels

Much effort has been put into place to understand the beneficial effect of nitrogen in stainless steels over the past years. A major step was the calculation of the atomic structure within the d-band of Fe-C and Fe-N carried out by [Rawers, 2003], [Gravriljuk & Berns, 1999] and [Mudali & Raj, 2004]. Therefore, nitrogen increases the state density on the Fermi surface whereas carbon leads to a decrease of state density. Consequently, a higher concentration of free electrons can be found in austenitic nitrogen alloyed steels - this result in a metallic character of interatomic bonds. This also explains the high ductility in HNS, even at high strengthening. Contrary, interatomic bonds in carbon austenites show a covalent characteristic. This is due to the localization of electrons at the atomic sites [Rawers, 2003]. The preference for different atoms to be nearest neighbors is defined as short range order and is mainly driven by the degree of metallic character of an intermetallic bond. A metallic interatomic bond supports a homogenous distribution as single interstitials, whereas a covalent bond results in clustering of atoms. These clusters can then potentially precipitate secondary phases such as carbides, nitrides etc. A cluster is to be realized as local accumulation of approx. 100 atoms [Berns, 2000]. The high thermodynamic stability of nitrogen stabilized austenites can also be led back on the hindered clustering of atoms [Rawers, 2003]. In summary, the electron configuration is therefore the main driver for an increased corrosion resistance. Due to nitrogen, the allocation of Cr-atoms within the lattice is homogenous so that Cr- clustering and formation of $M_{23}C_6$-carbides is reduced. Since nitrogen delays the precipitation of carbides as seen in fig. 6 & 7 , the likeliness of a local Cr depletion is limited.

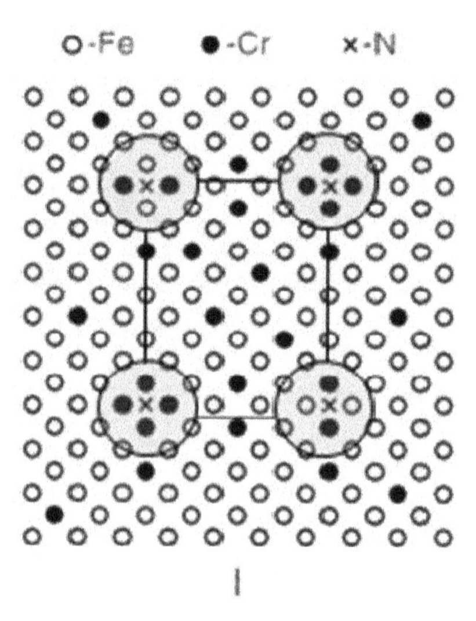

Fig. 6. Schematic of a short range order. Nitrogen increases the concentration on free electrons. Thereby forming a non-directional bonding and an equal distribution of the atom in crystal lattice. [Berns, 2000]

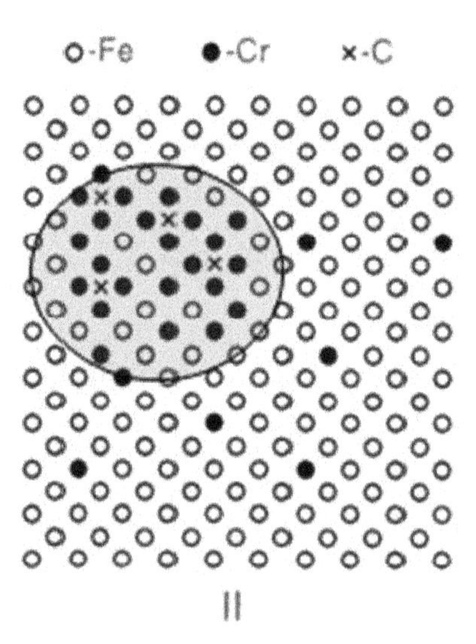

Fig. 7. Schematic of a cluster formation. Carbon decreases the concentration of free electrons. Thereby forming a directional bonding of non equal distribution of the atom in crystal lattice. [Berns, 2000]

3.2 Carbides

In dependency to the carbon content and the tempering time, austentic steels tend to precipitate $M_{23}C_6$-Carbides at the grain boundaries. Through this the ductility and corrosion resistance of the material significantly declines. However, the strength properties have no mentionable change. The susceptibility for intercrystalline corrosion clearly increases.

The precipitation behavior of this carbide can only be prevented through a quick quench in the critical temperature range. Fig.8 shows the location of the precipitation with relation to the alloy composition. Fine carbides are beneficial with regards to the corrosion resistance as the local chrome depletion is less in comparison to coarser carbides. The Cr depletion can be balanced out through an extended homogenisation (i.e. holding time) within the precipitation area.

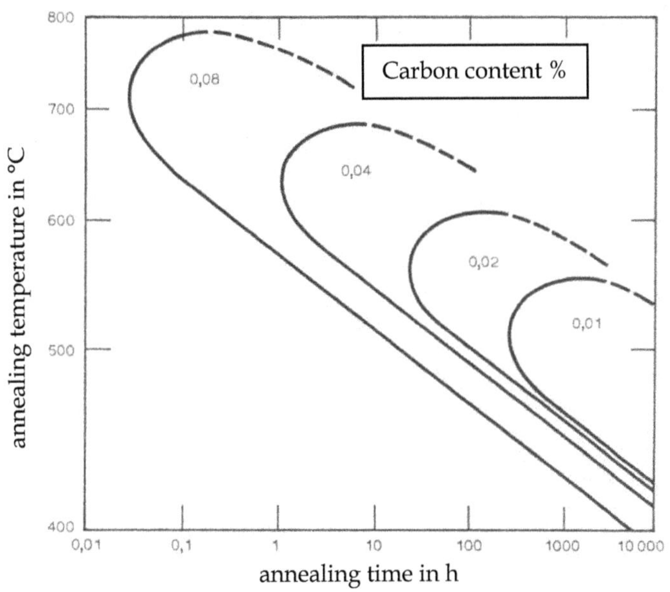

Fig. 8. Influence of the carbon content in location of grain decay in unstable austentic steels with circa 18 % Cr und 8 % Ni. Examination in Strauß-Test. [Thyssen, 1989]

For 12% Cr-steels, containing nitrogen it has been observed by [Pickering, 1988] that nitrogen lowers the martensite start temperature M_S; 1% nitrogen lowers M_S by 450 °C.

[Pickering, 1988] has investigated the influence of nitrogen on the carbide morphologies. The main type is as previously described the $M_{23}C_6$ type. In Nb-containing alloys, M_4X_3 have been observed where nitrogen can occupy interstitial dislocations. It also can be solutioned within M_6C.

HNS martensitic steels are also characterised with good high temperature strength and show an according hot forming behavior. Under circumstances these steels are found in thermo mechanical forging and rolling applications. The last forming step will effectively increase the dislocation density so that adequate nucleation for a desired precipitation exists. For example, the precipitation of carbides, nitrides as well as carbon nitride could be finely distributed. This could be of interest to the high temperature strength. The previously mentioned effects of fine carbides concerning the dissolution and corrosion resistance are also valid here.

3.3 Nitride and nitrogen perlite

In the case of the austenitic steels it should be considered, that in the temperature range of approx. 500- 900 °C and in connection with the alloy composition a precipitation of nitrides (Type Cr_2N) occurs. This nitrogen perlite identified microstructure raises significantly the susceptibility to cracking of the steel but can also support intergranular cracking. Depending on the alloy composition, the precipitation window for nitrogen perlite or other nitrides are adjusted to higher or lower temperatures. The figures 9 & 10 show exemplary an austenitic structure with beginning and advance nitrogen perlite precipitation. Clear to recognize at what speed that the precipitation occurs.

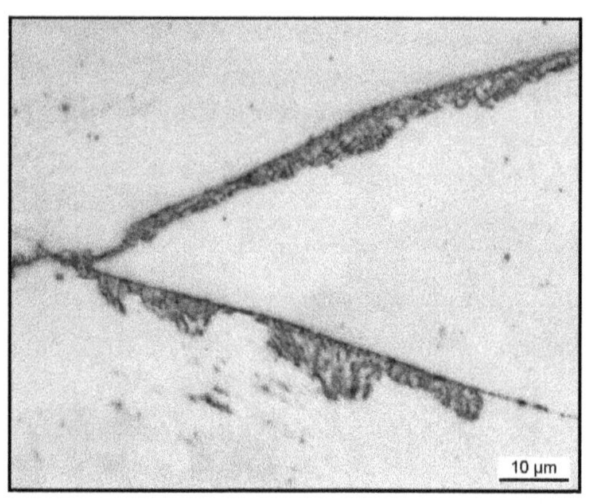

Fig. 9. Beginning of precipitation of nitrogen perlit cold worked austenitic structure 1.3816. 800 °C/15 min.

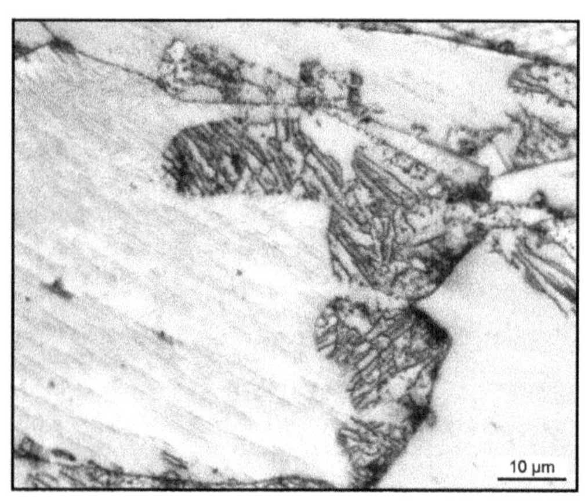

Fig. 10. Advanced precipitation of nitrogen perlit cold worked austenitic structure 1.3816. 800 °C/30 min.

For prevention of such brittle phases the precipitation area of the hot forming must be followed through fairly quick. The nitrogen level has obviously an impact on the precipitation kinetics of Cr_2N, see figure 11 for details. Best corrosion resistance can be achieved if all nitrogen is in solid solution, i.e. no nitrides are precipitated.

As seen in fig. 11 the precipitation depends on both, the alloy composition and holding time. The Cr_2N-formation has been reported by [Pickering, 1988] to be a major issue to high Cr and/or high Ni-alloys

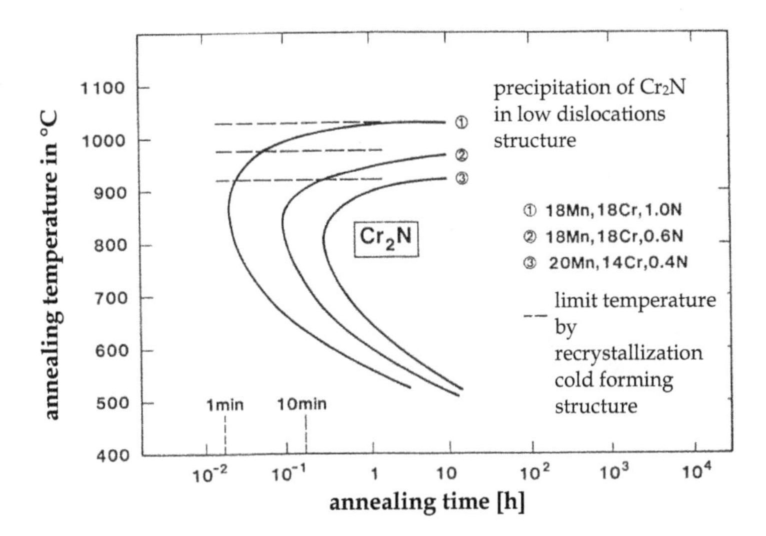

Fig. 11. Precipitation of Cr_2N in 18Mn18Cr at various nitrogen levels. [Uggowitzer, 1991]

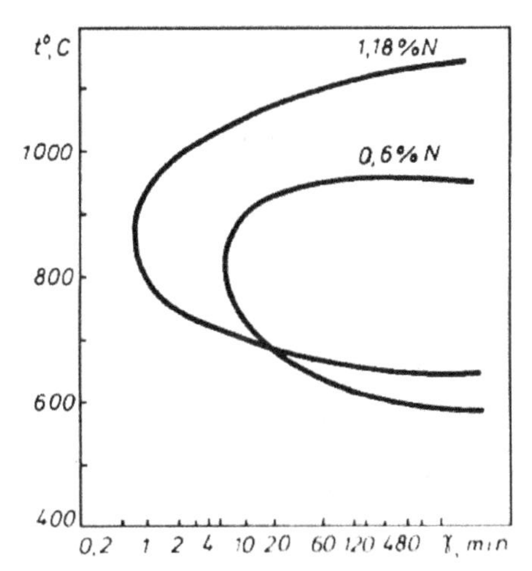

Fig. 12. TTT-diagram for the beginning of Cr_2N-precipitation at different nitrogen contents. Curve is based on X5CrMnN18-12. [Rashev et al., 2003]

Other intermetallic phases, such as Laves-phase, Z-phase and χ-phase have been investigated with regards to nitrogen alloying within literature. Generally, nitrogen seems to delay the formation of intermetallic phases. The underlying mechanism has been discussed controversially; the enhanced solubility of Chromium and Molybdenum due to nitrogen or its influence on Gibbs free energy for phase formation [Mudali & Rai, 2004]. The following table will provide an overview about the influence of nitrogen on some intermetallic phases [Mudali & Rai, 2004], [Heino et al., 1998].

Description	Influence of nitrogen...	Remark
σ	Suppresses formation of σ	
χ	Shifted to longer times Narrows temperature range for precipitation	M6C appears instead of χ
$M_{23}C_6$	Suppresses formation of $M_{23}C_6$	Can be replaced by M_6C with increasing nitrogen contents
Laves	Precipitation is shifted to higher temperature but accelerated	
R		See Laves

Table 3. Overview about the role of nitrogen alloying on the precipitation behavior of some intermetallic phases. [Mudali & Raj, 2004], [Heino et al., 1998]

4. Corrosion resistance of HNS

The role of nitrogen in stainless steel with regards to the corrosion resistance has been previously reported within literature [Pleva, 1991], [Truman, 1988], [Pedrazzoli &Speidel, 1991], [Dong et al., 2003], [Mudali & Rai, 2004]. This chapter will review today´s knowledge and present own data; corrosion fatigue and high temperature corrosion will not be covered.

4.1 General

It is well known that nitrogen in high alloyed steels improves the corrosion resistance; this is especially true for pitting and crevice corrosion. Additionally, nitrogen helps to prevent the alloy from stress corrosion cracking, though in an oblique way. Generally, the beneficial effect of nitrogen can be led back to an enrichment of nitrogen at the oxide/metal-interface and its influence on passivation. It has been reported that the enrichment increases with increasing potentials. However, these mechanisms have been discussed controversially [Mudali & Rai, 2004]:

- a formation of nitrides or mixed nitride layer, e.g. Ni_2Mo_3N
- enrichment of negatively charged N-ions, i.e. $N^{\delta-}$. These ions will lower the potential gradient of the passivation film and reject Cl^--ions
- formation of Cr_2N. A high local Cr-concentration should improve the corrosion resistance. However, this approach is unlikely as nitrogen does not change the matrix composition underneath the passivation film. No Cr depletion whatsoever has been reported.

Nitrogen does not show any influence on the thickness of the oxide layer, investigations of various alloys with different nitrogen contents have confirmed an average thickness of 12-22 Å [Mudali & Rai, 2004] within all varieties.

A synergism of nitrogen and molybdenum is suggested by many authors [Mudali & Rai, 2004], [Pickering, 1988], [Pedrazzoli & Speidel, 1991]. Molybdenum shifts the metal dissolution to higher potentials which will consequently lead to an increased enrichment of nitrogen at the metal/oxide-interface. In this case, nitrogen can lower the current density below the critical value for pitting corrosion [Mudali & Rai, 2004].

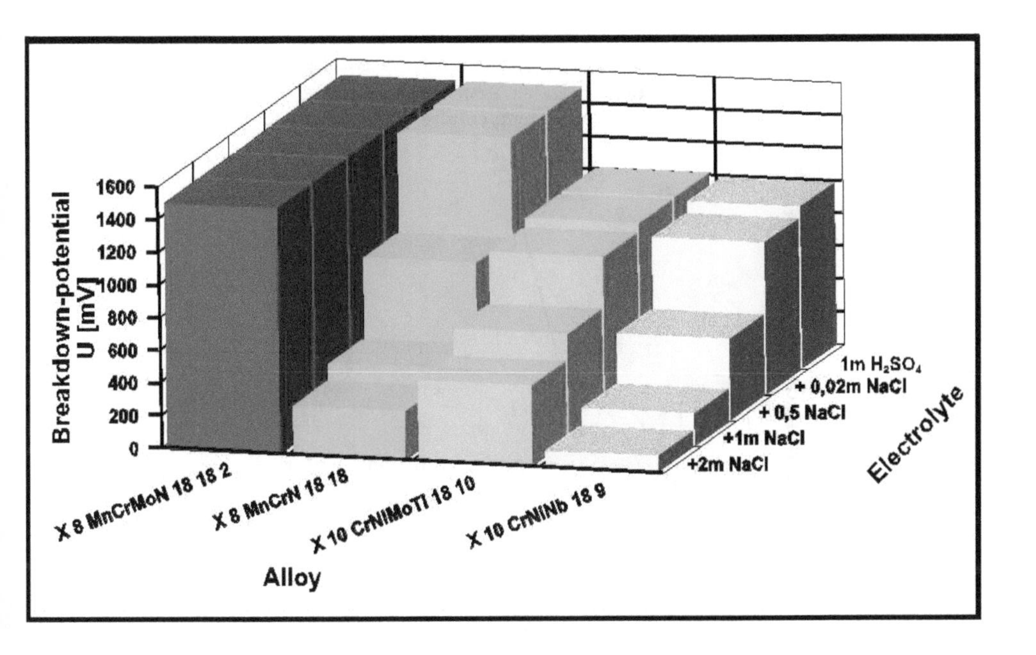

Fig. 13. Breakdown potential of HNS and commercial stainless steels in various electrolytes. [ETE-11]

It has been suggested that Molybdenum and nitrogen support the formation of highly mobile ions that interact with the passivation film. Addtionally, nitrogen seems to have a buffer effect by reacting as follows in oxidizing corrosive media:

$$[N]+4H^{+}+3e^{-}\rightarrow NH_{4}^{+} \quad (4)$$

The formation of NH_4^+ - ions helps to increase the pH value which results in an improved repassivation and reconditioning of the base material [Mudali & Rai, 2004]. It also has impact on the depronotation, which might explain the good performance in acids and halide containing liquids such as Cl^-, Br^- and I^- [Truman, 1988].

4.2 The role of nitrogen on pitting and crevice corrosion

Pitting corrosion is a very serious and harmful type of corrosion and is classified as local corrosion, characterized by small holes or pits. Usually, a repassivation cannot be achieved so that these pits can initiate cracks. This is the main reason why pitting often comes along with stress corrosion cracking. Pitting can be determined either by current-density-curves or a critical pitting temperature. It has been reported that nitrogen lowers the passivity in the current-density diagram. In austenitic steels, 1 % nitrogen improves the pitting potential by 600 mV [Pedrazzoli & Speidel, 1991]. Crevice corrosion follows generally the same principles; however the conditions are significantly tougher due to the geometric impact (electrolyte concentration in crevice). This will be covered at a later stage.

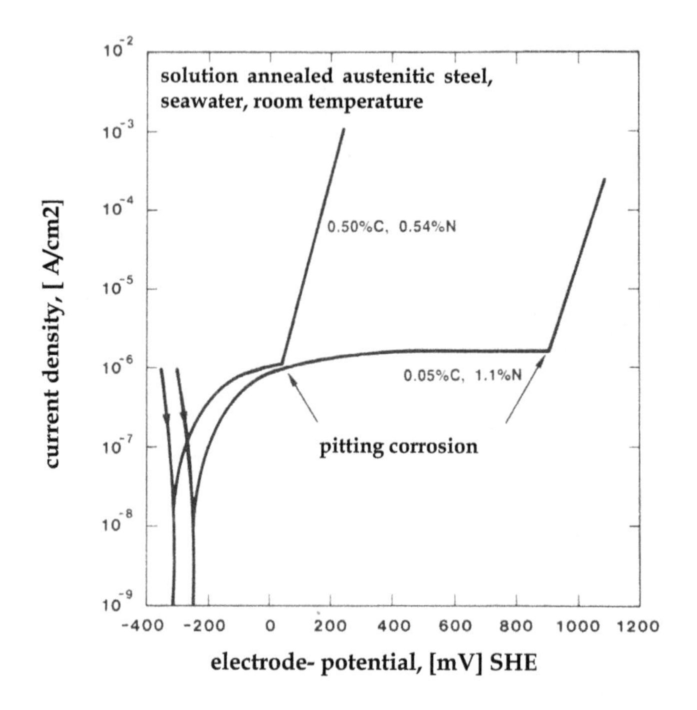

Fig. 14. Influence of nitrogen on Current-density depending of a given alloy. [Pedrazzoli & Speidel, 1991]

Pitting corrosion is a well-known corrosion problem for stainless steels. It can come along with sensitivation, i.e. a local Cr depletion can support pitting corrosion. Therefore, any segregation, welding joint, heat treatment etc. can have an impact on pitting corrosion.

The critical pitting temperature (CPT) is defined at what temperature pitting occurs. A common range for stainless steels is 10-100 °C and obviously depends on the alloy composition. [Pedrazzoli & Speidel, 1991] has reported that the critical temperature for crevice corrosion (CCT) is approx. 20 ° lower compared to pitting, see fig.16 & 17 for details.

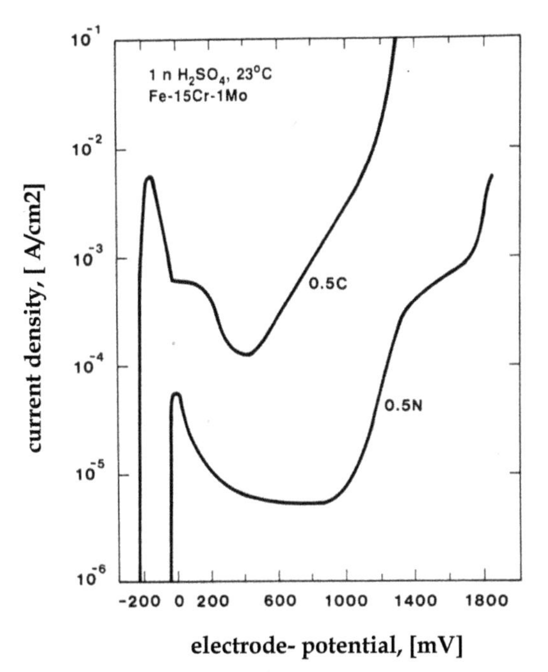

Fig. 15. Comparison of nitrogen and carbon on Current-density depending of a given alloy. [Pedrazzoli & Speidel, 1991]

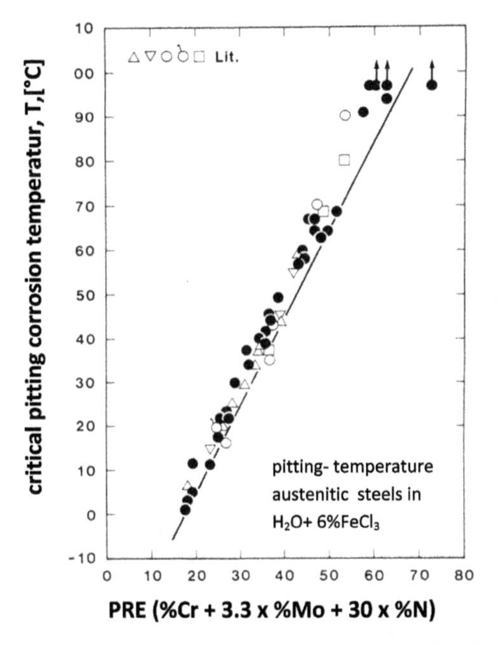

Fig. 16. Critical Pitting Temperature (CPT) as a function of PRE. [Pedrazzoli & Speidel, 1991]

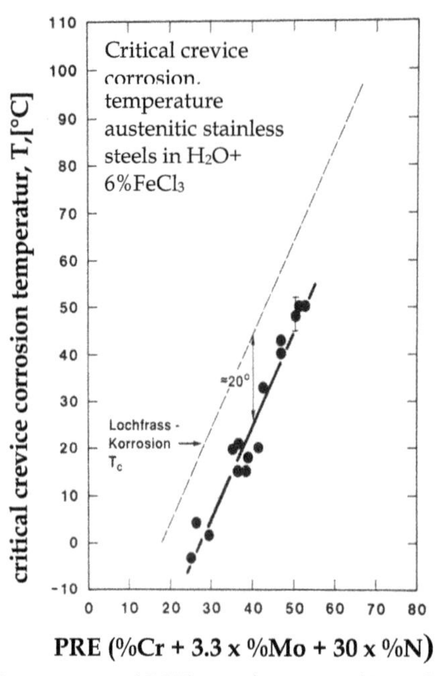

Fig. 17. Critical Crevice Temperature (CCT) as a function of PRE. [Pedrazzoli & Speidel, 1991]

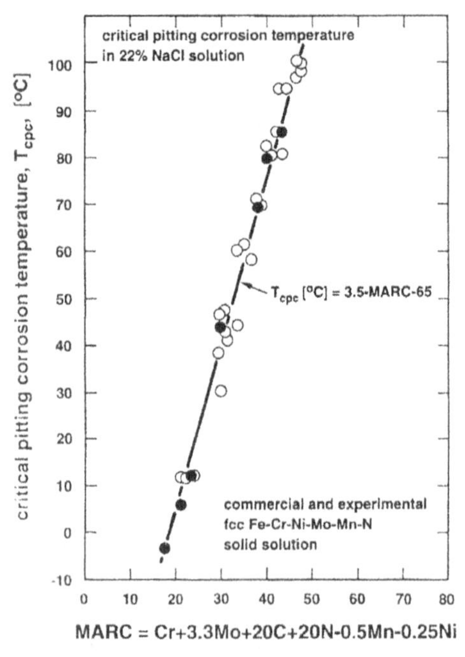

Fig. 18. MARC equation to rank various alloying element in regards to the alloy pitting resistance. [Speidel & Theng-Cui, 2003]

A commonly accepted ranking of alloys in terms of their pitting addiction is the Pitting resistance equivalent (PRE). It is defined as

$$PRE = Cr\,(\%) + 3{,}3\,Mo\,(\%) + x\,N\,(\%) \text{ whereas } x = 13 \ldots 30 \tag{5}$$

It has been suggested by [Pleva, 1991] to use x = 16 for steels containing Mo < 4,5 % and x = 30 for steels containing Mo 4,5 – 7,0 %.

The PRE does not take any other elements but Cr, Mo and N into account. [Speidel & Theng-Cui, 2003] has suggested a new figure to include also C, Mn and Ni into the equation and has defined MARC (**M**easure of **a**lloying for **r**esistance to **c**orrosion):

$$MARC = Cr\,(\%) + 3{,}3\,Mo\,(\%) + 20\,C\,(\%) + 20\,N\,(\%) - 0{,}5\,Mn\,(\%) - 0{,}25\,Ni\,(\%) \tag{6}$$

The MARC-equation is the first formula that considers carbon to be beneficial against pitting. [Bernauer & Speidel, 2003] has suggested a high carbon + nitrogen alloyed steel with improved pitting resistance. This is due to the higher thermodynamic stability of Cr-Mn-N-C systems compared to carbon-free Cr-Mn-N steels. However, both carbon and nitrogen must not form any precipitations but stay into solid solution.

A very global description of the influence of alloying elements on pitting potential was published by [Pedrazzoli & Speidel, 1991]. As seen in fig. 19, nitrogen and molybdenum have a significant impact on the potential shift.

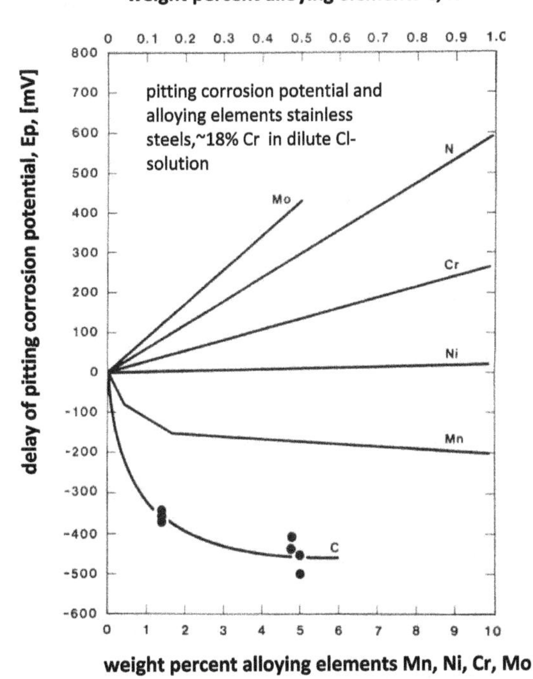

Fig. 19. Influence of various alloying elements on the pitting potential. [Pedrazzoli & Speidel, 1991]

Cold forming is supposed to have an impact on corrosion resistance; however the role of nitrogen in this case is not fully clear. Within Literature, the following has been reported [Mudali & Rai, 2004]:

- cold forming in stainless steels: no significant influence on pitting potential
- cold forming in nitrogen alloyed steels: a cold forming degree up to 20 % improves the critical pitting potential (CPP). A drop in CPP at higher deformation rates has been reported.

The improved pitting potential at low deformation rates i.e. below 20 % is due to the decreased tendency for twin formation. At higher deformation rates, deformation bands will appear which will be influenced by nitrogen (width and dislocation configuration). [Pleva, 1991] reports that the degree of cold working does not show any influence on the pitting corrosion. This has been also confirmed by own data on X8CrMnN 18-18 material, see fig.20 & 21 [ETE-11].

Various investigations have tried to explain the mechanism of nitrogen in terms of pitting. It has been agreed, that nitrogen stabilizes the γ-range in a very clear way. This is important to prevent δ-ferrite, esp. in Mo containing alloys. It also supports a homogenous, single-phase microstructure and avoids carbides to precipitate.

4.3 Ammonium theory

Nitrogen and Molybdenum obviously show a synergism in regards to pitting. Molybdenum seems to support the Cr_2O_3-formation by acting as an electron acceptor. This also leads of a depronotation of hydroxides. In addition, nitrogen reacts as follows:

$$[N]+4H^{+}+3e^{-}\rightarrow NH_4^{+} \tag{7}$$

The NH_4^+ - formation will increase the pH-value which support the repassivation. [Mudali & Rai, 2004] reports that NH_4^+ - ions have been confirmed by XPS within the passivation layer.

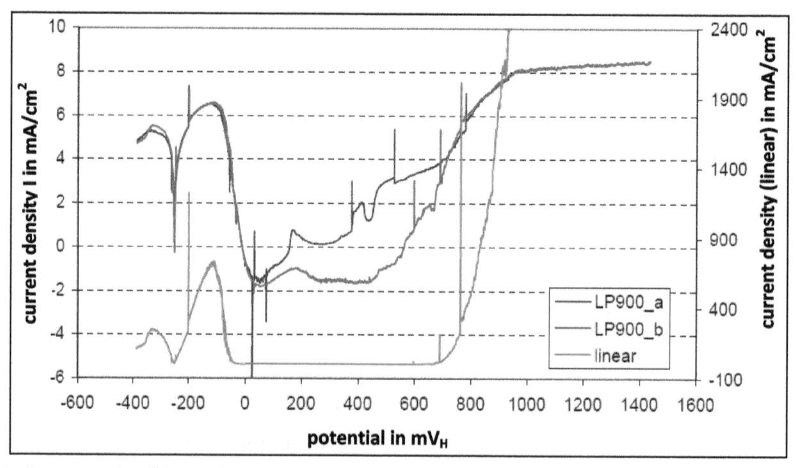

Fig. 20. Influence of cold working on current-density of X8CrMnN 18-18 (0 % cold work). UR: -250 / UL: 722 mV / ΔU: -972 mV / 1m H_2SO_4+ 0,5m NaCl [ETE-11]

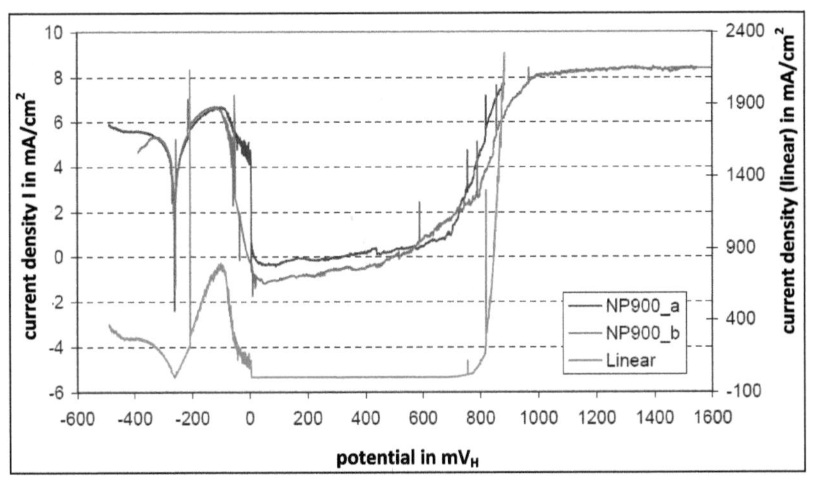

Fig. 21. Influence of cold working on current-density of X8CrMnN 18-18 (37 % cold work). UR: -250 / UL: 780 mV / ΔU: -1030 mV / 1m H_2SO_4+ 0,5m NaCl [ETE-11]

4.4 Surface enrichment theory

The surface enrichment theory is based on the general idea, that nitrogen is build into the lattice underneath the passive layer in solid solution. This nitrogen rich layer shall avoid dissolution of the substrate. Tentatively, there are chemical reactions with Cr and Mo who might change the local potentials as well. The formation of various N-rich phases has been reported, such as Cr_2N or Ni_2Mo_3N [Mudali & Rai, 2004], [Pickering, 1988]. Negatively charged N-ions, i.e. $N^{\delta-}$ are supposed to enrich at the metal/oxide interface. These ions will lower the potential gradient of the passivation film and reject Cl^- -ions.

4.5 Inhibitive nitrate formation theory

This theory covers the formation of pit growth inhibiting species. It is basically linked with the ammonia formation theory. The NH4+-formation in the pit tip appears to happen quicker than the OH—formation due to oxide reduction at the pit entrance.

$$[N]+4H^+ +3e^- \rightarrow NH_4^+ \tag{8}$$

The repassivation by NH4+ can be described as follows:

$$NH_4^+ + H_2O \rightarrow NH_4OH + H^+ \tag{9}$$

$$NH_4OH + H_2O \rightarrow NO_2^- + 7H^+ + 6e^- \tag{10}$$

$$NO_2^- + H_2O \rightarrow NO_3^- + 2H^+ + 2e^- \tag{11}$$

$$NH_4^+ + 2H_2O \rightarrow NO_2^- + 8H^+ + 6e^- \tag{12}$$

$$NH_4^+ + 3H_2O \rightarrow NO_3^- + 10H^+ + 8e^- \tag{13}$$

Crevice corrosion underlies basically the same principles than pitting; due to the geometry of the crevice the corrosion conditions are believed to be more challenging. [Pedrazzoli & Speidel, 1991] has reported that the critical temperature for crevice corrosion is approx. 20 ° lower compared to pitting.

4.6 The role of nitrogen on Intergranular corrosion (IGC)

IGC is mainly driven by the depletion of Cr at the grain boundaries and/or the precipitations of carbides, usually $M_{23}C_6$. Therefore, Carbon is supposed to be the main driver for IGC but also grain size, cold working and heat treatment have a significant influence on IGC.

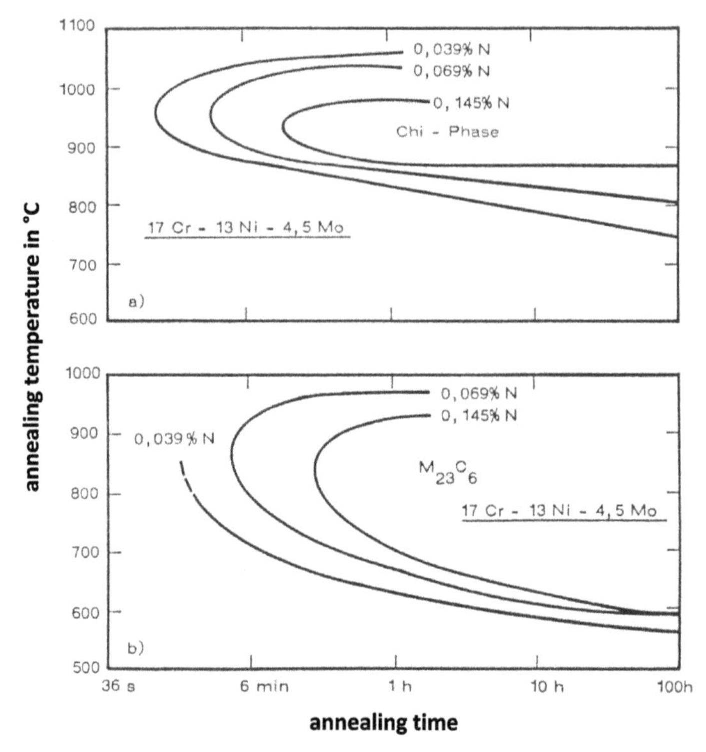

Fig. 22. Precipitation of M23C6 and χ-phase of a given alloy (17Cr-13Ni-4,5Mo) depending on annealing times and temperatures. [Gillessen et al., 1991]

As discussed in the previous chapter (microstructure), nitrogen tends to delay the $M_{23}C_6$ formation as it changes the Cr activity within the carbide. It also increases the passivity (i.e. lowering the current-density) and avoids the formation of α´- martensite at grain boundaries. However, this is only valid as long as nitrogen is in solid solution. It has been reported [Truman, 1988], [Dong et al., 2003] that an excess of nitrogen can lead to Cr_2N precipitations on the grain boundaries which can significantly decrease the intergranular corrosion resistance.

4.7 The role of nitrogen on Stress corrosion cracking (SCC)

It is commonly known that high alloyed steels are generally sensitive to SCC. The role of nitrogen against SCC has been discussed over the years but it appears that the positive effect of nitrogen is to be seen in a more oblique way. As previously discussed, nitrogen delays the carbide precipitation and avoids a local Cr depletion. Additionally, the crack growth velocity does not only depend on the actual Cr content, but also on the C content. The crack growth is much higher with increased C levels. It has been reported that nitrogen doesn´t have any influence on crack growth velocity at C > 0,5 % . The impact of carbon is therefore higher compared to that of nitrogen. [Pedrazzoli & Speidel, 1991]

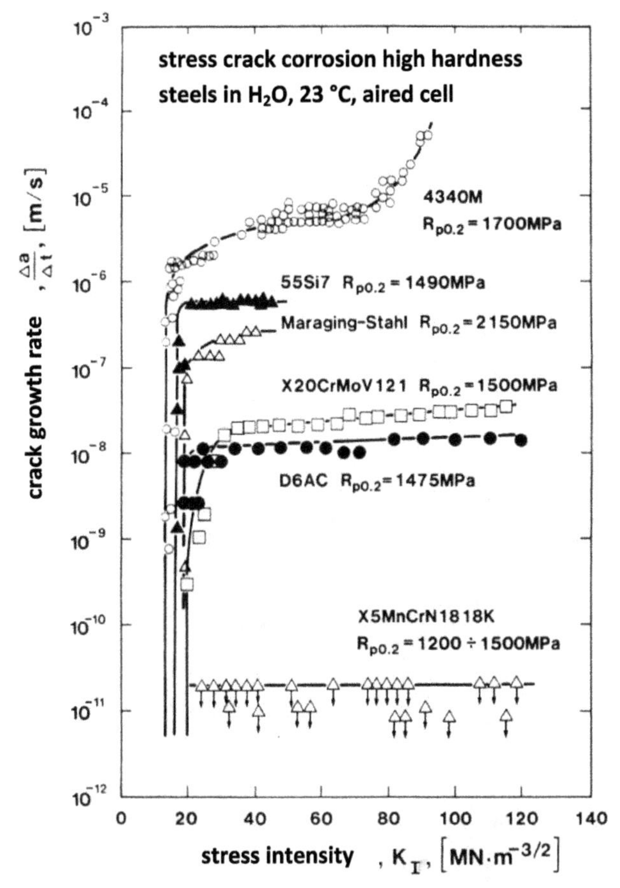

Fig. 23. Crack growth velocity for various alloys as a function of stress intensity. SCC for stainless steels, Water, 23 °C, ventilated. [Pedrazzoli & Speidel, 1991]

[Pickering, 1988] reports that the role of nitrogen is somehow inconsistent. In principle, nitrogen reduces the stacking fault energy which would be diametric to corrosion resistance. Low stacking fault energy in fcc-lattice is undesired in terms of SCC. Elevated nitrogen levels above 0,3 % are reported to be beneficial as the support of passivity is obviously higher than the influence on stacking fault energy. [Mudali & Raj, 2004] confirms that increased stacking fault energy does improve the SCC resistance. As nitrogen decreases the stacking fault energy it should detrimental to SCC. Carbon, due to the formation of wavy slip bands, should be theoretically beneficial. However, it appears that the role of nitrogen on SCC is fairly complex and depends on the alloy design and corrosion media.

The benefit of nitrogen alloying appears to be more oblique – the delay in $M_{23}C_6$ precipitation and improved pitting corrosion resistance has been recognized to be beneficial against SCC since pits are likely to initiate SCC.

5. Applications of HNS

High nitrogen steels are characterized through an interesting material profile that has led to a variety of demanding applications. The following will provide an overview about today´s use of HNS alloy, please note it reflects only pressurized alloys, i.e. PESR alloys. Any air melted nitrogen alloy steels will not be part of this chapter.

5.1 Martensitic steels

Since its development in the early 1990s, Cronidur 30 (X30CrMo 15-1 plus 0,4 % nitrogen) has been approved as high performace alloy for aerospace applications such as spindles, shafts and bearings. It is used as material for helicopter bearings, flap traps, fuel pumps etc. It excellent corrosion resistance combined with a high hardenability of 60 HRC makes it unique. The alloy has become an important material to industries as cutleries and knifes, general engineering, medicals (since it is free of nickel) and powder metallurgy.

Fig. 24. High precision bearing of Cronidur 30 [ETE-11].

Fig. 25. Extruder screw, Cronidur 30 [ETE-11].

A heat resistant alloy is based on X15CrMoV12-1 and contains 0,2 % nitrogen (trade name HNS 15). It precipitates fine V(C,N) and provides a good creep resistance and heat resistance up to 650 °C. Above this temperature the appearance of Laves-phase restricts its usage.

A Molybdenium-free version is known as HNS 28 and consists of X28Cr13 with 0,5 % nitrogen. Its purpose was the closed-die casting industry where a good polish and corrosion resistance is required.

5.2 Austenitic steels

A main driver for the development of HNS austenitic steel was the power generation in the 1980´s. A material for retaining rings was required that could resist the mechanical loads but also stress corrosion cracking. This has finally led to the introduction of X8CrMnN 18-18 also known as P900 or 18-18. This alloy combines superb mechanical properties, e.g. high ductility at elevated strength level with good corrosion resistance.

Further developments have also come up with Mn-stabilized austenites, e.g. X13CrMnMoN 18-14-3 (P2000). This alloy can achieve strength level (YS) beyond 2000 MPa, still with good ductility and corrosion resistance. One should consider heat treatment conditions and corresponding part dimensions with regards to the precipitation of Cr_2N.

Another market is powder metallurgy, i.e. thermal spraying and metal injection moulding (M.I.M.). The powders are gas atomized and very homogenous in terms of nitrogen and chemical composition. Main consumers are jewelry and general engineering.

Fig. 26. Generator shaft with retaining rings, X8CrMnN 18-18 [ETE-11].

Fig. 27. Retaining ring. X8CrMnN 18-18 [ETE-11].

6. References

AIF (2003) AIF-Abschlußbericht „ Untersuchungen zur wirtschaftlichen Warmumformung neuer hoch stickstofflegierter nichtrostender Stähle in Abhängigkeit vom Stickstoffgehalt, des Oberflächenzustandes und der Ofenatmosphäre", AiF-Vorhaben Nr. 13888N/II, 01.09.2003 bis 31.08.2007

Berns, H (2000). Stickstoffmartensit, Grundlage und Anwendung, HTM Härtereitechnische Mitteilungen, Ausgabe 1/2000, Bd. 55, Hansa Verlag, p.10

Bernauer, J., Speidel, M.O (2003). Effects of carbon in high-nitrogen corrosion resistant austenitic steels, High Nitrogen Steels 2003 Conference proceedings, Vdf Hochschulverlag AG ETH Zürich, Switzerland, 2003, pp. 159-168,

Dailly, R.; Hendry, A (1998). The Effect of Nitrogen on the mechanical behavior of cold-worked austenitic stainless steel rod, High Nitrogen Steels 1998 Conference proceedings, TransTechPublications Ltd, Switzerland, 1998, pp.427-435

Dong, H., Lin, Q., Rong, F., Su, J., Xin, C., Lang, Y., Kang, X (1998). Development and applications of nitrogen alloyed stainless steels in China, High Nitrogen Steels 2003 Conference proceedings, Vdf Hochschulverlag AG ETH Zürich, Switzerland, 2003, pp. 53-61

ETE (2011). Energietechnik Essen GmbH, internes Firmenarchiv, 2010

Gillessen, C., Heimann, W., Ladwein, T (1991). Entwicklung, Eigenschaften und Anwendung von konventionell erzeugten hochstickstoffhaltigen austenitischen Stählen, Ergebnisse der Werkstoffforschung, Verlag Thubal-Kain, Schweiz, Zürich, 1991, pp.167-187

Gavriljuk, V., Berns, H (1999). High nitrogen steels, Springer Berlin, 1. Auflage, 1999

Heino, S, Knutson-Wedel, M., Karlsson, B (1998). Precipitation in a high nitrogen superaustenitic stainless steel, High Nitrogen Steels 1998 Conference proceedings, TransTechPublications Ltd, Switzerland, 1998, pp. 143-148

Holzgruber, W (1988). Process technology for high nitrogen steels, High Nitrogen Steels 1988 Conference proceedings, The Institute of Metals, London, Brookfield, 1989, pp. 39-48

Mudali, U.K., Raj, B (2004). High nitrogen steels and stainless steels - manufacturing, properties and applications, New Delhi, Chennai, Mumbai, Kolkata, Narosa Publishing House, ASM International, 2004

Pedrazzoli, R., Speidel, M.O (1991). Korrosion und Spannungsrisskorrosion von stickstoffhaltigen Stählen, Ergebnisse der Werkstoffforschung, Verlag Thubal-Kain, Schweiz, Zürich, 1991, pp.103-121

Pickering, F.B. (1998). Some beneficial effects of nitrogen in steel, High Nitrogen Steels 1988 Conference proceedings, The Institute of Metals, London, Brookfield, 1989, pp.10-31

Pleva, J (1991). Korrosionsfeste stickstofflegierte Stähle - Eigenschaften und Erfahrungen, Ergebnisse der Werkstoffforschung, Verlag Thubal-Kain, Schweiz, Zürich, 1991, pp. 153-165

Rashev, T., Andreev, C., Manchev, M., Nenova, L. (2003). Creation and development of new high nitrogen steels in the Institute of Metal Science at Bulgarian Academy of Science, High Nitrogen Steels 2003 Conference proceedings, Vdf Hochschulverlag AG ETH Zürich, Switzerland, 2003, pp.241-257

Rawers, J (2003). Preliminary study into the stability of interstitial nitrogen and carbon in steels, High Nitrogen Steels 2003 Conference proceedings, Vdf Hochschulverlag AG ETH Zürich, Switzerland, 2003, pp. 273-280

Speidel, M.O., Theng-Cui, M (2003). High-Nitrogen austenitic stainless steels, High Nitrogen Steels 2003 Conference proceedings, Vdf Hochschulverlag AG ETH Zürich, Switzerland, 2003, pp.63-73

Thyssen Technische Berichte (1989). N.N.; Thyssen Technische Berichte, 15. Band 1989, Heft 1

Truman, J.E (1988). Effects of nitrogen alloying on corrosion behaviour of high alloy steels, High Nitrogen Steels 1988 Conference proceedings, The Institute of Metals, London, Brookfield, 1989, pp. 225-239

Uggowitzer, P. (1991). Uggowitzer, P.; Ultrahochfeste austenitische Stähle, Ergebnisse der Werkstoffforschung, Verlag Thubal-Kain, Schweiz, Zürich, 1991, pp. 87-101

3

Corrosion Resistance of Directionally Solidified Casting Zinc-Aluminum Matrix

Alicia Esther Ares[1,2,3], Liliana Mabel Gassa[1,4]
and Claudia Marcela Mendez[3]
[1] *Researcher of CIC, CONICET,*
[2]*Materials, Modeling and Metrology Program,*
Faculty of Sciences, National University of Misiones, Posadas,
[3]*Materials Laboratory, Faculty of Sciences, National University of Misiones, Posadas,*
[4]*INIFTA, National University of La Plata, Faculty of Exact Sciences, La Plata,*
Argentina

1. Introduction

Generally solidification leads to two types of grain morphologies: columnar and equiaxed. The origin of each one has been the subject of numerous theoretical and experimental researches in the field of metallurgy for many years. Columnar grains often grow from near the mold surface, where the thermal gradients are high, and the growth is preferentially oriented in a direction close to the heat flux. When the gradients are reduced near the center of the casting, equiaxed grains grow in all space directions leading to a material with more isotropic macroscopic mechanical properties and a more homogeneous composition field than with columnar structure. Depending on the application, one type of grain is preferred and thus favoured, e.g. equiaxed grains in car engines and columnar grains in turbine blades as reported by Reinhart et al, 2005 and McFadden et al., 2009.

Since the grain structure influences the properties of a casting, a great deal of effort has been devoted in the last decades to understand the mechanism behind the development of the macrostructure during solidification. Thus, equiaxed grains can nucleate and grow ahead of the columnar front causing an abrupt columnar to equiaxed transition (CET) whose prediction is of great interest for the evaluation and design of the mechanical properties of solidified products. As a consequence, it is critical for industrial applications to understand the physical mechanisms which control this transition during solidification (Spittle, 2006).

In order to realize the control of the columnar and equiaxed growth, it is necessary to understand the columnar to equiaxed transition (CET) mechanism during solidification, and make clear the CET transition condition. Fundamentally, it is necessary to have knowledge of the competition between nucleation and growth during solidification. Qualitatively, the CET occurs more easily when an alloy has a high solute concentration, low pouring temperature (for casting), low temperature gradient, high nucleation density in the melt and vigorous melt convection.

However, a quantitative understanding of the CET requires a thorough comprehension of all physical mechanisms involved.

In 1984, Hunt first developed an analytical model to describe steady-state columnar and equiaxed growth, and to qualitatively reveal the effects of alloy composition, nucleation density and cooling rate on the CET. On the other hand, he used a very simple empirical relationship to describe the variation of the undercooling with alloy composite and solidification rate. Cockcroft et al. (1994) used a more recent growth theory for the columnar and equiaxed growth but without considering high velocity non-equilibrium effects under rapid solidification. Recently, based on Hunt's CET model, Gäumann et al. (1997, 2001) developed a more comprehensive model by combining KGT model (Kurtz et. al., 1986) for directional solidification with LKT model (Lipton et. al., 1987) for the undercooling melt growth, with high velocity non-equilibrium effects to be taken into account. Gäumann et al. (2001) succeeded in applying their model to epitaxial laser metal forming of single crystal.

In previous research, the authors of this work carried out experiments in which the conditions of columnar to equiaxed transition (CET) in directional solidification of dendritic alloys were determined The alloy systems in this work include Pb-Sn (Ares & Schvezov, 2000), Al-Cu (Ares et. al., 2011), Al-Mg (Ares et. al., 2003), Al-Zn and Zn-Al alloys (Ares & Schvezov, 2007). These experiments permit to determine that the transition occurs gradually in a zone when the gradient in the liquid ahead of the columnar dendrites reaches critical and minimum values, being negative in most of the cases. The temperature gradients in the melt ahead of the columnar dendrites at the transition are in the range of -0.80 to 1.0 °C/cm for Pb–Sn, -11.41 to 2.80 °C/cm for Al–Cu, -4.20 to 0.67 °C/cm for Al–Si, -1.67 to 0.91 °C/cm for Al–Mg, -11.38 to 0.91 °C/cm for Al–Zn. Two interphases are defined; assumed to be macroscopically flat, which are the liquidus and solidus interphases. After the transition, the speed of the liquidus front accelerates much faster than the speed of the solidus front; with values of 0.004 to 0.01, 0.02 to 0.48, 0.12 to 0.89, 0.10 to 0.18 and 0.09 to 0.18 cm/s, respectively. Also, the average supercooling of 0.63 to 2.75 1C for Pb–Sn, 0.59 to 1.15 1C for Al–Cu, 0.67 to 1.25 1C for Al–Si, 0.69 to 1.15 1C for Al–Mg, 0.85 to 1.40 1C for the Al–Zn and was measured, which provides the driving force to surmount the energy barrier required to create a viable solid–liquid interface (Ares et al., 2005). A semi-empirical model to predict the columnar to equiaxed transition is developed based on experimental results obtained from measurements during solidification of lead–tin alloys directly upwards (Ares et al., 2002). The measurements include the solidification velocities of the liquidus and solidus fronts, and the temperature gradients along the sample in the three regions of liquid, mushy and solid. The experimental data was coupled with a numerical model for heat transfer. With the model, the predicted positions of the transition are in agreement with the experimental observations which show that the transition occurs when the temperature gradient reaches values below 1°C/cm and the velocity of the liquidus front increases to values around 0.01cm/s.

In addition, the thermal parameters, type of structure, grain size and dendritic spacing with the corrosion resistance of Zn-4wt%Al, Zn-16wt%Al and Zn-27wt%Al alloys were correlated (Ares et al., 2008). The polarization curves showed that the columnar structure is the most susceptible structure to corrosion, in the case of the alloy with only 4wt%of Al. The rest of the structures presented currents of peaks in the same order which were independent to the concentration of Al composition presenting in the alloy.

The biggest susceptibility to corrosion of the alloys with columnar structure can be observed by analyzing the values of Rct (charge transfer resistance) obtained using the electrochemical impedance spectroscopy (EIS) technique. In Zn-4wt%Al and Zn-27wt%Al, the corrosion susceptibility depends on the structure of the alloy. The alloy with 16wt%Al is less resistant to corrosion and their susceptibility to corrosion is independent of the structure. The alloy with 27wt%Al and the CET structure is the alloy which has the most corrosion resistance. When the critical temperature gradient becomes more negative, the Rct values increase. In the case of the correlation of Rct values and the structural parameters such as the grain sizes and secondary dendritic spacing, Rct values increase when the grain size and secondary dendritic spacing increase. But this does not happen for ZA16 alloy.

Composite materials obtained by solidification of alloys have made remarkable progress in their development and applications in automotive and aerospace industries in recent decades. Among them the most current applications are the zinc and aluminum base composite materials (Long et al., 1991; Rohatgi, 1991). It is well-known that the corrosion behavior of MMCs is based on many factors such as the composition of the alloy used, the type of reinforcement particles used, the reinforcement particle sizes and their distribution in the matrix, the technique used for the manufacture, and the nature of the interface between the matrix and reinforcement. A very slight change in any of these factors can seriously affect the corrosion behavior of the material.

In short, there is little research related to the study of mechanical and electrochemical properties of Zn-Al alloys as well as Zn-Al alloys MMCs containing SiC and Al_2O_3 particulations with different grain structures in the matrix. Also there is lack of fundamental study on the performance of Zn-Al alloys and their MMCs in corrosive environments when both solidification microstructure and type of particle distribution are in consideration. In the present research, Zn-Al-SiC and Zn-Al-Al_2O_3 composites are prepared and solidified by vertical directional solidification method. By means of voltammograms and electrochemical impedance spectroscopy, the corrosion resistances of Zn-Al matrix composite materials with different types of particles are obtained and analyzed and the results are compared.

2. Materials and methods

2.1 Alloys and metal matrix composites preparation

Zinc-Aluminum (ZA) alloys of different compositions were prepared from zinc (99.98 wt pct), aluminum (99.94 wt pct), and composites were prepared by adding SiC and Al_2O_3 particles to the alloys. The compositions of the alloys and composites prepared and directionally solidified are: Zn-27wt%Al, Zn-50wt%Al, Zn-27wt%Al + 8vol%SiC, Zn-27wt%Al + 15vol% SiC, Zn-50wt%Al + 8vol%SiC, Zn-50wt%Al + 15vol%SiC, Zn-27wt%Al + 8vol%Al_2O_3, Zn-27wt%Al + 15vol%Al_2O_3.

The chemical compositions of the commercially pure metals used to prepare the alloys are presented in Table 1. The molds were made from a 23 mm i.d. and 25 mm e.d. PYREX (Corning Glass Works, Corning, NY) tube, with a flat bottom, a cylindrical uniform section and a height of 200 mm. The sample was a cylinder 22 mm in diameter and 100 mm in height.

Chemical composition of Zn	
Element	Weight percent, wt%
Zn	99.98 ± 0.2
Fe	0.010 ± 0.01
Si	0.006 ± 0.0001
Pb	0.004 ± 0.001
Others	< 0.001 ± 0.0001
Chemical composition of Al	
Element	Weight percent, wt%
Al	99.94 ± 0.2
Fe	0.028 ± 0.0001
Si	0.033 ± 0.001
Pb	0.001 ± 0.0001
Others	< 0.001 ± 0.0001

Table 1. Chemical composition of the Zn and Al used to prepare the alloys.

2.2 Directional solidification

The alloy samples were melted and solidified directionally upwards in an experimental set-up described elsewhere (Ares et. al., 2007). It was designed in such a way that the heat was extracted only through the bottom promoting upward directional solidification to obtain the columnar-to-equiaxed transition (CET), see Figure 1 (a).

In order to reveal the macrostructure, after solidification the samples were cut in the axial direction, polished, and etched using concentrated hydrochloric acid for 3 seconds at room temperature for the zinc-aluminum alloys, followed by rinsing and wiping off the resulting black deposit. The microstructures were etched with a mixture containing chromic acid (50 g Cr_2O_3; 4 g Na_2SO_4 in 100 ml of water) for 10 seconds at room temperature (Vander Voort, 2007). Typical longitudinal macrostructure of different areas of the sample are shown in Figure 1 (b) to (d).

The position of the transition was located by visual observation and optical microscopy. The distance from the chill zone of the sample was measured with a ruler. It is noted in Figure 1 that the CET is not sharp, showing an area where some equiaxed grains co-exist with columnar grains. As was reported before, the size of the transition area is in the order of up to 10 mm (Ares et al., 2007, 2010). The grain structure was inspected by visual observation under Arcano® optical microscopy.

2.3 Corrosion tests

For the electrochemical tests, samples of 20 mm in length of each zone and for each concentration were prepared as test electrodes (see Figure 1), polished with sandpaper (from SiC #80 until #1200) and washed with distilled water and dried by natural flow of air.

All the electrochemical tests were conducted in 3wt% NaCl solution at room temperature using an IM6d Zahner®-Elektrik potentiostat coupled to a frequency analyzer system.

A conventional three-compartment glass electrochemical cell with its compartments separated by ceramic diaphragms was used. The test electrodes consisted of sections of the ZA ingots (see Figure 1) were positioned at the glass corrosion cell kit (leaving a rectangular area in contact with the electrolyte). The potential of the test electrode was measured against a saturated calomel reference electrode (0.242 V vs NHE), provided with a Luggin capillary tip. The Pt sheet was used as a counter electrode.

Voltammograms were run between preset cathodic (open circuit potential ≈ -1.500 V) and anodic (Es,a = -0.700 V) switching potentials at potential sweep rates (v), at 0.002 $V.s^{-1}$. Impedance spectra were obtained in the frequency range of 10^{-3} Hz and 10^{5} Hz at open circuit potential.

For comparison purposes, experiments using pure metals and aluminum-based alloys with different structures were conducted under the same experimental conditions. All the corrosion tests experiments were triplicate and the average values and graphical outputs are reported.

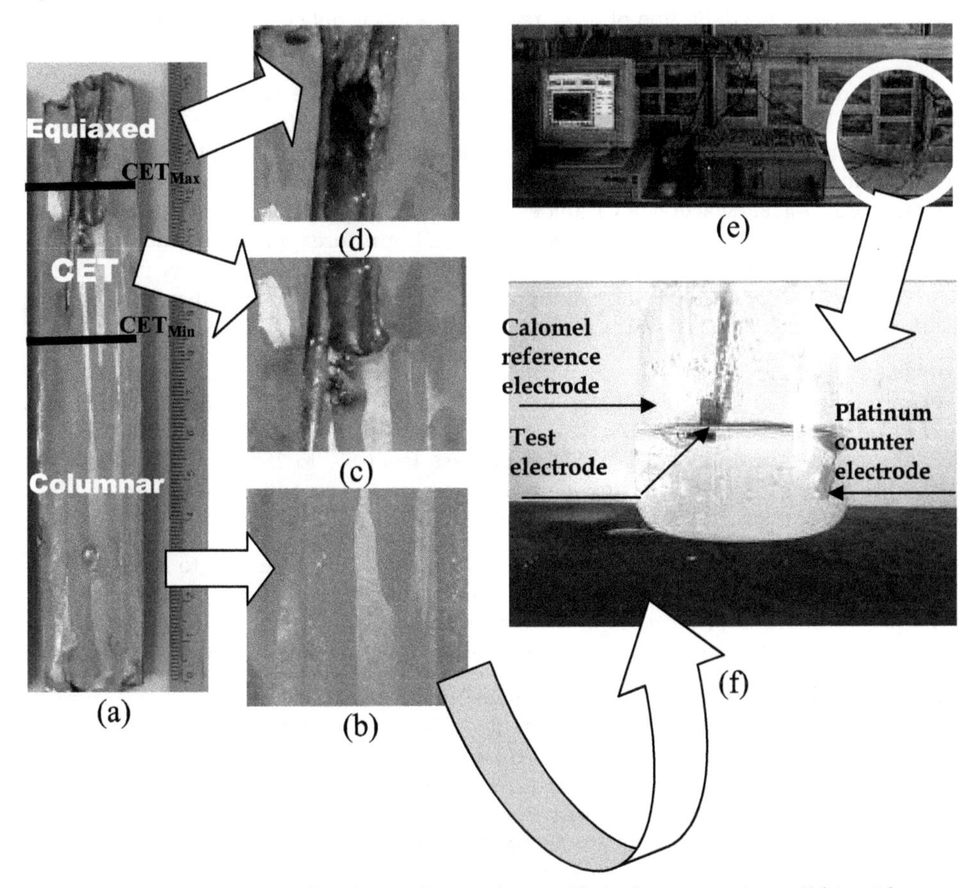

Fig. 1. Experimental device for electrochemical tests. (f) A glass corrosion cell kit with a platinum counter electrode and a sutured calomel reference electrode (SCE).

3. Results and discussion

3.1 Voltammetric data

During the anodic potential scanning, the voltammogram of equiaxed zinc shows that the current is practically zero until it reaches a potential of -1 V, where the current rises sharply, starting the active dissolution of metal (Figure 2 (a)). The negative potential scan shows a hysteresis loop, suggesting that this current increase was due to the start of a process of pitting, and two cathodic current peaks at about -1.2 V and - 1.3 V (called C_1 and C_2). These peaks could be associated with the reduction of $Zn(OH)_2$ and ZnO, respectively (Zhang, 1996). The composition of corrosion products formed on the zinc surface may be not uniformly distributed. The different compositions of the films formed can explain the difference in the outcomes reported in the literature.

The properties of corrosion products are a function of various material and environmental factors and thus vary essentially from situation to situation. For example, only one peak appears in the case of the columnar zinc (Figure 2 (b)). As the concentration of aluminum in the alloy increases, the definition of these reduction peaks is not clear, although the C_2 peak is dominant (Figure 2 (c)).

In the case of the alloys, the values of the anodic currents are similar for the same Al concentration, independently of the structure, and the most important difference is observed in the distribution of the cathodic current peaks, which indicates the different characteristics of the films formed during the anodic scan. These results can be attributed to the aggressive/depassivating action of Cl^- anions (Augustynski, 1978). At present, the mechanism of film formation is still uncertain. For the case of CET structure, profiles are more complex, because the proportion of one or other structure (columnar or equiaxed) can vary from sample to sample (Figure 2 (d)). Also, as the concentration of Al increases, the voltammetric profile of the different structures tends towards the response of pure aluminum (Figure 2 (e)).

Analyzing the response of the composites (Zn-27wt%Al + 8vol%SiC, Zn-27wt%Al + 15vol%SiC (Figure 2(f)), Zn-50wt%Al + 8vol%SiC, Zn-50wt%Al + 15vol%SiC (Figure 4 (g)), Zn-27wt%Al + 8vol%Al_2O_3, Zn-27wt%Al + 15vol%Al_2O_3 (Figure 2 (h))), we observed that when the volume percent of SiC particles increase from 8% to 15% in ZA27 and ZA50 matrix, the rate of dissolution of the alloy increases. In the case of the addition of Al_2O_3 particles to ZA 27 matrix the rate of dissolution is approximately the same.

This different distribution of the peaks in the voltammograms gives rise to surface layers with different corrosion products, as shown in the micrographs of Figure 3, where samples with higher proportion of particles in the matrix show the formation of a thicker layer of corrosion products. Also, it is observed the formation of pitting on the electrode surface.

Three distinctive features in the potentiodynamic curves can be clearly observed (Figure 4): (i) the potential at which the anodic current during the forward anodic bias increases sharply form the passive current level (breakdown or critical pitting potential E_p); (ii) a hysteresis loop (difference between forward and reverse scans) and (iii) the potential at which the hysteresis loop is completed during reverse polarization scan after localized corrosion propagation (repassivation potential E_r). Stable pits form at potentials noble to E_p and will grow at potentials noble to Er (Frankel, 1998). Also, for many years it has been recognized that Ep measurements are applicable to naturally occurring pit initiation on stainless alloys in chemical and marine environments (Wilde, 1972; Bilmes et al., 2005).

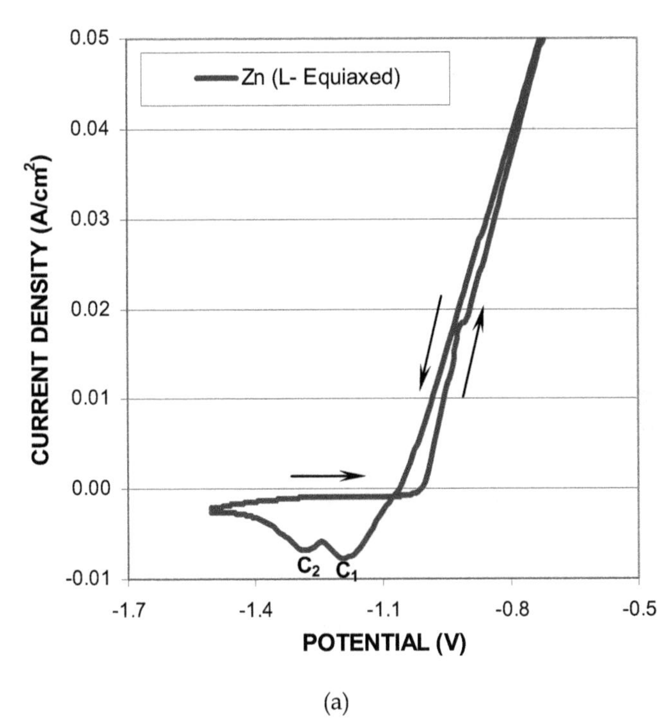

Zn (L- Equiaxed)
CURRENT DENSITY (A/cm²)
0.05
0.04
0.03
0.02
0.01
0.00
-0.01
C2
C1
-1.7
-1.4
-1.1
-0.8
-0.5
POTENTIAL (V)

(a)

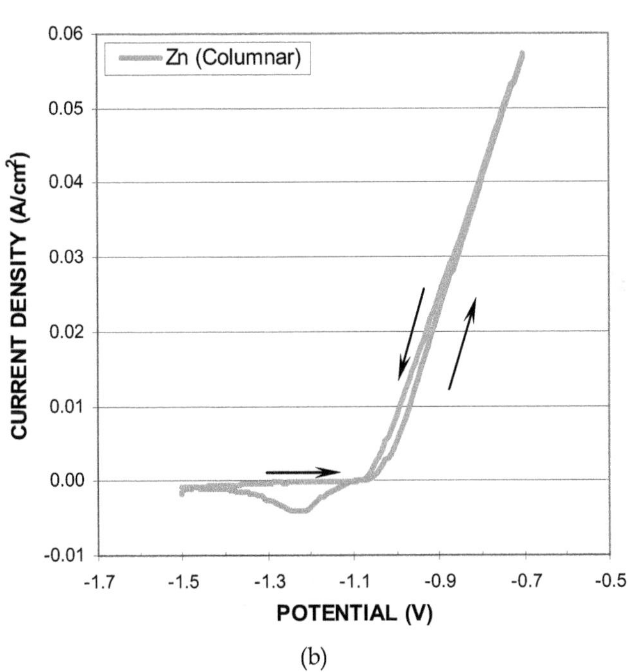

Zn (Columnar)
CURRENT DENSITY (A/cm²)
0.06
0.05
0.04
0.03
0.02
0.01
0.00
-0.01
-1.7
-1.5
-1.3
-1.1
-0.9
-0.7
-0.5
POTENTIAL (V)

(b)

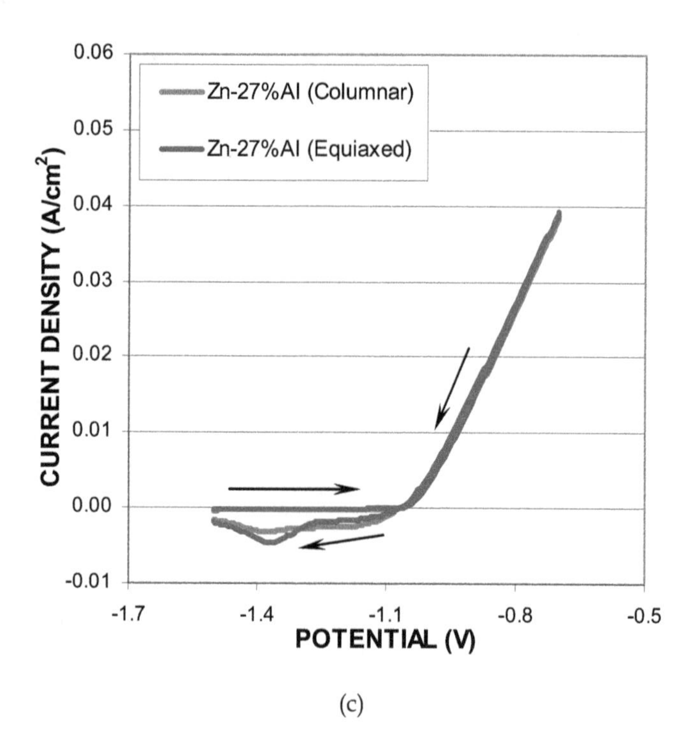
Zn-27%Al (Columnar)
Zn-27%Al (Equiaxed)
CURRENT DENSITY (A/cm²)
0.06
0.05
0.04
0.03
0.02
0.01
0.00
-0.01
-1.7
-1.4
-1.1
-0.8
-0.5
POTENTIAL (V)

(c)

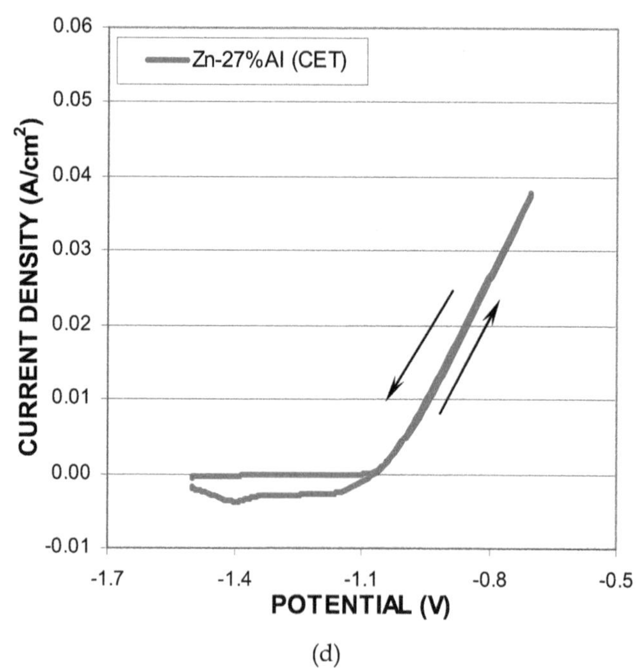
Zn-27%Al (CET)
CURRENT DENSITY (A/cm²)
0.06
0.05
0.04
0.03
0.02
0.01
0.00
-0.01
-1.7
-1.4
-1.1
-0.8
-0.5
POTENTIAL (V)

(d)

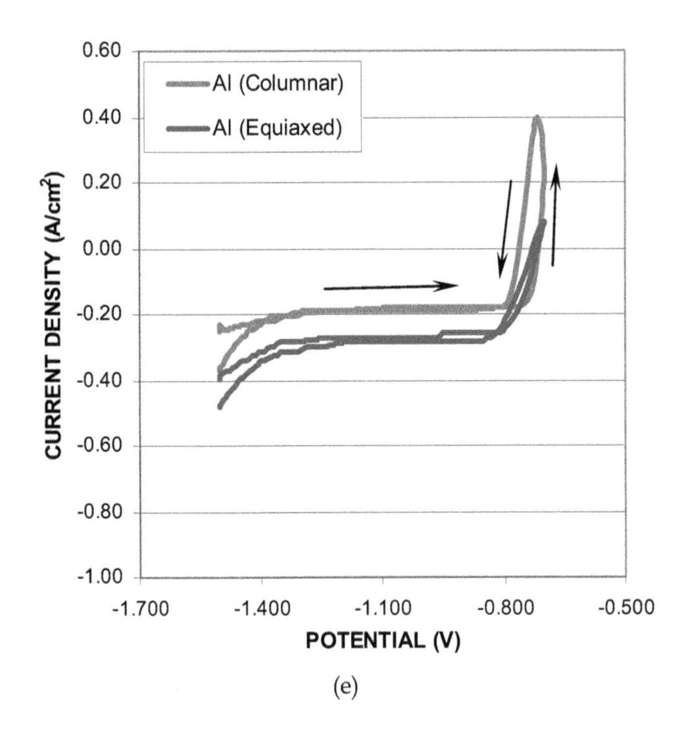
Al (Columnar)
Al (Equiaxed)
CURRENT DENSITY (A/cm2)
0.60
0.40
0.20
0.00
-0.20
-0.40
-0.60
-0.80
-1.00
-1.700
-1.400
-1.100
-0.800
-0.500
POTENTIAL (V)

(e)

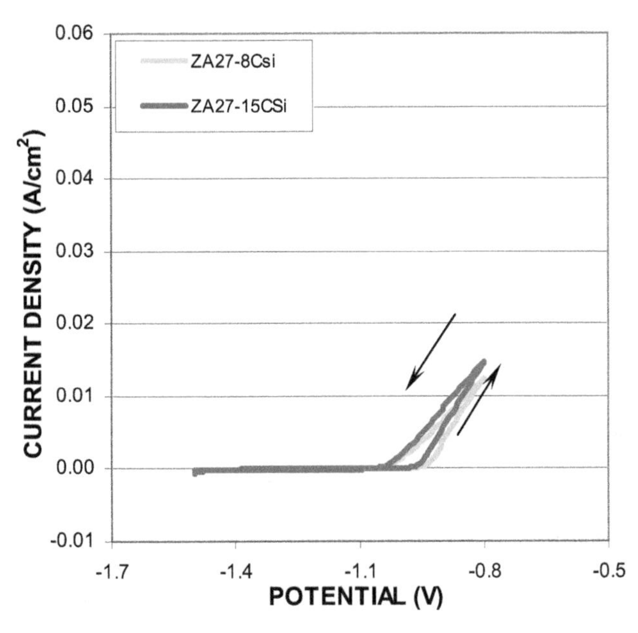
ZA27-8Csi
ZA27-15CSi
CURRENT DENSITY (A/cm2)
0.06
0.05
0.04
0.03
0.02
0.01
0.00
-0.01
-1.7
-1.4
-1.1
-0.8
-0.5
POTENTIAL (V)

(f)

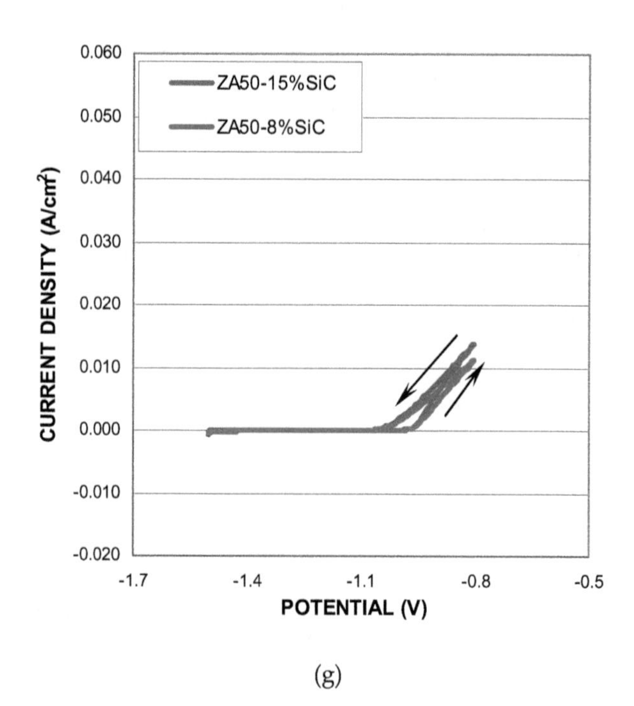

ZA50-15%SiC
ZA50-8%SiC
CURRENT DENSITY (A/cm²)
0.060
0.050
0.040
0.030
0.020
0.010
0.000
-0.010
-0.020
-1.7
-1.4
-1.1
-0.8
-0.5
POTENTIAL (V)

(g)

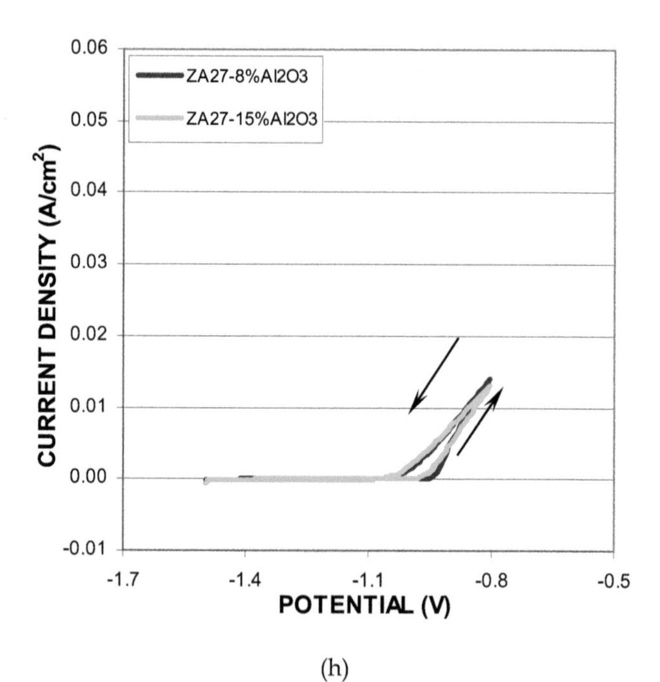

ZA27-8%Al2O3
ZA27-15%Al2O3
CURRENT DENSITY (A/cm²)
0.06
0.05
0.04
0.03
0.02
0.01
0.00
-0.01
-1.7
-1.4
-1.1
-0.8
-0.5
POTENTIAL (V)

(h)

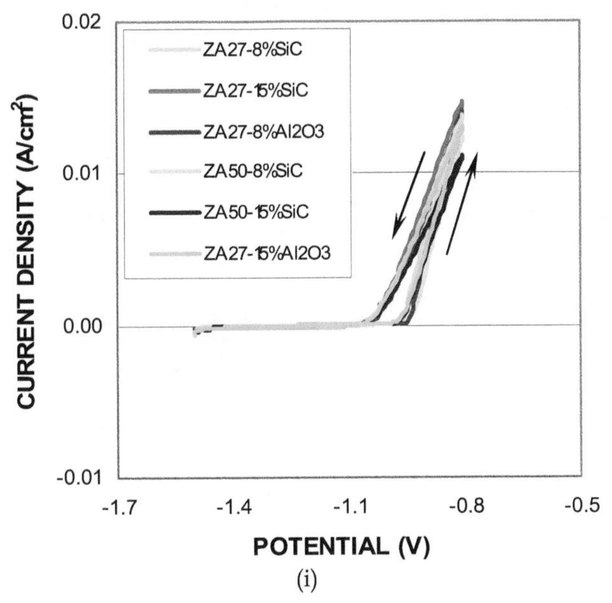

(i)

Fig. 2. Voltammograms of (a) pure Zinc with equiaxed structure, (b) pure Zinc with columnar structure (c) Zn-27wt%Al alloy with columnar and equiaxed structures, (d) Zn-27wt%Al alloy with CET structure, (e) pure Aluminum with columnar and equiaxed structure, (f) Zn-27wt%Al + 8vol%SiC, Zn-27wt%Al + 15vol%SiC, (g) Zn-50wt%Al + 8vol%SiC, Zn-50wt%Al + 15vol%SiC, (h) Zn-27wt%Al + 8vol%Al_2O_3, Zn-27wt%Al + 15vol%Al_2O_3 and (i) All types of composites.

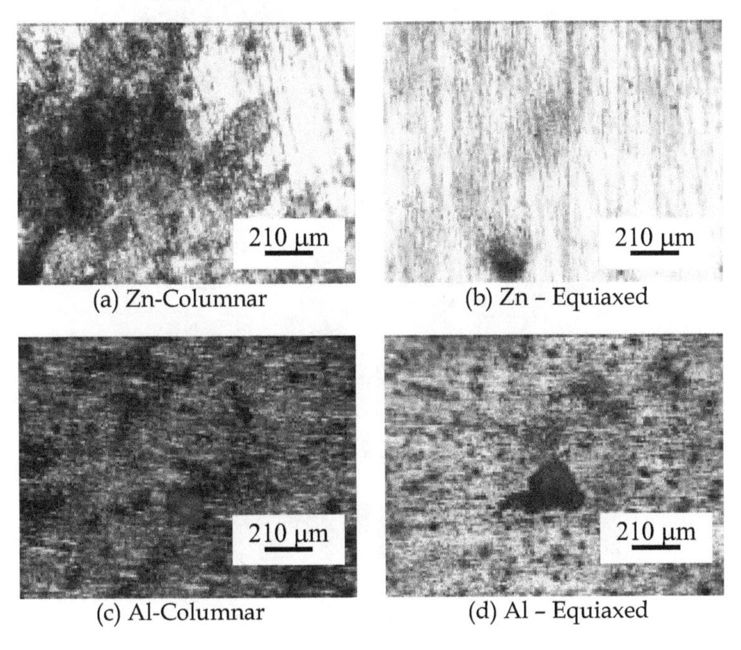

(a) Zn-Columnar (b) Zn – Equiaxed

(c) Al-Columnar (d) Al – Equiaxed

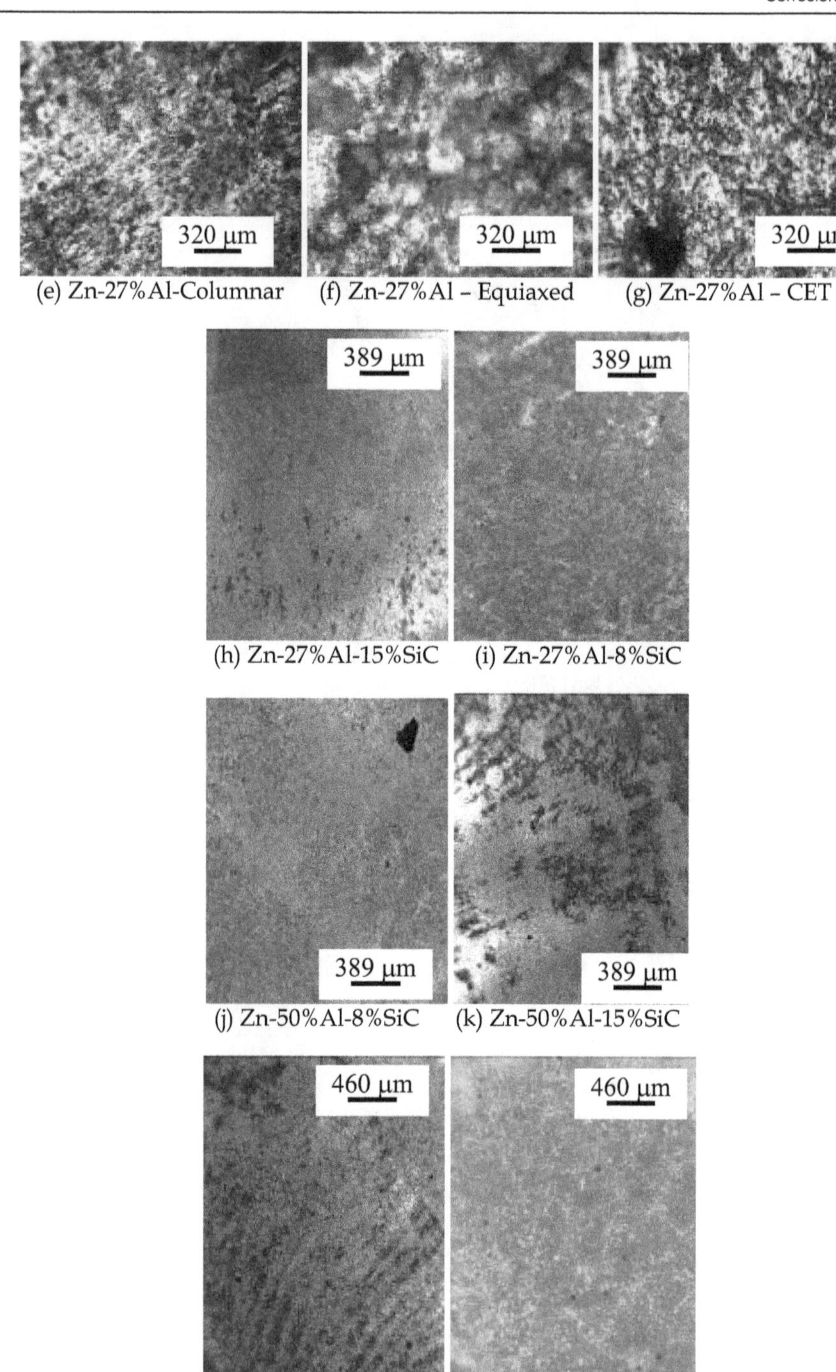

(e) Zn-27%Al-Columnar (f) Zn-27%Al - Equiaxed (g) Zn-27%Al - CET

(h) Zn-27%Al-15%SiC (i) Zn-27%Al-8%SiC

(j) Zn-50%Al-8%SiC (k) Zn-50%Al-15%SiC

(l) Zn-27%Al-15%Al_2O_3 (m) Zn-27%Al-8%Al_2O_3

Fig. 3. Micrographs of different alloy samples and structures.

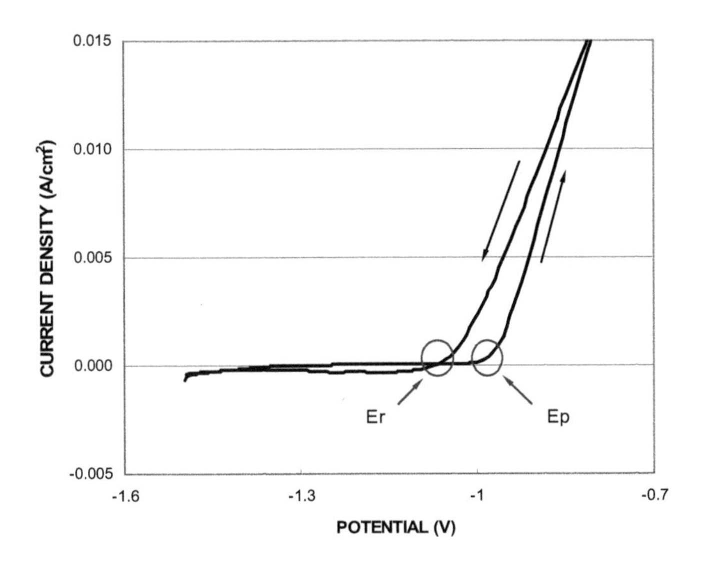

Fig. 4. Representative curve.

In all cases, the E / I response of the alloys shows the typical hysteresis indicates the phenomenon of pitting and found that the more susceptible are the composites than the alloys. The most susceptible are those containing neither SiC nor Al_2O_3 in the matrix, see Table 2 and Figure 5.

Alloy / Composite	Ep (V)	Er (V)	ΔEp-r (mV)	Ecorr (V)	ΔEr-corr (mV)
ZA50-15vol%SiC	-1.002	-1.093	91	-1.271	178
ZA50-8vol%SiC	-0.988	-1.082	94	-1.302	220
ZA27-8vol%Al_2O_3	-0.974	-1.086	112	-1.297	211
ZA27-15vol%Al_2O_3	-0.974	-1.079	105	-1.298	219
ZA27-15vol%SiC	-1.002	-1.086	84	-1.312	226
ZA27-8vol%SiC	-0.981	-1.107	126	-1.245	138
ZA27 CET	-1.02	-1.06	30	-1.102	42
ZA27 Col	-1.028	-1.052	24	-1.071	19
ZA27 Eq	-1.038	-1.068	40	-1.068	0

Table 2. The susceptibility to corrosion, ΔE, was measured as the difference between the potential of pitting, Ep, and the repassivation potential, Er.

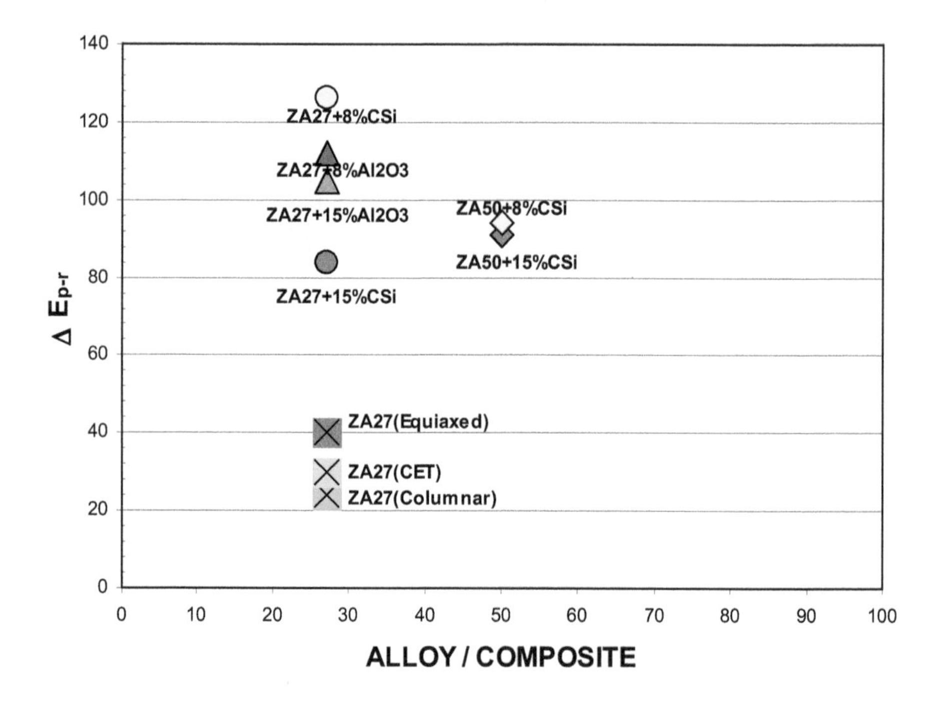

Fig. 5. ΔE_{p-r} as a function of concentration and alloy structure.

The susceptibility to corrosion was measured as the difference between the potential of pitting, Ep, and the repassivation potential, Er, as Δ Ep-r and the difference between the repassivation potential and the corrosion potential of each sample through Table 2 it is possible to observe that the values of repassivation potential for materials without particles in the matrix are near the corrosion potential, but not in the case of the the other samples.

3.2 Electrochemical impedance spectroscopy data

Impedance spectra are strongly dependent on the composition and structures of the alloys and composites. Figure 6 shows the experimental Nyquist diagrams for all the alloys and composites used. All the diagrams show one capacitive time constant at high frequencies and a non-well defined time constant at low frequencies, probably associated with diffusion processes also reported in the literature (Deslouis et al., 1984; Trabanelli at al., 1975). It can be seen that as the concentration of aluminum in the alloy increases, the second time constant approximates the response associated with a diffusion process in finite thickness, due to the formation of a more compact oxide.

In some cases, the shape of the Nyquist diagrams for CET structure in alloys resembles that of those with equiaxed structure and in others those with columnar structure, depending on the relative amount of each phase in the CET structure, which in turn depends on the region where the specimen was obtained.

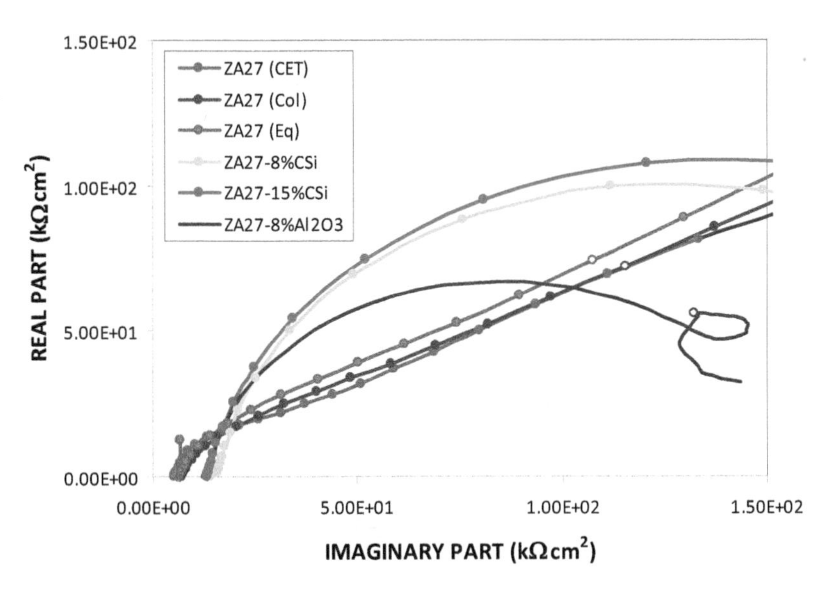

Fig. 6. (a) Nyquist Diagram for different samples.

The whole set of experimental impedance spectra can be discussed according to the following total transfer function.

$$Z_t(jw) = R_\Omega + Z \tag{1}$$

with:

$$\frac{1}{Z} = \frac{1}{R_{ct} + Z_W} + j_W.C \tag{2}$$

where R_Ω is the ohmic solution resistance, ω= 2πf; C_{dl} the capacitance of the electric double layer, R_{ct} the charge transfer resistance and Z_W the diffusion contributions in impedance spectra.

$Z_W = R_{DO}$ (jS) $^{-05}$ for semi-infinite diffusion contribution and $Z_W = R_{DO}$ (jS) $^{-05}$ coth (jS) $^{-05}$ is related to diffusion through a film of thickness d, formed on the electrode, where R_{DO} is the diffusion resistance and the parameter S= d2ω/D, where d and D are the diffusion thickness and diffusion coefficient related to the transport process (Fedrizzi et al., 1992).

The good agreement between experimental and simulated data according to the transfer function given in the analysis of Eqs. 1 and 2 using non-linear least square fit routines is shown in Figure 8 (a) at high frequencies. At low frequencies (less than 10-1 Hz) is not achieved a good fit with this model, since the impedance measurement does not give us enough information to define a new input capacitance, this occurs for samples ZA27-8% SiC, ZA50-8% SiC and ZA27-Al2O3. This process can represent by an equivalent circuit in Figure 8 (a).

For ZA27samples and those containing 15%SiC at high frequencies seems to be defined one second capacitive loop corresponding to corrosion processes controlled by precipitation and dissolution of ions Zn, see Figure 7 (b, c and d). The equivalent circuit corresponds to that showed in Figure 8 (b).

The values of C_{dl}, C_1 and R_{ct} determined from the optimum fit procedure are presented in Table 3.

The analysis of the impedance parameters associated with the time constant at low frequencies is difficult because in some cases the loop it is not complete. However, it was possible to calculate from by fitting an approximate value of diffusion coefficient $D \approx 10^{-10} - 10^{-12}$ cm^2/s.

High values of capacity confirm the formation of porous corrosion products, as can be seen in Figure 9. These high values of capacity may also be correlated with an increase in the area.

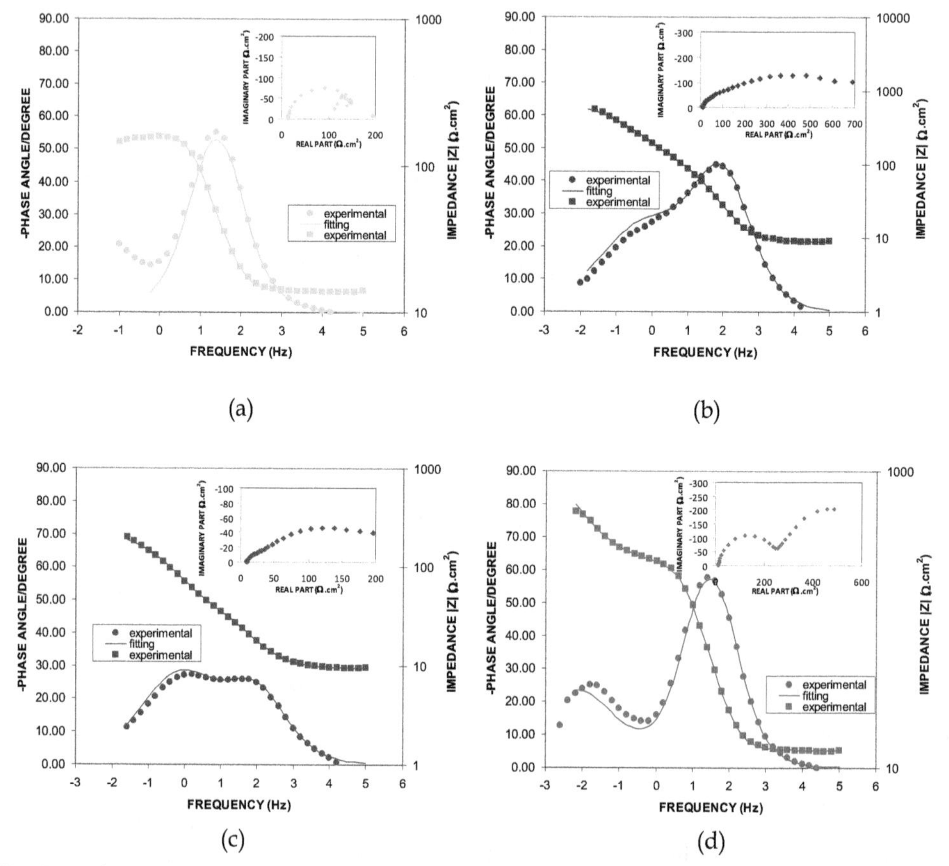

Fig. 7. Bode y Nyquist plot for – (a) ZA50-8%SiC (b) ZA27 L(CET) (c) ZA27 L(Columnar) (d) ZA27-15%SiC

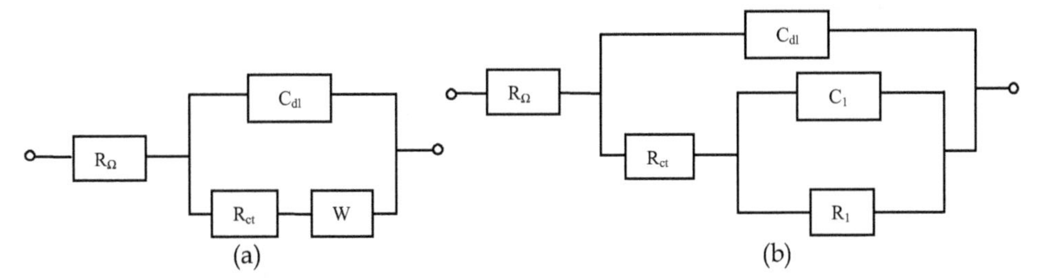

Fig. 8. Equivalent circuit of EIS for different samples.

The corrosion current can be related to the R_{ct} in the case of mixed control (Epelboin et al., 1972), where the polarization resistance technique fails, according to the following expression:

$$R_{ct} = b_a\, b_c / 2.303(b_a+b_c)I_{corr} \tag{3}$$

However, it is important to note that the Rct values are not directly related to the susceptibility to corrosion of the different alloys and composites. They are related to the rate of charge transfer reactions that give rise to the formation of a passive layer on the surface of the samples (the impedance measurements are at open circuit potential only). The protective characteristics of these passive films depend on the preparation conditions of the alloys, the distribution of elements in the alloy and the presence on the surface of active sites for adsorption of chloride ion.

Ions are formed during the anodic dissolution of alloy

$$Zn \rightarrow Zn^{+2} + 2\, e^-$$

which can react with hydroxyl ions

$$H_2O + \tfrac{1}{2}\, O_2 + 2\, e^- \rightarrow 2\, OH^-$$

and can be generated $Zn(OH)_2$

The ZA27 alloy with different structures is less resistant to corrosion and its susceptibility to corrosion is dependent of the structure. The ZA27 - 15%CSi composite has the highest corrosion resistance. Also, discriminating by type of composite materials, the MMCs with SiC are more corrosion resistant than those MMCs prepared with alumina particles.

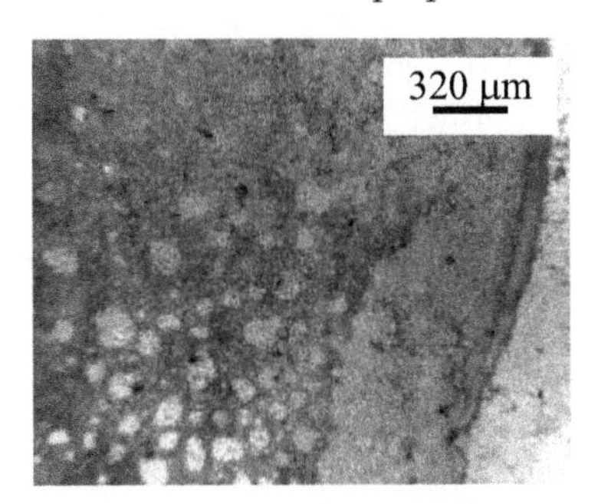

Fig. 9. Micrograph of ZA27 – 8%SiC sample after EIS test.

Alloy / Composite (wt pct)	R_Ω (Ω.cm^2)	C_{dl} (F/cm^2)	R_{ct} (Ω.cm^2)	R_1 (Ω.cm^2)	C_1 (F/cm^2)
Zn27-15%SiC	13.03	7.95e-5	1317.23	278.67	1.35e-03
Zn50-15%SiC	16.48	3.65e-5	------	19.33	5.18e-05
Zn-27Al (L-Columnar)	8.85	4.85e-5	94.78	615.29	7.51e-05
Zn-27%Al (L-CET)	9.41	1.04e-4	26.88	193.42	6.51e-04
Zn-27%Al (L-Equiaxed)	8.56	5.75e-5	87.01	304.59	6.05e-04
Zn-27%Al - 8%CSi	14.47	9.22e-5	228.77		
Zn-27%Al - 15%Al_2O_3	12.60	6.94e-5	151.78		
Zn-27%Al - 8%Al_2O_3	11.64	4.22 e-5	139.57		

Table 3. Principal parameters obtained from the EIS analysis.

4. Conclusions

The highest corrosion resistance or susceptibility to corrosion is a complex function of the alloys and composite composition, structure and the exposed surface, all of which determine the protective characteristics of the film that formed on the alloys.

Alloys with a higher aluminum content have a higher corrosion resistance, mainly due to the formation of a protective film.

Even at a higher Al concentration, the corrosion resistance depends on the structure of the alloys.

The results also indicate that the corrosion resistance of Zn-Al-SiC and Zn-Al-Al_2O_3 MMCs composites has demonstrated the improvement in comparison to ZA alloys in 3wt% NaCl solutions.

5. Acknowledgment

The authors would like to thank Consejo Nacional de Investigaciones Científicas y Técnicas (CONICET) for the financial support.

6. References

Ares, A.E.; Schvezov, C.E. (2000) Solidification Parameters During the Columnar-to-Equiaxed Transition in Lead-Tin Alloys. *Metallurgical and Materials Transactions A*, Vol. 31, 1611-1625, ISSN 1073-5623/83

Ares, A.E.; Gueijman, S.F.; Schvezov, C.E. (2010) Experimental Study of the Columnar-to-Equiaxed Transition During Directional Solidification of Zinc-Aluminum Alloys and Composites, *J. Crystal Growth*, Vol. 312, pp. 2154-2170, ISSN 0022-0248

Ares, A.E.; Caram, R.; Schvezov, C.E. (2003) Columnar-to-Equiaxed Transition Studies in Aluminum-Magnesium and Aluminum-Zinc Alloys, *Proceedings of Light Metals 2003*, ISBN, San Diego, California, United States, March of 2003

Ares, A.E.; Gueijman, S.F.; Caram, R.; C.E. Schvezov (2005) Analysis of Solidification Parameters During Solidification of Lead and Aluminum Base Alloys. *J. Crystal Growth*, Vol. 275, pp. 235-240, ISSN 0022-0248

Ares, A.E. ; Schvezov, C.E. (2007) Influence of Solidification Thermal Parameters on the Columnar to Equiaxed Transition of Al-Zn and Zn-Al Alloys. *Metallurgical and Materials Transactions A*, Vol. 38, pp. 1485-1499, ISSN 1073-5623/83

Ares, A.E.; Gueijman, S.F.; Schvezov, C.E. (2002) Semi-Empirical Modeling for Columnar and Equiaxed Growth of Alloys. *J. Crystal Growth*, Vol. 241, pp. 235-240, ISSN 0022-0248

Ares, A.E.; Gassa, L.M.; Gueijman, S.F.; Schvezov, C.E. (2008) Correlation Between Thermal Parameters, Structures, Dendritic Spacing and Corrosion Behavior of Zn-Al Alloys With Columnar to Equiaxed Transition. *J. Crystal Growth*, Vol. 310, pp. 1355-1361, ISSN 0022-0248

Ares, A.E.; Caram, R.; Schvezov, C.E. (2006) Relation between As-Cast Mechanical Properties, Microstructure and Solidification Conditions for Zn-Al Alloys, Proceedings of MCWASP International Conference Modeling of Casting, Welding and Advance Solidification Processes - XI, Opio, France, June of 2006

Ares, A.E.; Gatti, I. P.; Gueijman, S.F.; Schvezov, C.E. (2009) Mechanical Properties of Zinc-Aluminum Alloys versus Structural and Thermal Parameters, Proceedings of MCWASP International Conference Modeling of Casting, Welding and Advance Solidification Processes - XII, Vancouver, Canadá, June of 2009

Augustynski, J. (1978) Etude De La Rupture De Passivite De Certains Metaux Electrochimiquement Actifs, *Corrosion Sci.*, Vol. 13, pp. 955-965, ISSN 0010-938X

Bilmes, P.D., Llorente, C.L., Saire Huamán, L., Gassa, L.M., Gervasi, C.A. (2005) Microstructure and pitting corrosion of 13CrNiMo weld metals. *Corrosion Science*, Vol. 48, p.p. 3261–3270, ISSN 0010-938X

Cockcroft, S. L.; Rappaz, M.; Mitchell, A. (1994) An Examination of Some of the Manufacturing Problems of Large Single-Crystal Turbine Blades for Use in Land-Based Gas Turbines, In: *Materials for Advanced Power Engineering*, Coutsouradis, J. et al., pp. 1161-1175, ISBN 0-7923-3075-7, New York: Kluwer Inc.

Deslouis, C.; Duprat, M.; Tulet-Tournillon, C. (1984) The Cathodic Mass Transport Process During Zinc Corrosion in Neutral Aerated Sodium Sulphate Solutions. *J. Electroanal Chem.*, Vol. 181, pp. 119-136, ISSN 0022-0728

Epelboin, I.; Keddam, M.; Takenouti, H. (1972) Use of Impedance Measurements for the Determination of the Instant Rate of Metal Corrosion. *Journal of Applied Electrochemistry*, Vol. 2, pp. 71-79, ISSN 0021-891X

Frankel, G.S. (1998) Pitting Corrosion of Metals. *Journal of the Electrochemical Society*, Vol. 145, pp. 2186–2198, ISSN 1945-7111

Fedrizzi, L.; Ciaghi, L.; Bonora, P.L.; Fratesi, R.; Roventi, G. (1992) Corrosion Behaviour of Electrogalvanized Steel in Sodium Chloride and Ammonium Sulphate Solutions, a Study by E.I.S. *J. Appl. Electrochem*, Vol. 22, pp. 247-254, ISSN 0021-891X

Gäumann, M.; Trivedi, R.; Kurz, W. (1997) Nucleation Ahead of the Advancing Interface in Directional Solidification, *Mater. Sci. Eng.* A, Vol. 226-228, pp. 763-769, ISSN 0921-5093, ISSN 0921-5093

Gäumann, M.; Bezençon, C.; Canalis, P. et al. (2001) Single-Crystal Laser Deposition of Superalloys: Processing-Microstructure Maps. *Acta Mater.*, Vol. 49, pp.1051-1062, ISSN 1359-6454

Hunt, J. D. (1984) Steady State Columnar and Equiaxed Growth of Dendrites and Eutectic. *Mater. Sci. Eng.*, Vol. 65, pp. 75-83, ISSN 0921-5093

Kurz, W.; Giovanola, B.; Trivedi, R. (1986) Theory of Microstructural Development During Rapid Solidification. *Acta Metall. Mater.*, Vol. 34, pp. 823-830, ISSN 0956-7151

Lipton, J.; Kurz,W.; Trivedi, R. (1987) Rapid Dendrite Growth in Undercooled Alloys. *Acta Metall.*, Vol. 35, pp.957-964, ISSN 0956-7151

Long, T.T.; Nishimura, T.; Aisaka, T.; Morita, M. (1991) Wear Resistance of Al-Si Alloys and Aluminum Matrix Composites. *Materials Transactions JIM*, Vol. 32, N° 2, pp. 181-188, ISSN 0916-1821

McFadden S.; Browne D.J.; Gandin C.A. (2009) A Comparison of Columnar-to-Equiaxed Transition Prediction Methods Using Simulation of the Growing Columnar Front. *Metall Mater Trans A*, Vol. 40, pp. 662-672, 1073-5623

Reinhart G.; Mangelinck-Noël N.; Nguyen-Thi H.; Schenk T.; Gastaldi J.; Billia B.; Pino P.; Härtwig J.;Baruchel J. (2005) Investigation of Columnar–Equiaxed Transition and Equiaxed Growth of Aluminium Based Alloys by X-ray Radiography. *Mater Sci Eng A*, Vol. 413–414, pp. 384–388, ISSN 0921-5093

Rohatgi, P. (1991) Cast Aluminum-Matrix Composites for Automotive Applications. JOM, Vol. 43, pp. 10–15, ISSN 1047-4838

Spittle J.A. (2006) Columnar to Equiaxed Grain Transition in as Solidified Alloys. *International Materials Reviews,* Vol.51, No.4, pp. 247–269, ISSN 0950-6608

Trabanelli, G.; Zucchi, F.; Brunoro, G.; Gilli, G. (1975) Characterization of the Corrosion or Anodic Oxidation Products on Zinc. *Electrodeposition Surf. Treat.*, Vol. 3, pp. 129-138, ISSN 0300-9416

Vander Voort, G.F. (June 2007) *Metallography Principles and Practice* (Fourth Edition) , ASM International, ISBN-10: 0-87170-672-5, New York, United States

Wilde, B.E. (1972) Critical Appraisal of some Popular Laboratory Electrochemical Tests for Predicting the Localized Corrosion Resistance of Stainless Alloys in Sea Water. *Corrosion*, Vol. 28, pp. 283–291, ISSN 00109312

Zhang, X. G. (1996) *Corrosion and Electrochemistry of Zinc* (First Edition), Plenum Press, ISBN 0-306-45334-7, New York and London

4

Corrosion Resistance of Pb-Free and Novel Nano-Composite Solders in Electronic Packaging

L.C. Tsao
*Department of Materials Engineering,
National Pingtung University of Science & Technology, Neipu, Pingtung,
Taiwan*

1. Introduction

Tin-lead (Sn-Pb) alloys for metal interconnections were first used about 2000 years ago. Recently, the use of alloys has become essential for the interconnection and packaging of virtually all electronic products and circuits. Sn-Pb solder alloys have been widely used in the modern electronics industry because of their low melting points, good wettability, good corrosion resistance, low cost, reasonable electrical conductivity, and satisfactory mechanical properties. However, due to health concerns, recent legislation, and market pressures [1], the electronic industry is moving toward green manufacturing as a global trend. In the area of packaging, mainly driven by European RoHS (Reduction of Hazardous Substances), lead was banned effective July 1, 2006, except in some exempt items. In addition, Pb and Pb-containing compounds, as cited by the Environmental Protection Agency (EPA) of the US, are listed among the top 17 chemicals posing the greatest threat to human life and the environment [2] because of lead's toxicity [3]. In the electronics industry, the lead generated by the disposal of electronic assemblies is considered hazardous to the environment. Therefore, developing viable alternative Pb-free solders for electronic assemblies is of principal importance.

2. Lead-free solder systems

Although several commercial and experimental Pb-free solder alloys are available as replacements for Sn-Pb solders, the following families of solders are of particular interest and are the prevailing choices of industry [4]: eutectic Sn-Ag, eutectic Sn-Cu, eutectic Sn-Zn, eutectic Bi-Sn, and Sn–In, as shown in Table 1. Since the properties of the binary Pb-free solders cannot fully meet the requirements for applications in electronic packaging, additional alloying elements are added to improve the performance of these alloys. Thus, ternary and even quaternary Pb-free solders have been developed [5-7], such as Sn-Ag-Cu, Sn-Ag-Bi, and Sn-Zn-Bi solder. However, the knowledge base on Sn-Pb solders gained by experience is not directly applicable to lead-free solders. In other words, the reliability of Pb-free solder joints in consumer products is attracting more interest and concern from both academia and technologists[8-10].

All system	Eutectic composition (wt.%)	Melting point or range (°C)
Sn-In	Sn52In	118(e)
Sn-Bi	Sn58Bi	138(e)
Sn-Zn	Sn9Zn	198.5(e)
Sn-Ag	Sn3.5Ag	221(e)
Sn-Cu	Sn0.7Cu	227(e)
Sn-Ag-Bi	Sn3.5Ag3Bi	206-213
Sn-Ag-Cu	Sn3.8Ag0.7Cu	217(e)
	Sn3.50.5Cu	218
Sn-Zn-Bi	Sn8Zn3Bi	189-199

Table 1. Data showing the enhancement of the mechanical properties of Pb-free solders[9, 10].

3. Nano-composite solders

As electronic devices continue to become lighter and thinner, they require much smaller solder joints and fine-pitch interconnections for microelectronic packaging. For example, portable electronic devices, such as portable computers and mobile phones, have become thinner and smaller while adding more complicated functions. The miniaturization of these electronic devices demands better solder-joint reliability. Hence, in all chip connection and ball grid array (BGA) technologies, solder interconnection through flip-chip assembly has been proven to offer the highest density of input/output (I/O) connections in a limited space. To meet the insatiable appetite for ever-finer I/O pitches and ever-higher I/O densities, C4 (controlled collapse chip connection) technology was developed by IBM in the mid 1960s, and this technology was applied to future microelectronic packaging. According to the International Technology Roadmap for Semiconductors (ITRS), the pad pitch may fall below 20 μm by the year 2016 [11]. In some flip chip packages, solder balls of 20μm in size are used to connect the pads on the chip and the print circuit board (Fig. 1). Furthermore, Thru-Silicon-Via (TSV) technologies are also lurking on the horizon as the next-generation higher-density chip connection technology, and they also require fine-pitch Pb-free solder interconnections.

The conventional solder technology may not guarantee the required performance at such pitches due to characteristics such as higher diffusivity and softening [12]. In order to solve these problems, efforts have been made to develop new Pb-free solders with a low melting point, good mechanical properties, better microstructure properties, and high creep resistance. Recently, Pb-free solders doped with nano-sized, nonreacting, noncoarsening oxide dispersoids have been identified as potential materials that could provide higher microstructure stability and better mechanical properties than the conventional solders [13-24]. Tsao et al. [14-16] studied the influence of reinforcing TiO_2 and Al_2O_3 nanoparticles on microstructural development and hardness of eutectic Sn-Ag-Cu solders. In their work, microhardness measurements revealed that the addition of TiO_2 and Al_2O_3 nanoparticles is helpful in enhancing the overall strength of the eutectic solder. Shen et al. [17] controlled the formation of bulk Ag_3Sn plate in Sn-Ag-Cu solder by adding ZrO_2 nanoparticles to reduce the amount of undercooling during solidification and thereby suppress the growth of bulk Ag_3Sn plates. Zhong and Gupta [18] successfully prepared a nano-Al_2O_3 reinforced nano-composite solder by mechanically intermixing nano-Al_2O_3 particles into Sn0.7Cu Pb-free solder, and this composite solder shows improved mechanical properties. The best tensile strength realized for the composite, which contains 1.5 wt.% alumina, far exceeds the strength of the eutectic Sn–Pb solder. Many authors have studied the effect of adding single-walled carbon nanotubes [19] or multi-walled carbon nanotubes [20, 21] on the mechanical properties of nano-composite solders. The data on the enhancement of the mechanical properties of nano-composite solders collected from some of the literature are listed in Table 2 [13, 14, 16, 22, 23]. Here, it should be stressed that although the addition of nanoparticles into solder matrices can improve the creep behavior[24], the effects on the corrosion resistance and mechanical properties of the nano-composite solders cannot be ignored.

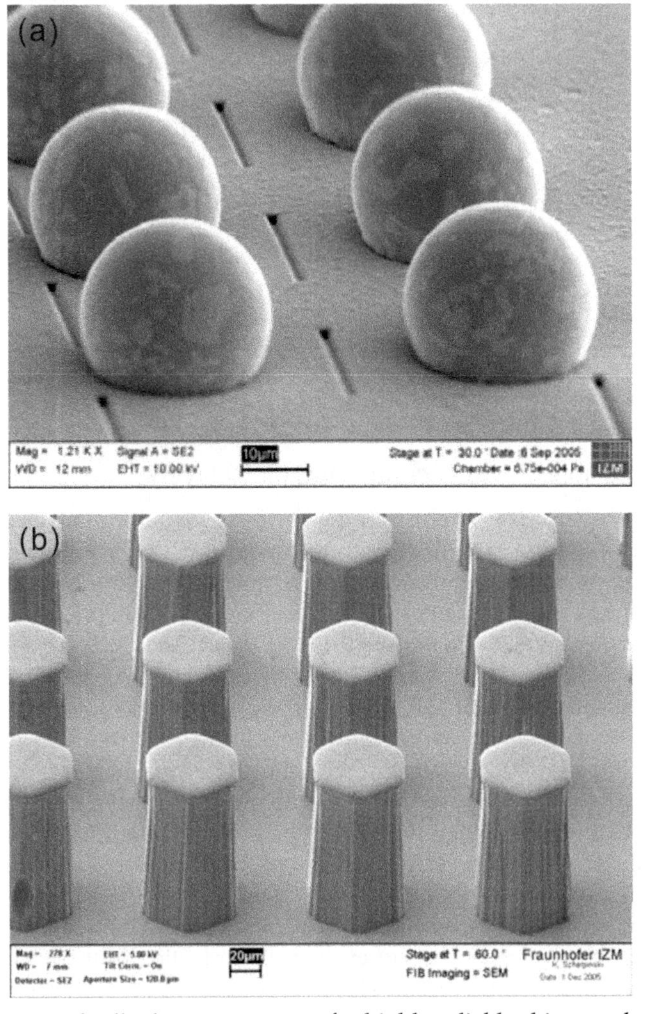

Fig. 1. Micro bump and pillar bump structures for highly reliable chip-to-substrate interconnects: (a) SnAg microbump (20 µm diameter), and (b) Cu pillarbump (height: 80 µm) [11]

Solder matrix	Reinforcement nanoparticles	Mechanical properties			
		0.2%YS (MPa)	UTS (MPa)	Elongation (%)	References
Sn4In4.1Ag0.5Cu	Nil	56±6	60±8	37±7	[22]
	1.0 vol.% Al_2O_3	72±6	75±6	21±3	
	3.0 vol.% Al_2O_3	73±3	77±6	11±3	
	5.0 vol.% Al_2O_3	74±3	76±2	10±0	
Sn3.5Ag0.7Cu	Nil	31±2	35±1	41±8	[23]
	0.01wt.% MWCNTs	36±2	47±1	36±2	
	0.04wt.% MWCNTs	36±4	46±6	37±2	
	0.07wt.% MWCNTs	33±3	43±5	35±4	
Sn3.5Ag0.5Cu	Nil	45.96±1.14	54.34±1.42	49.2±1.3	[16]
	0.25 wt.% Al_2O_3	48.81±1.23	60.20±1.84	47.3±0.8	
	0.5 wt.% Al_2O_3	52.56±1.56	62.44±1.76	44.0±1.2	
	1.0 wt.% Al_2O_3	57.22±1.8	68.05±1.63	43.5±2.1	
	1.5 wt.% Al_2O_3	61.45±2.3	70.05±2.06	32.5±3.2	
Sn3.5Ag0.25Cu	Nil	53.2	55.7	48.6	[14]
	0.25 wt.% TiO_2	59.5	61.5	40.5	
	0.5 wt.% TiO_2	67.6	69.1	32.1	
	1.0 wt.% TiO_2	69.3	70.1	25.2	

Table 2. The data showing the enhancement of the mechanical properties of nano-composite solders[13, 14, 16, 22, 23].

4. The interfacial intermetallic compound (IMC) layers

In connected metals, all the common base materials, coatings, and metallizations, such as Cu, Ni, Ag, and Au, form intermetallic compounds (IMC) with Sn, which is the major element in Sn solders. Cu is the material most frequently used for leads and pads on flip chip substrates and printed wiring boards. It is now known that in the solder/Cu interfacial reaction, Sn reacts rapidly with Cu to form Cu_3Sn (ε-phase) and Cu_6Sn_5 (η-phase) [25]. Other metal substrate/solder interfacial reactions form IMCs, such as Ag_3Sn[26] (Sn solder/Ag), Sn-Ni [27] (Sn solder/Ni), Ag-In[28] (In solder/Ag) and Cu-In IMC[29] (In solder/Cu). These intermetallic compounds are generally more brittle than the base metal, which can have an adverse impact on the solder joint reliability. Excessive thickness may also decrease solder joint ductility and strength [30-34]. Recently, we found that a great number of nano-Ag_3Sn particles form on the Cu_6Sn_5 IMC when the solders contain Ag_3Sn precipitate phase after a Pb-free Sn3.5Ag0.5Cu (SAC) nano-composite solder/Cu substrate interface reaction[30, 31]. These nanoparticles apparently decrease the surface energy and hinder the growth of the Cu_6Sn_5 IMC layer during soldering and aging. All these results indicate that Gibbs absorption theory can be used to explain the formation of these nanoparticles and their effects on the surface energy of the IMC. Many studies have reported that nano-sized, nonreacting, noncoarsening oxide dispersoid particles, such as TiO_2 [30-32], Al_2O_3 [33], Y_2O_3[34], CNTs [35], and ZrO_2[36] can affect the growth rate of interfacial IMC.

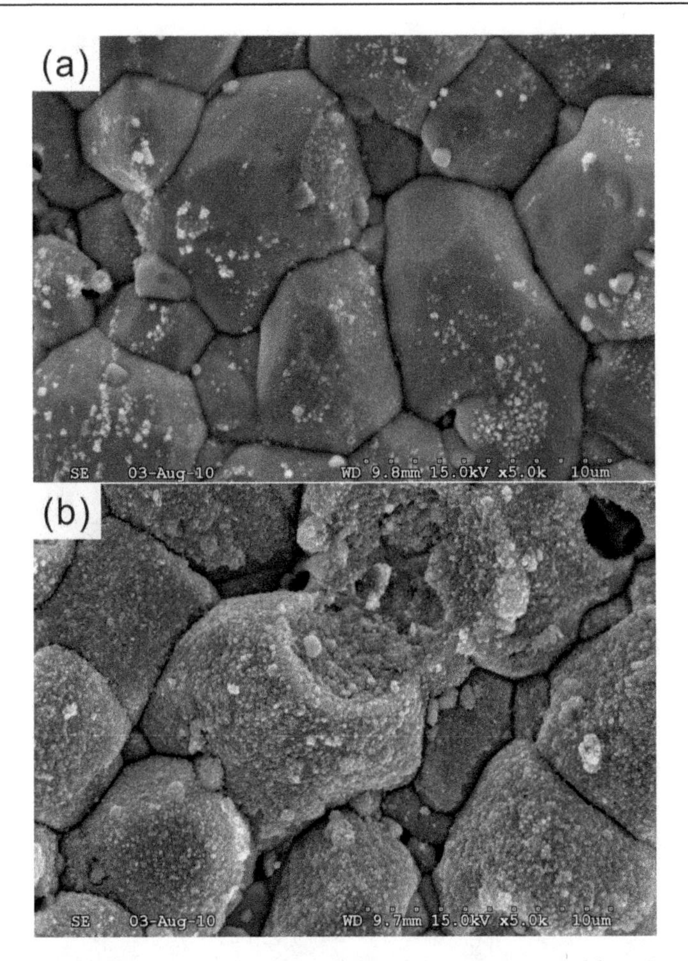

Fig. 2. Top view of the IMC at the interfaces of the nano-composite solder joints on Cu substrate after aging for 7 days at 175°C: (a) SAC and SAC- TiO_2 [31].

5. Corrosion behavior of Pb-free solder joints

The diversity of materials, drive toward miniaturization, and globalization have significantly contributed to the corrosion of microelectronic devices [37]. However, the key point is that solder joints are often exposed to corrosive environments that can accelerate the corrosion process. Although corrosion resistance is an important parameter in choosing solder alloys, the corrosion behavior of Sn-Pb solder joints was rarely of interest because the oxide that forms on the tin-lead alloy is relatively stable. Mori et al. showed that both Pb-rich and Sn-rich phases dissolve when the Sn-Pb solder alloy is immersed in corrosive solution, and the corrosion rate is slower than that of the Sn-Ag solder [38, 39]. Compared to traditional Sn-Pb solders, Sn-Ag-Cu solders are easily corroded in corrosive environments due to their special structures (as shown in Fig. 3). The presence of Ag_3Sn in Sn-Ag-Cu solders accelerates the dissolution of tin from the solder matrix into a corrosive medium

Fig. 3. Surface morphology changes of solder balls after the salt spray test for 96 hrs: (a) Sn-Pb solder, and (b) SAC solder[39].

because of the galvanic corrosion mechanism [39]. When corrosion occurs in the solder joints, it may change the microstructure of corroded regions and provide crack initiation sites, thereby decreasing the mechanical properties of the joints. Lin and Lee have investigated both Sn-Pb and Sn-Ag-Cu solder alloy wafer-level packages, with and without pretreatment by 5% NaCl salt spray, with thermal cycling to failure. The salt spray test did not reduce the characteristic lifetime of the Sn-Pb solder joints, but it did reduce the lifetime of the Sn-Ag-Cu solder joints by over 43% (Fig.4). The characteristic lifetime cycle number was 1384 for the as-assembled and non-salt spray treated components, but it was only 786 for the components which were treated in 5 wt.% NaCl salt spray for 96 h. In addition, the presence of multiple corrosion sites per solder joint poses an additional risk factor to the structural stability of the joint, for corrosion sites are all potential crack initiation sites.

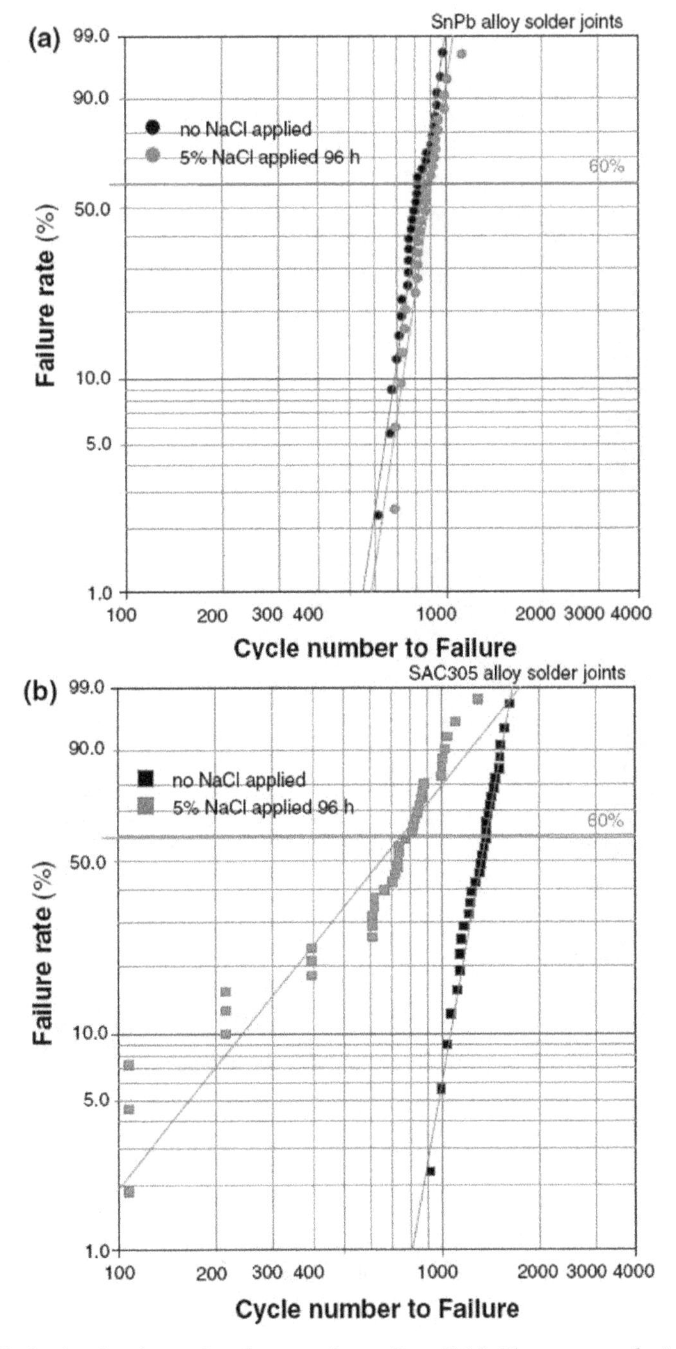

Fig. 4. Weibull plot for the thermal cycling results on 5 wt.% NaCl aqueous solution (salt spray) treated WLCSP: (a) Sn-Pb solder alloy samples and (b) SAC305 solder alloy samples [40].

Unlike Sn-Pb joints, which have a dual phase structure and block the path of corrosion due to the existence of phase boundaries, the SAC305 joint is basically pure Sn with coarse islands of Ag_3Sn and Cu_6Sn_5 intermetallic precipitate (Fig. 5). A corrosion crack can propagate and lead to additional corrosion along the way, without interruption from the Sn phase structure. Although both materials show strong resistance to corrosion, the localized nature of the corroded area at critical locations causes significant degradation in Sn-Ag-Cu solder joints[40].

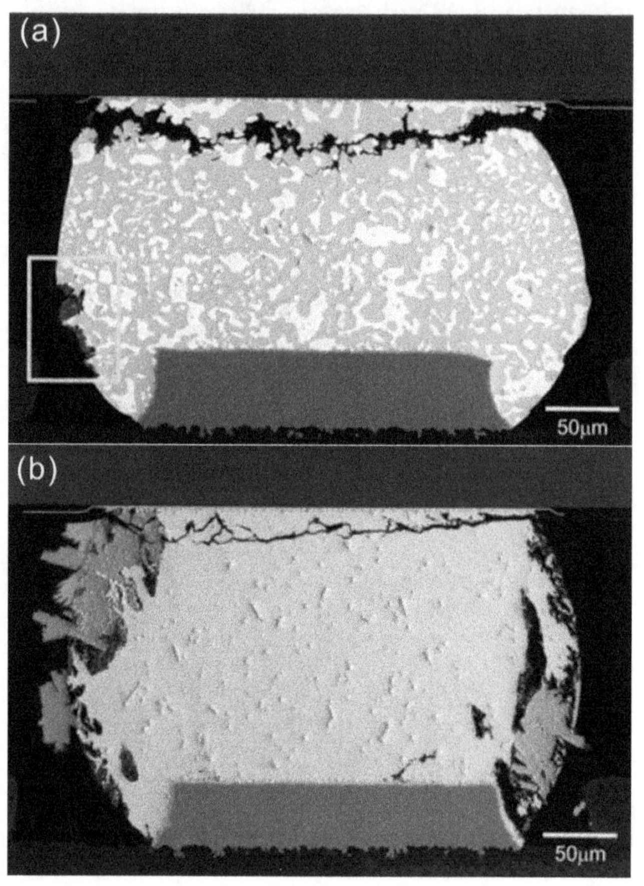

Fig. 5. Cross-section SEM microstructure after salt spray treatment and then thermal cycling: (a) Sn-Pb, and (b) SAC305 solder joint [40].

6. Galvanic corrosion of soldering

Corrosion of solder alloys, in the presence of a suitable electrolyte can occur either due to the potential difference between the major phases in the alloy or galvanic coupling between one or more phases of the alloy and other parts of the microelectronics device. Some metals that are frequently used in microelectronics are Cu, Au, Ag, Ni and Pd. The standard emf for these metals and metals used in solder alloys are listed in Table 3[4]. Especially, advanced packaging technologies make the solder alloy susceptible to corrosion problems

[41]. Thus, in the electronics industry, corrosion has become a significant factor in recent years because of the extremely complex systems that have been developed and the increasing demand on their reliability [42, 43]. For example, using Cu and Sn metals allows fine-pitch interconnections to be fabricated at relatively low cost. These features make Cu-Sn based SLID bonding very appealing for 3D stacked applications (Fig. 6) [44].

Metals used in solder	Metals used in microelectronics				
	Au	Ag	Cu	Ni	Pd
Sn	1.636	0.935	0.473	-0.114	1.123
Pb	1.626	0.925	0.463	-0.124	1.113
In	1.842	1.141	0.679	0.092	1.329
Zn	2.263	1.562	1.10	0.513	1.75

Table 3. Δemf values for metals commonly used in microelectronics[4].

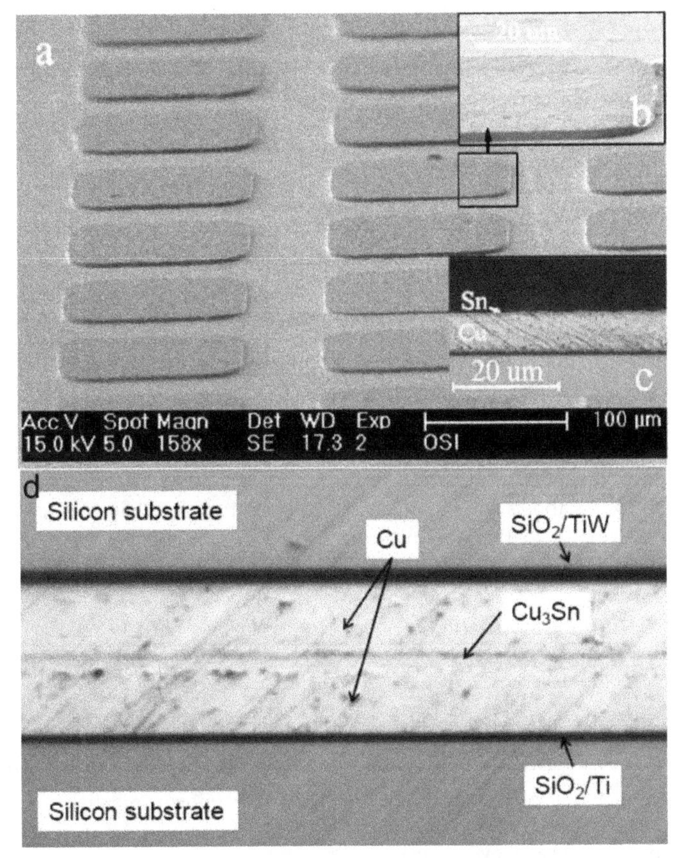

Fig. 6. Electroplated pads of 5 µm Cu and 200 nm Sn: (a) and (b) SEM image with different magnification; (c) Cross-section view under optical microscope; and (d) Cross-section view of a fluxless bonded Cu/Sn Interconnect [44].

The joining of materials with solders generally results in a multi-layer structure in which IMC are formed between substrate and solders. Such a structure in a flip chip package is a galvanic couple. The galvanic corrosion behavior of the solder bump structures have a great effect upon reliability[45]. For instance, the galvanic current densities of the Sn solder with respect to the IMC Cu_6Sn_5 and Cu_3Sn, and base Cu have been investigated (Fig. 7). It appears that Sn solder has a greater galvanic current density and thus is very subject to corrosion, and it is especially so in coupling with the formation of Cu_3Sn layers than with Cu_6Sn_5 layers. The galvanic current densities of the Sn37Pb solders of Cu_3Sn, Cu, and Cu_6Sn_5 are about 38, 16, and 5 ($\mu A/cm^2$), respectively.

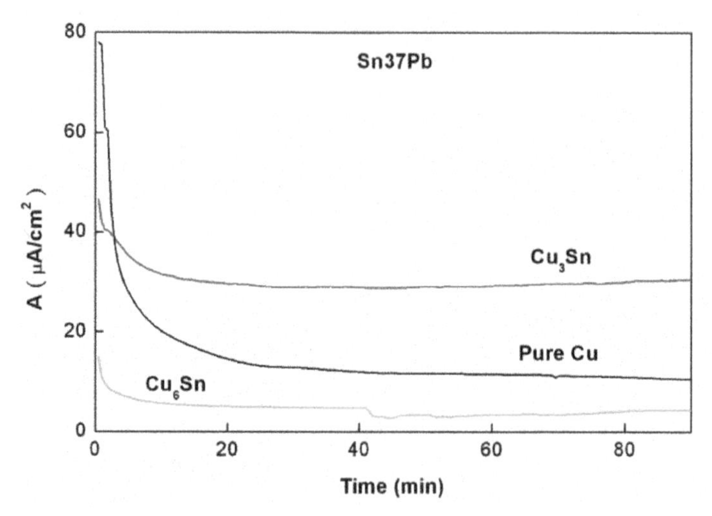

Fig. 7. The galvanic current densities of the solder with respect to intermetallic compounds Cu_6Sn_5 and Cu_3Sn, and Cu substrate, in a 3.5 wt.% solution [45].

Increasing the copper content, which reacts with Sn to form IMC, significantly improves the corrosion resistance of solders and increases the corrosion current density (I_{corr}), as shown in Fig. 8, 9 and Table 4. At above 460 mV_{SCE}, the passivation current densities of all specimens are around $10^{-1}A/cm^2$, with the declining sequence of Sn37Pb ≥ Cu_6Sn_5 > Cu_3Sn > Cu.

Specimens	Φ_{corr} (mV_{SCE})	Φ_b (mV_{SCE})	$\Delta\Phi$ (mV)	I_{corr} ($\mu A/cm^2$)	I_p (mA/cm^2)
Sn37Pb	-584.4	-303.0	281	6.48	67.7
Cu_6Sn_5	-457.7	-45.0	412	2.61	56.9
Cu_3Sn	-309.0	-8.9	300	48.17	18.3
Cu	-192.1	236	428	391.6	6.5

$\Phi_{corr.}$: corrosion potential; $I_{corr.}$: corrosion current density; Φ_b: breakdown potential;
$\Delta\Phi = \Phi_{corr.} - \Phi_b$, Φ_p: passivation range of solder alloy;
I_p: passivation current density at above 460 mV_{SCE}.

Table 4. Corrosion properties in a 3.5 wt.% NaCl solution for the Sn37Pb solder, Cu_6Sn_5 IMC, Cu_3Sn IMC and pure Cu samples [45].

It can be seen that the galvanic corrosion behavior of Cu_3Sn is generally greater than that of Cu_6Sn_5 for the flip chip package in a 3.5 wt. % NaCl solution environment. This indicates that the formation of IMC Cu_3Sn and Cu_6Sn_5 layers causes many problems with corrosion behavior and reliability.

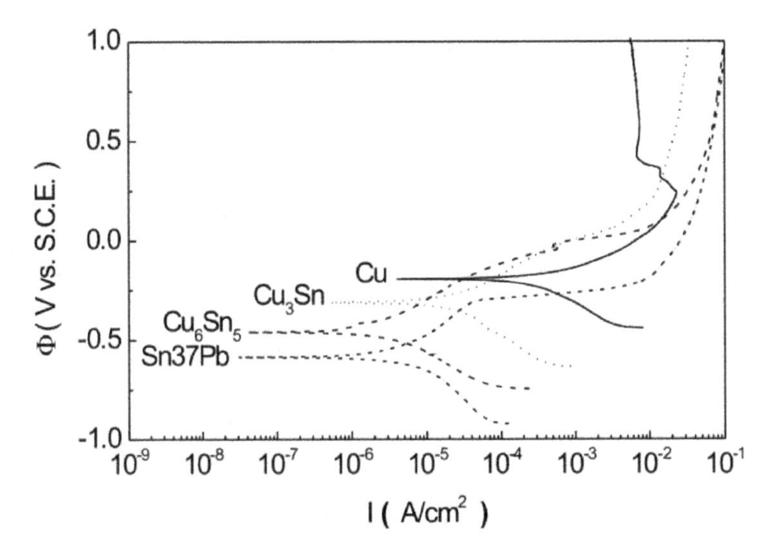

Fig. 8. The potentiodynamic polarization curves of Sn37Pb solder, Cu_6Sn_5 IMC, Cu_3Sn IMC, and pure Cu samples in a 3.5 wt.% NaCl solution [45].

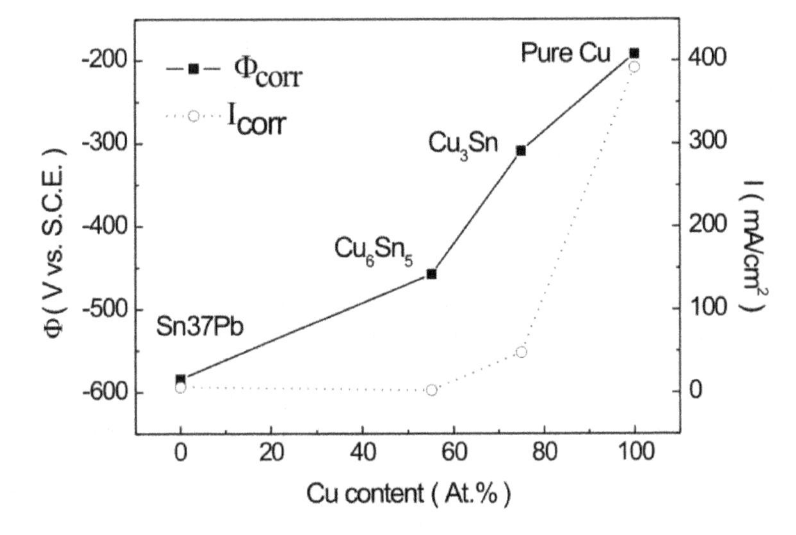

Fig. 9. Effect of Cu content on both Φ_{corr} and I_{corr} during polarization of the Sn37Pb solder, Cu6Sn5, Cu3Sn, and Cu substrate in 3.5 wt.% NaCl solution[45].

7. Corrosion behavior of Pb-free solder

Both the particular design of the electronic system, and the manner in which it is mounted in a substrate or printed wiring board, the solder connection can be exposed to the atmosphere. The solder is thus not only exposed to air, but also moisture and other corrosives such as chlorine and sulfur compounds. The ability of the solder to be able to withstand corrosion property is therefore relevant to the long-term reliability of solder joints [4]. In addition, solder alloys are electrically connected with other metallic components in the electronic device. Some metals that are frequently used in microelectronics are Cu, Au, Ag, Ni and Pd. Therefore, there is also the potential for galvanically induced corrosion of the solder, which could exacerbate any atmospheric corrosion that might be occurring. However, the properties of these lead free alloys in corrosive environments has not been widely reported, though it is of importance in many automotive, aerospace, maritime and defence applications [46]. Some researchers have studied the corrosion behaviour of Sn–Zn–X solders [47, 48] and Sn–Zn–Ag–Al–XGa [49], but few [50, 51] have studied the corrosion properties of Sn–Ag, Sn–Cu and Sn–Ag–Cu solders. Zinc is both metallurgically and chemically active. The presence of Zn in the solder alloy results in poor corrosion resistance, which is an important problem to address before practical application of this material [49]. Hence, the electrochemical corrosion behaviour of Pb-free Sn-Zn binary solder and Sn-Zn-X (X=Bi, Ag and Al) solder alloys have been investigated in NaCl solution by potentiodynamic polarization techniques[52-55]. Lin et al. [47-49] have investigated the corrosion behaviour of Sn–Zn–Al, Sn–Zn–Al–In and Sn–Zn–Ag–Al–XGa solders in 3.5% NaCl solution. They found that Sn–Zn–Al alloy [47] undergoes more active corrosion than Sn–37Pb alloy. Furthermore, they found that 5In–9(5Al–Zn)–YSn and 10In–9(5Al–Zn)–Sn alloys exhibit electrochemical passivation behaviour, and the polarization behaviours of these two alloys are similar to that of 9(5Al–Zn)–Sn alloy. Sn-Ag-M (M=In, Bi) solders exhibit poor corrosion behaviour as compared to that of Sn-Pb eutectic solder (0.1M NaCl solution)[56]. In contrast, increasing the copper content (from 0.8 to 6.7 at.%) enhances the corrosion resistance of Sn-Ag solder alloys, which exhibit improved passivity behaviour as compared to Sn-Pb eutectic solder. EPMA results indicate that the Ag_3Sn IMC is retained after the polarization test. Hence, the Ag_3Sn is more noble than the β-Sn phase. The pit formation on the surface of Sn–Ag–M alloys is due to the dissolution of the tin-rich phase. Wu et al. [51] has studied the corrosion behaviors of five solders in salt and acid solutions by means of polarization and EIS measurements. The Sn3.5Ag0.5Cu solder has the best corrosion resistivity due to the high content of noble or immune elements (Ag and Cu) and theorized stable structure, whereas the Sn9Zn and Sn8Zn3Bi solder have the worst corrosion behavior. Nevertheless, the four Pb-free solders exhibit acceptable corrosion properties, since there is not much difference in key corrosion parameters between them and the Sn37Pb solder. The corrosion data of the solders in 3.5 wt.% NaCl solutions are listed in Tables 5 [46, 51]. Lin and Mohanty et al. [46-49] studied the corrosion properties of Sn–Zn–X and Sn–Zn–Ag–Al-XGa in NaCl solution, and their results showed that the corrosion product on the surface could be SnO, SnO_2, $SnCl_2$ and ZnO, etc., depending on the applied potential. Li et al. [46] confirms that the corrosion product on the Sn–Pb and lead free solders is tin oxide chloride hydroxide ($Sn_3O(OH)_2Cl_2$).

Solder	Scanning rate	Ecorr (mV)	*Icorr* (A/cm²)	E_p (mV)	I_p (mA/cm²)	References
Sn37Pb	1 mV/s	-588	1.905×10^{-6}	-201	4.989	[51]
Sn9Zn		-940	2.691×10^{-5}	-326	2.938	
Sn8Zn3Bi		-1291	1.380×10^{-5}	9	8.035	
Sn3.5Ag0.5Cu		-605	5.370×10^{-7}	-236	4.083	
Sn3.5Ag0.5Cu9In		-578	7.413×10^{-6}	-158	1.524	
Sn0.7Cu	30 mV/s	-688	1.78×10^{-7}	-	0.74	[46]
Sn3.5Ag		-705	4.9×10^{-7}	-	0.49	
Sn3.8Ag0.7Cu		-727	0.89×10^{-7}	-	1.07	

Ecorr – corrosion potential, Ip – passivation current density, Icorr – corrosion current density, Ep – passive potential.

Table 5. Experimental data of the testing solders under polarization in 3.5 wt.% NaCl solution.

8. Corrosion behavior of Pb-free nano-composite solder joints

This author has recently worked on the development of nano-composite solders in microelectronic packaging by applying two methods of fabrication: mechanical mixing of inert nano-particles (Fig. 10) and precipitation of nano-IMC in the solder matrix (Fig. 11) [57]. The average size of the nominally spherical nano-Al_2O_3 particles was 100 nm in diameter.

Notably, the addition of nano-particles decreased the size of dendrite β-Sn grains, the needle-like Ag_3Sn grains, and Ag_3Sn phase located between the average spacing. When 1 wt% was added, the superfine spherical nano-Ag_3Sn grains were about 0.16 ± 0.06μm in length and 0.15 ± 0.05 μm in diameter, and the average spacing between them was a significant improvement (0.14 ± 0.05μm), significantly smaller than the sizes found in the SAC composite solder. However, large Ag_3Sn IMCs were not observed in the Pb-free SAC solder. Another, author reported that the effects of nano-TiO_2 particles on the interfacial microstructures and bonding strength of Sn3.5Ag0.5Cu nano-composite solder joints in ball grid array (BGA) packages with immersion Sn surface finishes [58]. It is clearly shown in Fig. 12a, b that the discontinuous Cu_6Sn_5 IMC layer grows with a rough scallop shape (Mark A), and wicker-Cu_6Sn_5 IMC forms on the rough scallop-shaped Cu_6Sn_5 IMC layer (Mark B) and grows into the SAC solder matrix. However, the addition of a small percentage of nano-TiO_2 particles alters the Pb-free Sn3.5Ag0.5Cu composite solder/pad interface morphology after reflowing, as shown in the SEM micrographs in Fig. 12c, d. Only the continuous scallop-shaped Cu_6Sn_5 IMC layer was detected at the interface. However, the wicker-Cu_6Sn_5 IMC disappeared at the interface with the Cu pads. In addition, the number of Ag_3Sn IMC forms increased in the eutectic area when the content of nano- TiO_2 particles was increased to 0.25–1 wt%. It is interesting that the smallest thickness of the IMC layer was achieved with the addition of 1 wt% of nano-TiO_2 particles. The thickness of the Cu_6Sn_5 IMC layer was reduced by 51%. The results indicate that the growth of the Cu_6Sn_5 IMC layer at the solder/pad interfaces of Sn3.5Ag0.5Cu is depressed through the small addition of nano-TiO_2 particles[58]. With the addition of 0.5–1 wt% nano-TiO_2 particles, fracture occurred in all of the solder joints as cracks propagated through the Sn3.5Ag0.5Cu composite solder balls, which ruptured mostly along the submicro Ag_3Sn IMC and solder matrix, as shown in Fig. 13a, b. This phenomenon is similar to that occurring in Pb-free Sn0.7Cu composite solder BGA packages[59].

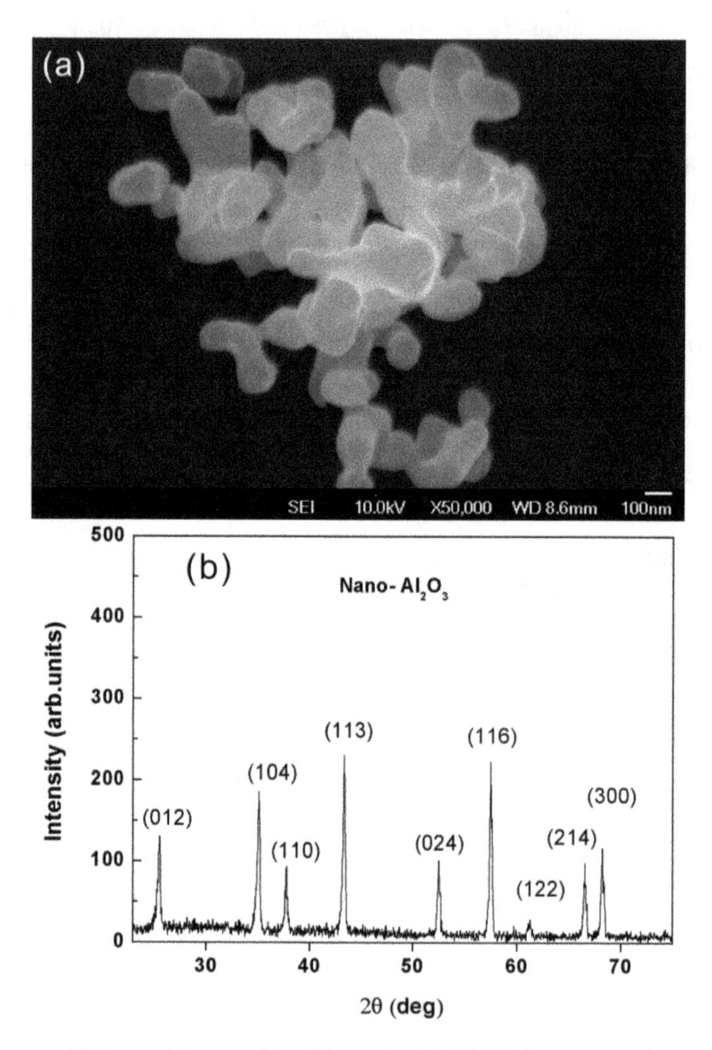

Fig. 10. The nano-Al_2O_3 particles used in this study: (a) FE-SEM micrograph, and (b) X-ray diffraction spectrum[57].

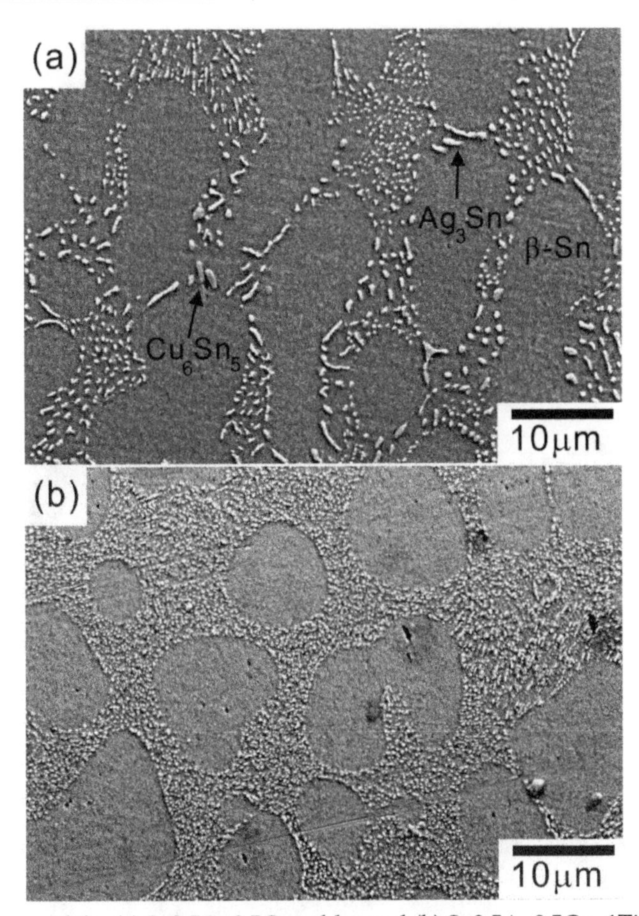

Fig. 11. SEM image of the (a) Sn3.5Ag0.5Cu solder and (b) Sn3.5Ag0.5Cu -1TiO_2 nano-composite solder[57].

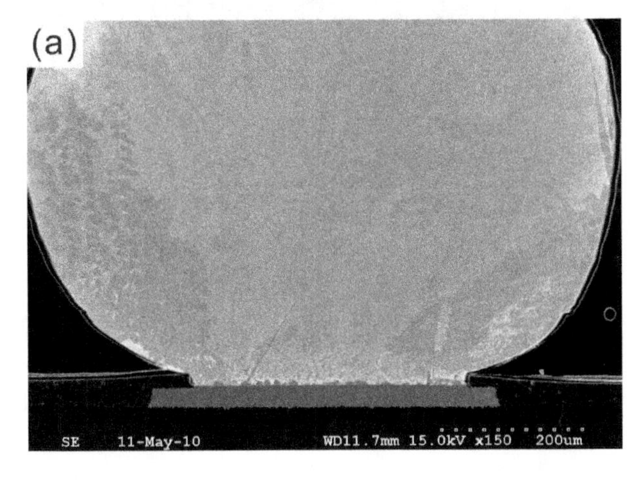

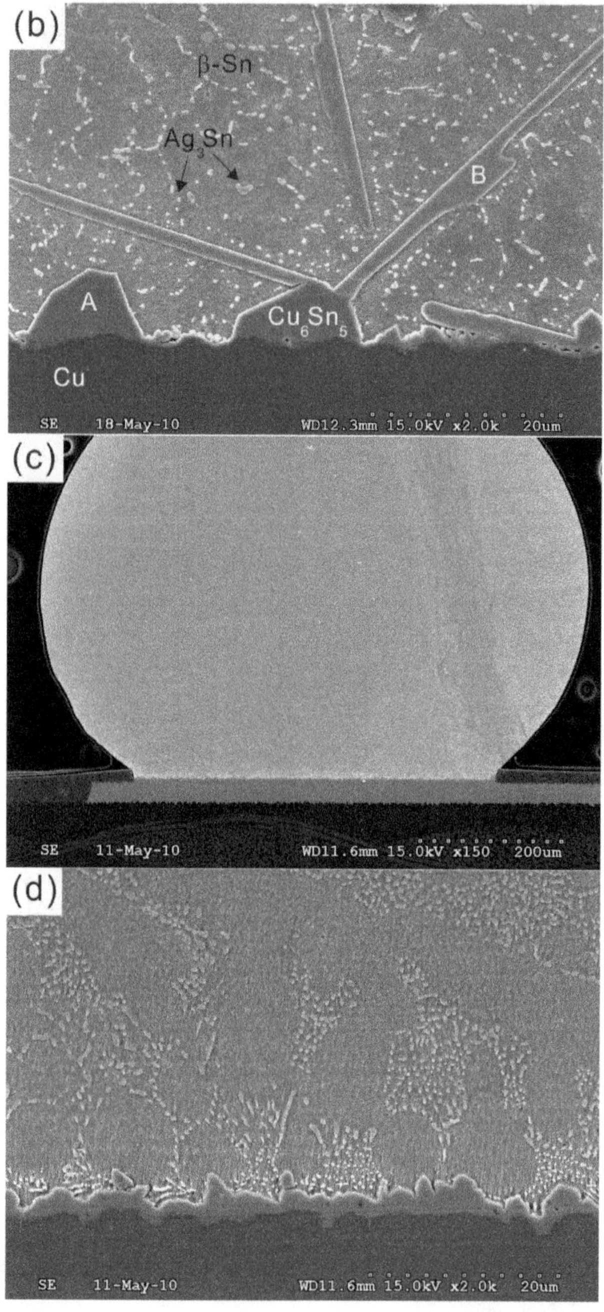

Fig. 12. Morphology of intermetallic compounds formed at the interfaces of the as-reflowed solder joints: (a) Sn3.5Ag0.5Cu, (b) (a) magnifications ; (c) d Sn3.5Ag0.5Cu-0.75TiO_2; (d) (c) magnifications [58].

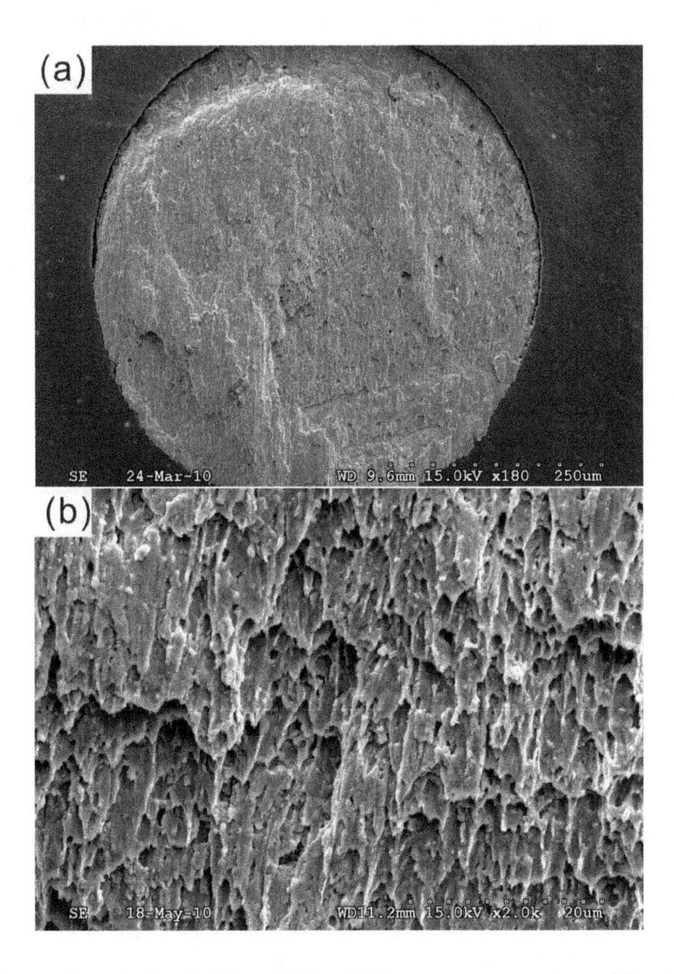

Fig. 13. Fractography of the Sn3.5Ag0.5Cu-1TiO_2 nano-composite solder joints in BGA packages after ball shear tests [58].

To achieve high reliability, solder materials must have high resistance to corrosive conditions such as moisture, air pollutants from industry, and oceanic environments[54]. Although corrosion of solder alloys is not currently a major problem for electronic devices used in normal environments, it may be a problem when they are used in harsh environments, such as oceanic environments. However, there is a lack of information regarding the corrosion resistance of nano-composite solders in corrosive environments.

Figure 14 shows the polarization curves of the Sn3.5Ag0.5Cu solder and the Sn3.5Ag0.5Cu nano-composite solder in 3.5 wt.% NaCl solution[60]. From the polarization curves, the corrosion potential (Φcorr), the breakdown potential (Φ_b), and the dynamic corrosion current density (*I*corr) have been determined (Table 6). The width of the passive region on the anodic polarization curves ($\Delta\Phi$ = Φ_b - Φcorr) in Table 6 indicates the pitting resistibility or the stability of the passive film on the Sn3.5Ag0.5Cu composite alloy surface. The corrosion potential (*Φ*corr) of the Sn3.5Ag0.5Cu nano-composite solder is slightly more passive than that of the Sn3.5Ag0.5Cu solder. This implies that a finer grain size produces more grain boundaries, which act as corrosion barriers. On the other hand, the breakdown potential (Φ_b) of the Sn3.5Ag0.5Cu nano-composite solders becomes much more passive with the addition of oxide nanoparticles. As Table 6 also indicates, the Sn3.5Ag0.5Cu solders possess a higher pitting tendency (smaller $\Delta\Phi$ value) than the Sn3.5Ag0.5Cu nano-composite solders. Rosalbino et al. reported that the pit formation at the surface of Sn–Ag–M alloys is due to the dissolution of the tin-rich phase [56]. In addition, the corrosion current densities were obtained by using the TAFEL extrapolation method. The corrosion current densities of the Sn3.5Ag0.5Cu solders and Sn3.5Ag0.5Cu nano-composite solders were very similar.

Solder	*Φ*corr (mV_{SCE})	*Φ*b (mV_{SCE})	$\Delta\Phi$ (mV)	*I*corr (μA/cm²)
Sn3.5Ag0.5Cu	-662.1	-284.1	378	0.36
Sn3.5Ag0.5Cu-0.5TiO2	-651.4	-95.1	556	0.27
Sn3.5Ag0.5Cu-0.5Al2O3	-642.1	-146.2	496	0.40

Φcorr : corrosion potential; Icorr: corrosion current density; Φb : breakdown potential;
ΔΦ = Φcorr - Φb.

Table 6. Corrosion properties in a 3.5 wt.% NaCl solution for the nano-composite solder [60].

Many studies have reported that the corrosion behavior of alloys depends on the second phase distribution, shown to be Mg alloy[61, 62] and Al alloy[63]. In the Sn3.5Ag0.Cu nano-composite solder alloys, the microstructure had finer β-Sn grains, a large amount of Ag_3Sn particles, and a small amount of oxidize nanoparticles. This leads to improvement of the corrosion behavior of the Sn3.5Ag0.5Cu nano-composite solder, such as greater corrosion resistance, the lower pitting tendency, and the smaller corrosion current density, respectively.

The corrosion products of Sn3.5Ag0.5Cu and Sn3.5Ag0.5Cu nano-composite solder have similar microstructures (Fig. 15). The corrosion products of Sn3.5Ag0.5Cu solder after polarization have a larger flake-like shape (Mark a) and small mushroom-like shape, and are loosely distributed on the surface, with different orientations (Fig.15a). On the other hand, the corrosion products of Sn3.5Ag0.5Cu nano-composite solder after polarization tests have only a flake-like shape, as shown in Fig. 15b (Mark a). Table 7 shows the surface element concentrations of solder corrosion products from EDS. According to the EDS analysis, the corrosion products of Sn3.5Ag0.5Cu and Sn3.5Ag0.5Cu nano-composite solder contain mainly Sn, O, and Cl (Fig.16). It can be seen that the corrosion products of the Sn3.5Ag0.5Cu solders and Sn3.5Ag0.5Cu nano-composite solders have slightly different compositions.

Solder		Surface element concentration (wt.%)				
		Sn	Ag	Cu	Cl	O
Sn3.5Ag0.5Cu	All area	69.26	4.05	0.37	13.40	12.60
	flake	72.97	0.43	-	15.71	9.58
	mushroom	74.87	1.29	-	17.97	5.76
Sn3.5Ag0.5Cu-0.5TiO$_2$	All area	68.7	3.82	0.68	13.39	13.41
	flake	64.22	0.50		15.03	20.25

Table 7. Surface element concentration of different solders after potentiodynamic polarization tests[60].

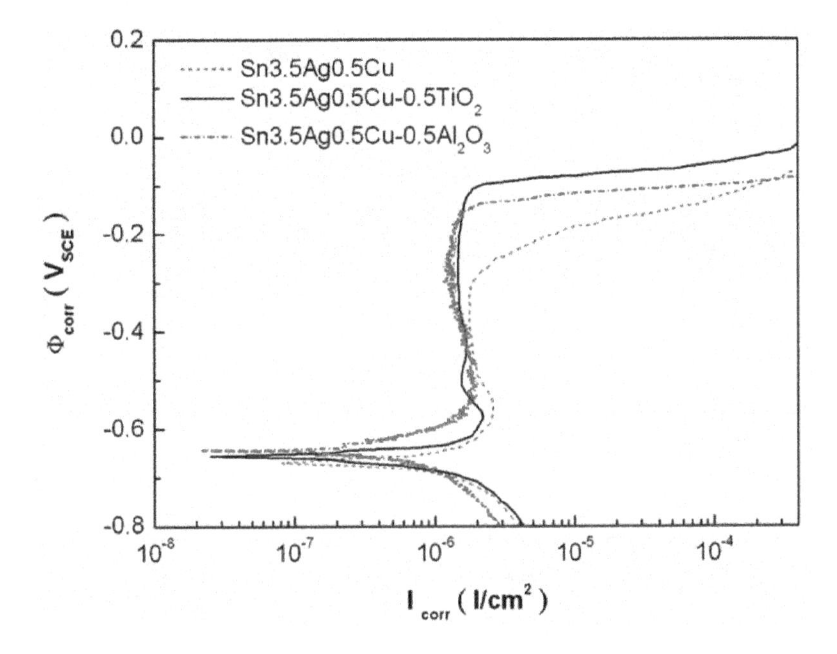

Fig. 14. The potentiodynamic polarization curves of the nano-composite solder in a 3.5wt.% NaCl solution: (a) Sn3.5Ag0.5Cu solder; (b) Sn3.5Ag0.5Cu-0.5TiO$_2$; and (c) Sn3.5Ag0.5Cu-0.5Al$_2$O$_3$ [60].

Fig. 15. Microstructure of the corrosion products on different solders after polarization tests (a) Sn3.5Ag0.5Cu solder, (b) Sn3.5Ag0.5Cu nano-composite solder[60].

During polarization testing in NaCl solution, the only possible cathodic reaction is oxygen reduction [49, 64]:

$$O_2 + 4e^- + 2H_2 \rightarrow 4OH^- \tag{1}$$

When the current density reaches about 10 mA/cm², many hydrogen bubbles evolve from the cathode due to the hydrogen evolution on the cathode:

$$2H_2O + 2e^- \rightarrow H_2 + 2OH^- \tag{2}$$

The reactions on the anode are quite complicated. Some possible anodic reactions have been reported in the literature [46, 64-66], as displayed below:

$$Sn + 2OH^- - 2e^- = Sn(OH)_2 \tag{3}$$

$$Sn + 2OH^- - 2e^- = SnO + H_2O \quad (4)$$

$$Sn + 4H_2O = Sn(OH)_4 + 4e^- + 4H^+ \quad (5)$$

The dehydration of $Sn(OH)_2$ and $Sn(OH)_4$ into SnO and SnO_2, respectively, has also been reported [46, 64,65]:

$$Sn(OH)_2 + 2OH^- + 2e^- = Sn(OH)_4 \quad (6)$$

$$Sn(OH)_2 + 2OH^- = SnO + H_2O \quad (7)$$

$$SnO + H_2O + 2OH^- + 2e^- = Sn(OH)_4 \quad (8)$$

However, Yu et al., after investigating the corrosion properties of Sn9Zn and Sn8Zn3Bi solder in NaCl solution, postulated the formation of a tin oxyhydroxychloride according to the following reaction[66]:

$$3Sn + 4OH^- + 2Cl^- = Sn_3O(OH)_2Cl_2 + H_2O \quad (9)$$

In addition, Li et al.[46] studied the corrosion properties of Sn-Ag, Sn–Ag–Cu, Sn–Cu, and SnPb solder in 3.5wt.% NaCl solution with different scanning rates, and their results showed that the corrosion product on the surface was tin oxide chloride hydroxide ($Sn_3O(OH)_2Cl_2$). In our case, the presence of such a surface layer, instead of a tin oxychloride layer, cannot be ruled out due to the detection limits of energy-dispersive spectroscopy. In order to understand the reaction during the corrosion products, XRD has been used to analyse the corrosion products on the surface after the polarization tests (Fig. 17). The results show that all the Sn3.5Ag0.5Cu and Sn3.5Ag0.5Cu solder materials have the same corrosion product, $Sn_3O(OH)_2Cl_2$, which is a complex oxide chloride hydroxide of tin[67]. This further confirms that the corrosion product on the Sn3.5Ag0.5Cu composite solders is $Sn_3O(OH)_2Cl_2$.

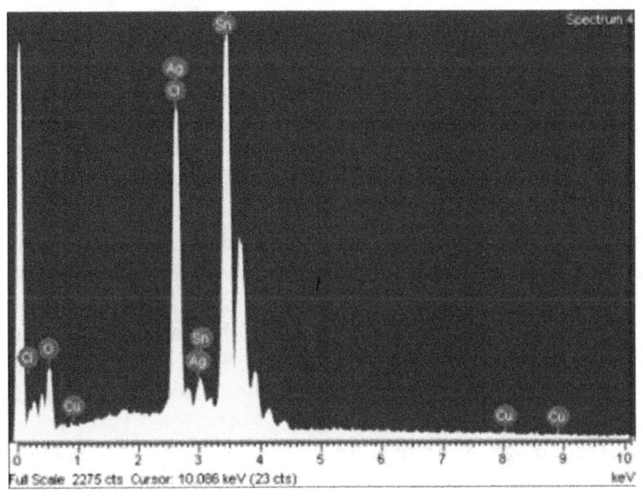

Fig. 16. EDS analysis of corrosion product of the Sn3.5Ag0.5Cu nano-composite solder after polarization tests[60].

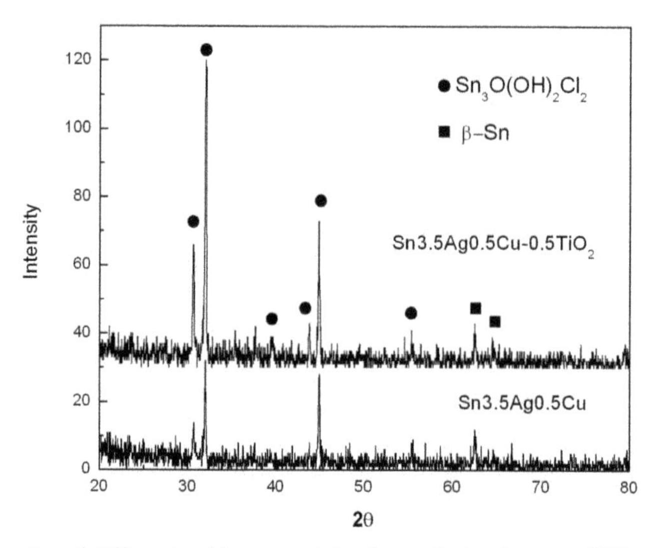

Fig. 17. XRD spectra of different solder materials after polarization tests[60].

9. References

[1] K. N. Tu, K. Zeng, Tin–lead (Sn–Pb) solder reaction in flip chip technology, Mater. Sci. Eng. R 34 (1) (2001) 1–58.

[2] E. P. Wood, K.L. Nimmo, In search of new lead-free electronic solders, J. Electron. Mater. 23 (8) (1994) 709–713.

[3] S. Jin, D. R. Frear, J.W. Morris Jr. Foreword, J. Electron. Mater. 23 (8) (1994) 709-713.

[4] M. Abtew, G. Selvaduray, Lead-free Solders in Microelectronics, Mater. Sci. Eng. R 27 (2000) 95-141.

[5] I. E. Anderson, J. C. Foley, B. A. Cook, J. Harringa, R. L. Terpstra, O. Unal, Alloying effects in near-eutectic Sn–Ag–Cu solder alloys for improved microstructural stability, J. Electron. Mater. 30 (9) (2001) 1050–1059.

[6] M. McCormack, S. Jin, G.W. Kammlott, H. S. Chen, New Pb-free solder alloy with superior mechanical properties, Appl. Phys. Lett. 63 (1) (1993) 15–17.

[7] A. Z. Miric, A. Grusd, Lead-free alloys, Soldering, Surf. Mount Technol. 10 (1) (1998) 19–25.

[8] N. C. Lee, Lead-free Soldering. In Daniel, L. and Wong, C. P. (eds.), Materials for Advanced Packaging, Springer Science Business Media (New York, 2009), pp. 181-218.

[9] K. N. Tu, A.M. Gusak, M. Li, Physics and materials challenges for lead-free solders, J. Appl. Phys. 93 (2003)1335–1353.

[10] H. T. Ma, J. C. Suhling, A review of mechanical properties of lead-free solder for electronic packaging, J. Mater. Sci. 44(2009) 1141–1158.

[11] International Technology Roadmap for Semiconductors—Assembly and Packaging, 2009 ed. (http://www.itrs.net/Links/2009ITRS/2009Chapters_2009Tables/2009_Assembly.pdf).

[12] E. C. C. Yeth, W. J. Choi, K. N. Tu, Current Crowding Induced Electromigration Failure in Flip Chip Technology, Appl. Phys. Lett. 80 (2002) 580-582.

[13] J. Shen, Y. C. Chan, Research advances in nano-composite solders, Microelectronics Reliability 49 (2009) 223–234.

[14] L. C. Tsao, S. Y. Chang, Effects of Nano-TiO_2 additions on thermal analysis, microstructure and tensile properties of Sn3.5Ag0.25Cu solder, Mater. Des. 31(2010) 990–993.

[15] L. C. Tsao, S.Y. Chang, C.I. Lee, W.H. Sun, C.H. Huang, Effects of nano-Al_2O_3 additions on microstructure development and hardness of Sn3.5Ag0.5Cu solder, Mater. Des. 31, (2010) 4831–4835.

[16] T. H. Chuang, M. W. Wu, S. Y. Chang, S. F. Ping, L. C. Tsao, Strengthening mechanism of nano-Al_2O_3 particles reinforced Sn3.5Ag0.5Cu lead-free solder, J. Mater. Sci.: Mater. Electron. 22 (2011)1021–1027.

[17] J. Shen, Y.C. Liu, D. J. Wang, H.X. Gao, Nano ZrO_2 particulate-reinforced lead-free solder composite, J. Mater. Sci. Technol. 22 (2006) 529-532.

[18] X. L. Zhong, M. Gupta, Development of lead-free Sn–0.7Cu/Al_2O_3 nanocomposite solders with superior strength, J. Phys. D: Appl. Phys. 41 (2008) 095403–095409.

[19] K. Mohan Kumar, V. Kripesh, Andrew A.O. Tay, Single-wall carbon nanotube (SWCNT) functionalized Sn–Ag–Cu lead-free composite solders, J. Alloy Compd. 450 (2008) 229–237.

[20] S. M. L. Nai, J. Wei, M. Gupta, Improving the performance of lead-free solder reinforced with multi-walled carbon nanotubes, Mater. Sci. Eng. A 423 (2006) 166–169.

[21] S. M. L. Nai, J. Wei, M. Gupta, Influence of ceramic reinforcements on the wettability and mechanical properties of novel lead-free solder composites, Thin Solid Films 504 (2006) 401–404.

[22] Z. X. Li, M. Gupta, High strength lead-free composite solder materials using nano-Al_2O_3 as reinforcement. Adv. Eng. Mater. 7 (2005)1049–1054.

[23] S. M. L. Nai, J. Wei, M. Gupta, Effect of carbon nanotubes on the shear strength and electrical resistivity of a lead-free solder. J. Electron. Mater. 37(2008) 515–522.

[24] H. Mughrabi, Plastic deformation and fracture of materials. Berlin: Springer-Verlag;. 1993. p. 315–322

[25] L. C. Tsao, 10th International Conference on Electronic Packaging. Technology & High Density Packaging (ICEPT-HDP 2009) (2009)1164 -1166.

[26] R. W. Wu, L. C. Tsao, S. Y. Chang, C. C. Jain and R. S. Chen, Interfacial reactions between liquid Sn3.5Ag0.5Cu solders and Ag substrates, J. Mater. Sci.: Mater. Electron. 22(8) (2011) 1181-1187.

[27] Görlich, D. Baither, G. Schmitz, Reaction kinetics of Ni/Sn soldering reaction, Acta Materialia 58(9) (2010) 3187-3197.

[28] T. H. Chuang, Y. T. Huang, L. C. Tsao, $AgIn_2$/Ag_2In transformations in an In-49Sn/Ag soldered joint under thermal aging, J. Electron. Mate. 30 (2001) 945-950.

[29] D. G. Kim, C. Y. Lee, S. B. Jung, Interfacial reactions and intermetallic compound growth between indium and copper, J. Mater. Sci.: Mater. Electron. 5(2) (2004) 95-98.

[30] L. C. Tsao, Evolution of nano-Ag_3Sn particle formation on Cu-Sn intermetallic compounds of Sn3.5Ag0.5Cu composite solder/Cu during soldering, J. Alloy Compd.509 (2011) 2326–2333.

[31] L. C. Tsao, Suppressing effect of 0.5 wt.% nano-TiO_2 addition into Sn–3.5Ag–0.5Cu solder alloy on the intermetallic growth with Cu substrate during isothermal aging, J. Alloy Compd. 509 (2011) 8441– 8448.
[32] L. C. Tsao, C.P. Chu, S. F. Peng, Study of interfacial reactions between Sn3.5Ag0.5Cu composite alloys and Cu substrate, Microelectron. Eng. 88 (2011) 2964-2969.
[33] S. Y. Chang, L. C. Tsao, M. W. Wu and C. W. Chen, The morphology and kinetic evolution of intermetallic compounds at Sn–Ag–Cu solder/Cu and Sn–Ag–Cu-0.5Al_2O_3 composite solder/Cu interface during soldering reaction, J. Mater. Sci.: Mater. Electron. DOI: 10.1007/s10854-011-0476-9.
[34] X.Y. Liu, M.L. Huang, C. M. L. Wu, Lai Wang, Effect of Y_2O_3 particles on microstructure formation and shear properties of Sn-58Bi solder, J Mater Sci: Mater Electron, J Mater Sci: Mater Electron 21 (2010) 1046–1054.
[35] S. M. L. Nai, J. Wei, M. Gupta, Interfacial Intermetallic Growth and Shear Strength of Lead-Free Composite Solder Joints, J. Alloy Compd. 473 (2009) 100–106.
[36] A. K. Gain, T. Fouzder, Y. C. Chan, Winco K. C. Yung, Microstructure, kinetic analysis and hardness of Sn–Ag–Cu–1 wt% nano-ZrO_2 composite solder on OSP-Cu pads, J. Alloy Compd. 509 (2011) 3319–3325.
[37] R. Ambat, P. Møller, A review of Corrosion and environmental effects on electronics, The Technical University of Denmark, DMS vintermøde proceedings, (2006).
[38] M. Mori, K. Miura, Corrosion of Tin Alloys in ulfuric and Nitric Acids, Corrosion Science, 44 (2002)887-898.
[39] F. Song, S. W. Ricky Lee, Corrosion of Sn-Ag-Cu Lead-free Solders and the Corresponding Effects on Board Level Solder Joint Reliability, 2006 Electronic Components and Technology Conference, (2006) 891-889.
[40] B. Liu, T. K. Lee, K. C. Liu, Impact of 5% NaCl Salt Spray Pretreatment on the Long-Term Reliability of Wafer-Level Packages with Sn-Pb and Sn-Ag-Cu Solder Interconnects , J. Electron. Mater. 40, 2011,doi: 10.1007/s11664-011-1705-y.
[41] Morten S. Jellesen et al., Corrosion in Electronics (Paper presented at 2008 Eurocorr Conference, Edinburgh, Scotland, 7–11 September 2008).
[42] R. Baboian, "Electronics," Corrosion Tests and Standards: Applications and Interpretation, ed. R. Baboian (West Conshohocken, PA: ASTM, 1996), www .corrosionsource. com/events/intercorr/baboian.htm).
[43] V. Chidambaram, J. Hald, R. Ambat, J. Hattel,A Corrosion Investigation of Solder Candidates for High-temperature Applications, JOM , 61 (2009) 59-65.
[44] H. Liu, K. Wang, K. Aasmundtveit, N. Hoivik, Intermetallic Cu_3Sn as Oxidation Barrier for Fluxless Cu-Sn Bonding, 2010 Electronic Components and Technology Conference, 853-857.
[45] L. C. Tsao, Corrosion Characterization of Sn37Pb Solders and With Cu Substrate Soldering Reaction in 3.5wt.% NaCl Solution, 2009 International Conference on Electronic Packaging Technology & High Density Packaging (ICEPT-HDP), (2009) 1164-1166.
[46] D. Li, P. P. C. C. Liu, Corrosion characterization of tin–lead and lead free solders in 3.5 wt.% NaCl solution, Corr. Sci. 50 (2008) 995–1004.
[47] K. L. Lin, T.P. Liu, The electrochemical corrosion behaviour of. Pb-Free Al–Zn–Sn solders in NaCl solution, Mater. Chem. Phys. 56 (1998) 171-176.

[48] K. L. Lin, F.C. Chung, T.P. Liu, The potentiodynamic polarization behavior of Pb-free XIn-9(5Al-Zn)-YSn solders, Mater. Chem. Phys. 53 (1998) 55-59.

[49] U.S. Mohanty, K.L. Lin, The effect of alloying element gallium on the polarisation characteristics of Pb-free Sn–Zn–Ag–Al–XGa solders in NaCl solution, Corros. Sci. 48 (2006) 662–678.

[50] D. Q. Yu, W. Jillek, E. Schmitt, Electrochemical migration of Sn–Pb and lead free solder alloys under distilled water, J. Mater. Sci. :Mater. Electron. 17 (2006) 219–227.

[51] B. Y. Wu, Y.C. Chan, M. O. Alam, Electrochemical corrosion study of Pb-free solders, J. Mater. Res. 21 (2006) 62-70.

[52] A. Ahmido, A. Sabbar, H. Zouihri, K. Dakhsi, F. Guedira, M. Serghini-Idrissi, S. El Hajjaji ,Effect of bismuth and silver on the corrosion behavior of Sn–9Zn alloy in NaCl 3wt.% solution, Mater. Sci. Eng. B 176 (2011) 1032– 1036.

[53] U. S. Mohanty, K. L. Lin, Electrochemical corrosion behaviour of Pb-free Sn-8.5Zn-0.05Al-XGa and Sn-3Ag-0.5Cu alloys in chloride containing aqueous solution, Corrosion Science 50 (2008) 2437–2443.

[54] J. Hu, T. Luo, A. Hu, M. Li, D. Mao, Electrochemical Corrosion Behaviors of Sn-9Zn-3Bi-xCr Solder in 3.5% NaCl Solution, J. Electron. Mater. 40, (2011) 1556-1562.

[55] T. C. Chang, J. W. Wang, M. C. Wang , M. H. Hon, Solderability of Sn-9Zn-0.5Ag-1In lead-free solder on Cu substrate Part 1. Thermal properties, microstructure, corrosion and oxidation resistance, J. Alloy Compd. 422 (2006) 239–243.

[56] F. Rosalbino, E. Angelini, G. Zanicchi, R. Marazza, Corrosion behaviour assessment of lead-free Sn-Ag-M (M = In, Bi, Cu) solder alloys, Mater. Chem. Phys. 109 (2008) 386–391.

[57] T. H. Chuang, M. W. Wu , S. Y. Chang, S. F. Ping, L. C. Tsao, Strengthening mechanism of nano-Al_2O_3 particles reinforced Sn3.5Ag0.5Cu lead-free solder, J Mater Sci: Mater Electron 22 (2011) 1021–1027.

[58] J. C. Leong , L. C. Tsao , C. J. Fang,C. P. Chu, Effect of nano-TiO_2 addition on the microstructure and bonding strengths of Sn3.5Ag0.5Cu composite solder BGA packages with immersion Sn surface finish, J Mater Sci: Mater Electron, (2011)1443-1449.

[59] L. C. Tsao, M. W. Wu, S. Y. Chang, Effect of TiO_2 nanoparticles on the microstructure and bonding strengths of Sn0.7Cu composite solder BGA packages with immersion Sn surface finish, J Mater Sci: Mater Electron DOI 10.1007/s10854-011-0471-1.

[60] L. C. Tsao, T. T. Lo, S. F. Peng, S. Y. Chang, Electrochemical behavior of a new Sn3.5Ag0.5Cu composite solder, 11th International Conference on Electronic Packaging Technology & High Density Packaging, (2010), 1013-1017.

[61] A. Pardo, M. C. Merino, A. E. Coy, R. Arrabal, F. Viejo, E. Matykina, Corrosion Behavior of Magnesium/Aluminum Alloys in 3.5 wt% NaCl, Corr. Sci., 50 (2008)823-834.

[62] G. Ben-Hamu, A. Eliezer, E. M.Gutman. Electrochemical Behavior of Magnesium Alloys strained in Buffer Solutions, Electrochim. Acta, 52 (2006)304-313.

[63] Y. Liu, Y. F. Cheng, Role of Second Phase Particles in Pitting Corrosion of 3003 Al Alloy in NaCl Solution, Mater. Corr., 61 (2010)211-217.

[64] S. D. Kapusta, N. Hackerman, Anodic passivation of tin in slightly alkaline solutions, Electrochim. Acta 25 (1980) 1625–1639.

[65] Q. V. Bui, N. D. Nam, B.I. Noh, A. Kar, J. G. Kim, S.B. Jung, Effect of Ag addition on the corrosion properties of Sn-based solder alloys, Mater. Corr. 61 (2010) 30-33.

[66] D. Q. Yu, C. M. L. Wu, L. Wang, The Electrochemical Corrosion Behavior of Sn-9Zn and Sn-8Zn-3Bi Lead-Free Solder Alloys in. NaCl Solution, 16th International Corrosion Congress, Beijing, P.R. China, (2005)19–24.
[67] F. Rosalbino, E. Angelini, G. Zanicchi, R. Carlini, R. Marazza, Electrochemical Corrosion Study of Sn–3Ag–3Cu Solder Alloy in NaCl solution, Electrochim. Acta, 54 (2009) 7231-7237.

5

Tribocorrosion: Material Behavior Under Combined Conditions of Corrosion and Mechanical Loading

Pierre Ponthiaux[1], François Wenger[1] and Jean-Pierre Celis[2]
[1]*Ecole Centrale Paris, Dept. LGPM, Châtenay-Malabry,*
[2]*Katholieke Universiteit Leuven, Dept. MTM, Leuven,*
[1]*France*
[2]*Belgium*

1. Introduction

1.1 Definition of tribocorrosion

Tribocorrosion can be defined as the study of the influence of environmental factors (chemical and/or electrochemical) on the tribological behavior of surfaces. In other words, the process leading to the degradation of a metallic and/or non-metallic material resulting from a mechanical contact (sliding, friction, impact, ...) combined to a corrosive action of the surrounding environment.

The origin of tribocorrosion is closely related to the presence of a passive film on material surfaces subject to wear and the modifications of these surfaces by friction or any other form of mechanical loading. In very general terms, the passive film (mainly oxide) is considered to be snatched in the contact area.

Oxide particles, referred to as 'debris", are released from the contacting materials. Then, the debris can be removed from the contact zone or on the contrary trapped in it. In the case of removal, the debris dissolve chemically or are dragged out by a hydraulic flow along the material surface. In this case, the tribocorrosion mechanism is based on a repeated tearing off of the oxide after each contact and eventually a removal of some of the underlying material depending on the intensity of mechanical stress acting on the contacting materials. The major concern is then to quantify and eventually to model the kinetics of repassivation as accurately as possible. This type of tribocorrosion process can be classified as an oxidative wear mechanism as, for example, the 'mild oxidative wear model' (Quinn, 1992). In the case of debris trapping, one has to consider that under appropriate hydrodynamical, chemical, and thermal contact conditions and relative speed of the two contacting bodies, the debris will remain temporarily in the contact zone mainly as colloids with a diameter usually in the range of a few hundred nanometers. Two cases may then be distinguished: (a) the debris accelerates the wear in comparison to the case of debris-free contacts is accelerated by an abrasive effect, or (b) the debris slows down the wear compared to the case where the contact zone is free of any debris, resulting in a protective effect.

Tribocorrosion may take place in practice in a large number of very different tribological systems consisting of mechanical devices containing metallic parts that are in contact with counterparts and exhibiting a relative movement placed in an environment revealing itself to be corrosive to at least one of the contacting materials. A non-limitative list of examples might contain machinery pumps, bearings, gears, ropes, electrical connectors, hinges, microelectromechanical systems (MEMs), and orthopedic implants like hip and knee implants.

1.2 Synergism between mechanical and chemical loading

To understand the importance and complexity of the phenomena taking place under tribocorrosion, one has to consider that the corrosiveness of a medium (liquid or gas) towards a material is highly dependent on the mechanical stresses that act onto a material, particularly at its surface exposed to that environment.

In tribocorrosion, five mechanisms may explain the synergism noticed between mechanical and chemical factors acting on contacting materials, namely:

1. the debris can speed up or reduce wear compared to what happens in the same environment where debris does not exist like e.g. in sliding contacts polarized at a large cathodic potential,
2. a galvanic coupling is established between the worn and unworn areas. It accelerates the anodic dissolution in the area where the metal is depassivated,
3. a galvanic coupling may be established between the two contacting counterparts,
4. an accumulation of dissolved species may take place in the liquid surrounding the contact. This may render the medium chemically or electrochemically more aggressive,
5. the mechanical loading in the contact area and its nearby zone may causes a work hardening of the materials. This work hardening can alter the kinetics of corrosion and/or repassivation processes.

A synergistic effect occurs in tribocorrosion when the mechanical process affects the corrosion process acting in a tribological system or vice versa. In these cases, the wear, W, found on a given component in a tribological system subjected to a mechanical loading in a given corrosive environment, will be very different and often much greater than the sum of the mechanical wear, W_{mo}, measured as a material loss under a given mechanical load in the absence of a corrosive environment, and the material loss induced by corrosion, W_{co}, in the absence of any mechanical contact (see Equation 1):

$$W \neq W_{mo} + W_{co} \qquad (1)$$

This result is partly explained by the fact that the corrosion resistance in the case of a metal depends on the presence at its surface of reaction layers, sometimes only a few atom layers thick, resulting from an interaction between the material and the surrounding environment. Such layers can be classified as oxides, solid precipitates, adsorbed layers, or passive surface films. Some of them like dense oxide layers, precipitates, or passive films play a protective role by isolating the underlying metal from a direct contact with a surrounding corrosive environment. This is particularly true in the case of stainless steels and other alloys containing chromium. Their passive surface film formed in ambient air or in contact with an aqueous solution is a few nanometers thick but gives them a high resistance to corrosion. The sliding of a hard counterbody material on such a surface is likely to damage that passive film what is known as a "depassivation" process by which the bare material is exposed to the corrosive environment. Various but essentially electrochemical processes can then compete on these bare surfaces, namely:

- the dissolution of the metal in the corrosive medium,
- the formation of a new compound that may contribute to the breakdown process, and
- the restoration of the protective film known as "repassivation" process.

1.3 Complexity of the tribocorrosion process

The following examples taken from literature illustrate quite well the numerous parameters and interactions that govern the tribocorrosion process. Lemaire & Le Calvar, 2000, described the wear of a cobalt-based alloy coating generally referred to as "stellite 6" applied on the gripper latch arms of the control rods command mechanisms in pressurized water reactors (PWR). The downwards movement of the control rods is controlled by gripper latch arms of which the protruding teeth are coated with Stellite 6. The teeth block the movement once they come in contact with the control bar at the circular grooves lining their surface. At each blocking step, there is a contact between teeth and inner part of the grooves at a moderate pressure estimated at 150 MPa. Subsequently a sliding takes place over a distance of approximately 0.1 mm before the control rods come to rest. In the middle of the primary cooling circuit stellite 6 does not undergo any significant corrosion in the absence of any mechanical stress, thanks to the protective action of the passive film on stellite 6 consisting of chromium oxides. However under field operating conditions where impact and sliding of the teeth on the control bar take place, corrosion is evident. The wear observed on the teeth was found not to depend only on the number of blocking steps as would be the case in absence of corrosion. But the wear was found to depend also on the time interval between two successive blocking steps. The wear rate for a given number of blocking steps appeared to increase with the latter.

A plausible hypothesis to interpret this behavior is to consider that between two successive blocking steps corrosion takes place on parts of the surface where the passive film was mechanically damaged in the preceding step. The wear progress is correlated with the time interval between successive blocking steps by the following simple empirical equation (Bom Soon Lee et al., 1999):

$$I(t) = I_0 \left(\frac{t}{t_0} \right)^{-n} + I_p \tag{2}$$

in which I(t) is the evolution of the dissolution current of a metal with time starting at the time the metal becomes depassivated due to a mechanical action and extending during the film restoration where dissolution and repassivation are competitive surface processes. The parameters I_0 and t_0 are constants, while I_p is the passivation current under steady state, and n has a value between 0.3 and 1.

However, the effect of sliding on the electrochemical reactivity of the surface of a metal is not always confined to the partial destruction of surface layers. Other phenomena resulting directly or indirectly from contacts between parts can influence the corrosion behavior of their surfaces. For example, under reciprocating sliding conditions at small displacement amplitude, known as fretting (Carton et al., 1995; Godet et al, 1991) cracks may appear at the rim of the contact zone even after only a small number of contact events. Another parameter to be considered in tribocorrosion is the stirring of the corrosive environment along the surfaces of contacting parts caused by their relative movement. It affects tribocorrosion since such a stirring modifies the transport kinetics of chemical species that are generated in the vicinity of the surfaces due to the corrosion-related reactions.

The effect of the environment (mechanical or physico-chemical) on the crack propagation is evident. This returns us, somehow, to the notion of stress corrosion with the nuance that the cracks are induced at a mechanical loading which is not constant in the fretting test under consideration.

In addition one has to consider the possible role of strain hardening and/or structural transformations induced by the sliding action on the electrochemical reactivity of the surface, speeding up or slowing down some reactions. The resulting material transformation may end up in the most stable phase of the material considered being a supersaturated solid solution obtained by the gradual dissolution of pre-existing precipitates. It may also become a nano-crystalline network of a few tens of nanometers in average size that contains a high density of dislocations with no preferred orientation. This structure is very hard but also very fragile. Its intrinsic reactivity with the environment differs necessarily from the one of the original surface. Moreover, starting with the formation of a network of micro-cracks on it, wear particles (known as 'debris') are generated.

In the case of tribocorrosion, it is important to consider the galvanic coupling that might result from the heterogeneity of the electrochemical state of non-rubbed surfaces, and rubbed surfaces undergoing a strain hardening and on which the surface layers are altered. This galvanic coupling causes the polarization of non-rubbed and rubbed areas, and modifies the kinetics of reactions in these areas.

One has also to consider the existence of a third body that consists of wear particles as visible in Figure 1. At first, this third body may interact with the environment to form oxides or hydroxides. If they are ejected out of the contact zone, they become strictly speaking 'debris' and contribute to a material loss. If they remain in the contact zone, they can modify the mechanical response of the system by favoring a sliding action between counterparts, by acting as an abrasive agent promoting the so-called "abrasive wear", or by affecting the reactivity of material surfaces.

In return, a corrosion process can modify the surface states of materials and in consequence the contact conditions. In that way corrosion can affect the sliding conditions (coefficient of friction, wear regime, ...). The interaction between friction and corrosion therefore induces a complex phenomenon of synergy.

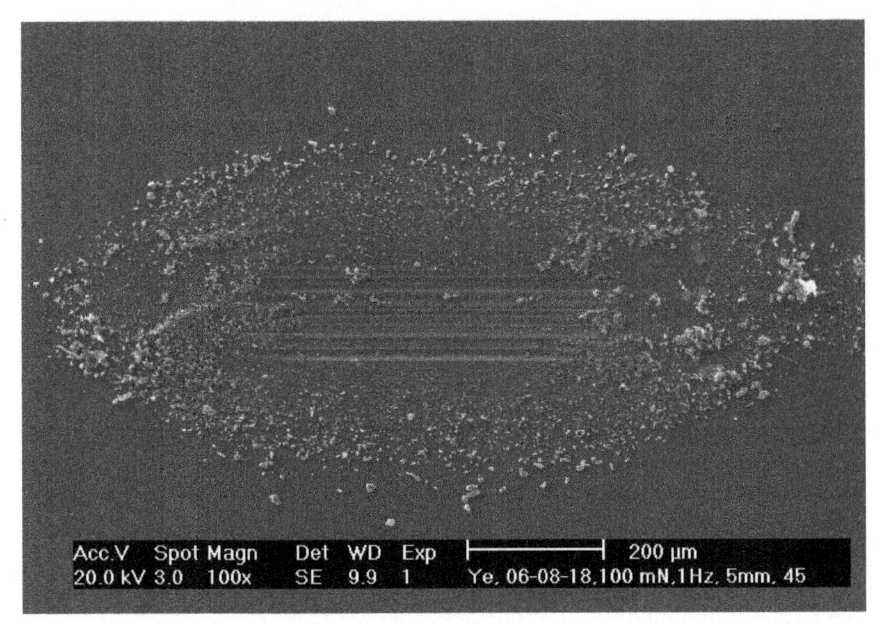

Fig. 1. Example of a third body resulting from a tribological process during reciprocating sliding. Some abrasive grooves are also visible in the central area of the sliding track.

2. The *in situ* study of tribocorrosion processes by electrochemical techniques

When performing a tribocorrosion test, one has to implement not only the following traditional concepts of tribological testing, namely:

- the representativeness of the contact (e.g. sphere-on-flat, sphere, cylinder-on-flat, flat-on-flat,...),
- the choice of the type of relative movement (continuous or reciprocating),
- the applied forces and the relative velocities, accelerations or displacements,
- the influence of the characteristics of the tribometer used on the wear induced, and
- the similarity of the wear mechanisms active in the laboratory test and in the field,

but he has also to take into account and to control simultaneously a large number of testing parameters, like:

- mechanical ones (e.g. contact frequency, noise, vibrations, surface roughness and residual stress ...),
- metallurgical ones (e.g. composition and structural state of contacting materials, microstructure, surface film composition and structure ...), and
- environmental ones (e.g. composition of the corrosive environment, pH, aggressiveness, viscosity, ionic conductivity, temperature, solid particles in suspension, stagnant or stirred,..).

These parameters determine the electrochemical reactivity of the surfaces and in consequence influence the contact conditions (wear regime, existence of a third body, friction ...).

Such tests therefore need to be instrumented to control and/or to record the contact conditions like the normal or tangential force, the relative displacement, velocity, acceleration and contact frequency ...). They need also to be instrumented with electrochemical techniques enabling the control and/or the recording of electrochemical parameters like the polarization of the contacting materials.

The choice of electrochemical techniques that can be implemented in a tribocorrosion test and the development of relevant models for the interpretation of the tribocorrosion mechanism are determined by the mechanical contact conditions being continuous or reciprocating. Electrochemical measurements can be performed with both types of tribometers. However, to be implemented under conditions that allow the interpretation of results, some methods require stationary electrochemical conditions, at least prior to starting up the measurements. In the case of continuous sliding, a quasi-stationary electrochemical surface state can often be reached, and all the electrochemical techniques available for corrosion studies (polarization curves, impedance spectroscopy, electrochemical noise,...), can be used. On the contrary, when reciprocating contact conditions prevail, the interpretations of experimental results are more complex due to the non-stationary electrochemical conditions. Measuring techniques suitable for the recording of current or potential transients will be used preferentially (Mischler et al., 1997; Rosset, 1999).

2.1 Open circuit potential measurements

Under sliding conditions, the measured open circuit potential is a mean value which depends on the electromotive force induced by the surface heterogeneity resulting from the coexistence of non-rubbed and rubbed areas which are in different electrochemical states, and on the areas of these zones and their spatial distribution that determines the non-uniform distribution of potential over the whole surface (Oltra et al, 1986). When applying for example a continuous sliding, this open circuit potential responds to the mechanical loading imposed, as shown in Table 1 and Figure 2.

Fn (N)	0	2.2	2.2	4.2
V (cm s-1)	0	0.5	0.8	0.8
Eoc (V vs SCE)	-0.15	-0.29	-0.32	-0.42

Table 1. Variations of the open-circuit potential E_{OC} of 316L stainless steel in artificial sea water as a function of varying tribological contact conditions. S.C.E.: saturated calomel electrode. F_n: normal force. V: sliding speed.

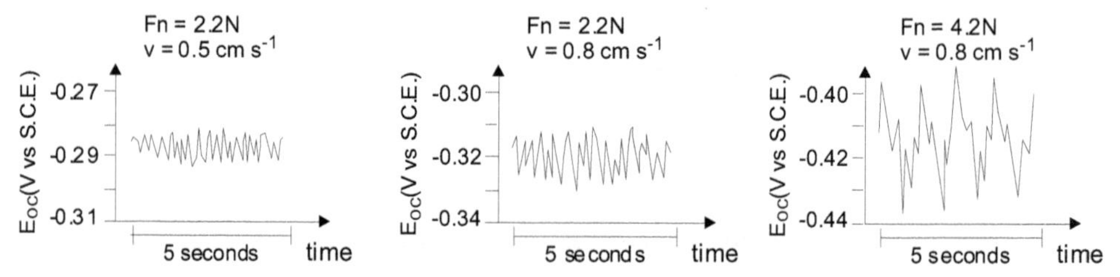

Fig. 2. Electrochemical noise at the open circuit potential E_{OC} of 316L stainless steel in artificial sea water for different tribological contact conditions. S.C.E.: saturated calomel electrode. F_n: normal force. V: sliding speed.

Table 1 gives the evolution of the mean value of the open circuit potential of a 316 L stainless steel in contact with alumina immersed in artificial seawater for different normal forces and velocities (Ponthiaux et al., 1997). Alumina is taken since it has a high electrical resistance and is chemically inert in the liquid used. At a zero normal force and speed, the measured open circuit potential corresponds to the passive state of the entire stainless steel surface. When the normal force or the speed increases, the open circuit potential shifts towards lower potential values. This shift can be explained by considering the following phenomena:

- on the rubbed area, the surface layer on materials can be partially destroyed (also called "depassivation") and that rubbed area acquires a potential that corresponds to the one of an "active material" undergoing dissolution or on which reduction reactions can take place, and
- an increase of the normal force or velocity tends to increase the area of active material.

In Figure 2, the rapid fluctuations of the open circuit potential arising during sliding are represented schematically. They constitute an 'electrochemical noise' representing the electrochemical response to the rapid and stochastic fluctuations of the new bare metal surface generated in the real contact area by sliding friction. This noise can provide information on the mechanism of friction as well as on the mechanism of electrochemical reactions involved in the depassivation - repassivation process (Déforge et al.). A more detailed analysis of open circuit potential measurements under sliding requires a more precise knowledge of the local surface state of contacting materials. Experimentally, microelectrodes can be used to determine potential values of rubbed and non-rubbed areas. Such techniques have already been used to study localized corrosion, and models have been proposed (Lillard et al., 1995). Note that the interpretation of local potential measurements or the development of theoretical models describing the potential distribution, can only be obtained with realistic assumptions on the reaction kinetics of reactions occurring at rubbed and non-rubbed surface areas. Such Information can only be obtained by using complementary methods.

2.2 Potentiodynamic polarization measurements

Potentiodynamic polarization curves obtained at increasing and decreasing potential scan in absence of any sliding is schematically shown in Figure 3. In the case the current measured originates from the whole surface of the tested sample that might be considered as being uniform.

Under sliding conditions, the currents measured during potentiodynamic polarization are in a first approach the sum of two components, namely the current originating from the rubbed area, and the one linked to the non-rubbed area. Under such conditions, the maximum dissolution current, I_M, varies with the mean contact pressure and sliding speed. However, these two test parameters do not necessarily affect in the same way the electrochemical behavior of the alloy:

- at constant speed an increase of the mean contact pressure may cause a decrease of the maximum dissolution current, I_M. This reflects then a slowing down of the dissolution of the alloy which can be due to the strain hardening of the material in the rubbed area. Moreover, the potential, E_M, at which this maximum current is noticed, remains almost unaffected. This indicates that the dissolution process is the same on rubbed and non-rubbed materials,

- at a fixed mean pressure, an increase of the sliding speed may cause, in turn, an increase of the maximum dissolution current. This increase of I_M is then a consequence of the increased bare surface area generated per unit of time. It indicates in that case that the dissolution of the bare material is faster than the non-rubbed material.

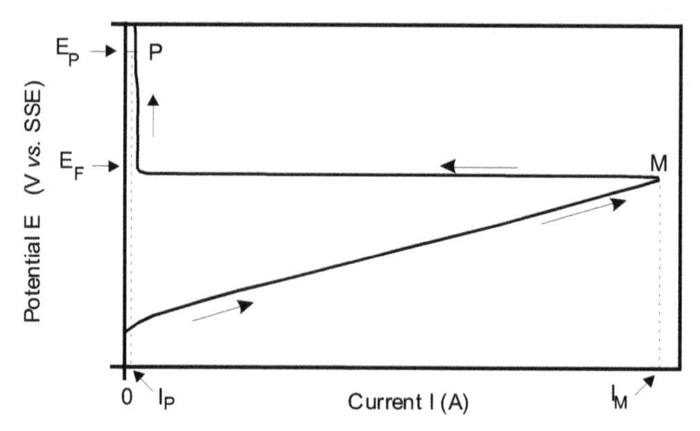

Fig. 3. Schematic potentiodynamic polarization curve recorded on a passivating metallic material (Fe-30%Ni alloy) in absence of any sliding: E_F = Flade potential, E_M = potential at maximum dissolution current I_M, I_P = passivation current.

The possible evolution of the passivation current, Ip, under sliding friction with the mean contact pressure, P_M, and the sliding velocity is schematically shown in Figure 4. That passivation current is increasing with these two testing parameters. That increase can be related to an increase in bare surface area due to:

- an increasing contact area with increasing normal force, and
- an increasing depassivated surface area generated per unit time with increasing sliding speed.

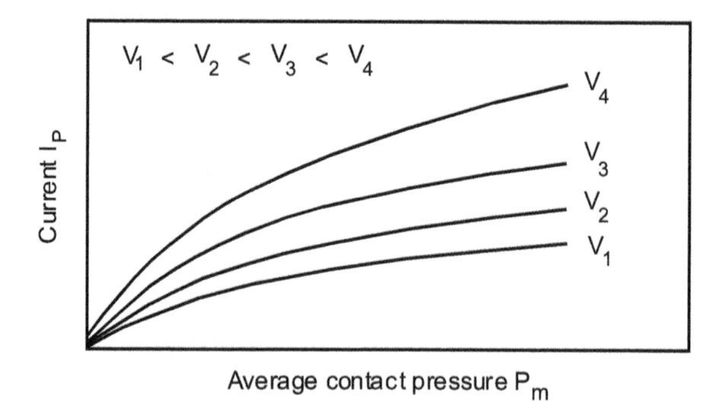

Fig. 4. Evolution of the passive current, I_p, with the mean contact pressure, P_m, at increasing sliding speeds. Example of a Fe-31%Ni alloy disc immersed in 0.5 M sulfuric acid and polarized at a fixed passive potential, under continuous sliding against alumina.

The extent of the passivation range is in general affected by the sliding. In some cases the Flade potential, E_F, shifts towards more anodic values at increasing contact pressure. This means that the stability of the passive film decreases. When the potential E_F is reached, the entire surface becomes active and dissolves. However, under these experimental conditions, the disruption of the passive film is only a local disturbance in the sliding track in a relatively small area that represents generally not more than 10% of the total sample area. Therefore sliding does not only affect the electrochemical state of the rubbed surface, but also the state of the whole surrounding sample surface.

Finally it should be stressed that the coefficient of friction varies with the applied potential during a potentiodynamic polarization. Changes in the value of the coefficient of friction reveal a possible change in the surface state of the materials in the sliding track (Ponthiaux et al., 1999).

The in-depth interpretation of the polarization curves frequently faces difficulties related to the non-uniform distributions of current and potential on the sample surface. This non-uniformity originates from the intrinsic effect of the sliding that causes an heterogeneity of the electrochemical surface reactivity, combined with the ohmic drop in the electrolyte. A full exploitation of the polarization curves in terms of local behavior is possible only if one can model the current and potential distributions under sliding conditions. This brings back to the same approach as in the case of the interpretation of open circuit potential measurements. Note that the effect of non-uniform distributions on the interpretation of polarization curves was already investigated in the absence of any sliding (Law & Newman, 1979; Ponthiaux et al., 1995; Tiedemann et al., 1973).

2.3 Repassivation current transients

Methods based on analysis of potential or current transients (Ponthiaux et al., 1995) are particularly well suited to study reciprocating sliding tests (Mischler et al., 1997). These methods are used to study between successive contact events, the rebuild of damaged surface layers (oxide, passive film ...). Under imposed polarization e.g. in the passivation range, at each stop-start event, a transient variation of current is noticed with time (see Figure 5). The charge corresponding to this transition can be attributed to the re-growth of a uniform film in the damaged area.

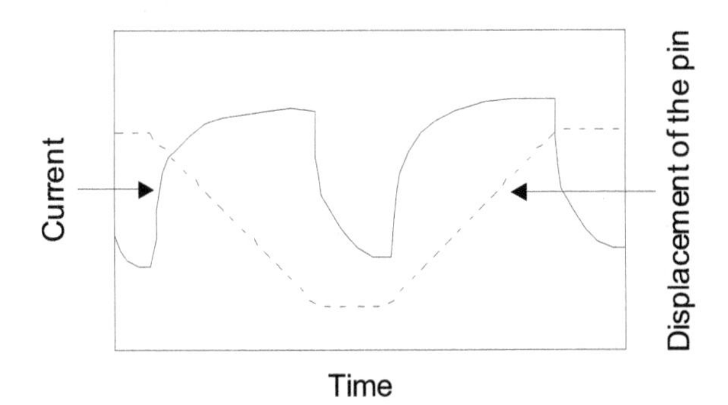

Fig. 5. Schematic evolution of the current transients obtained under reciprocating sliding.

In order to study in more details the mechanism and kinetics of a re-growth of the passive film, the "potential jump" method can be used. However, such a method cannot be applied on metals such as aluminum, for which the passive film can not be reduced by a cathodic polarization, or for some steels for which the contamination of the surface by reduction products can affect the initial current increase at the potential jump.

2.4 Electrochemical impedance measurements

This method requires quasi-stationary electrochemical conditions of currents and potentials. The impedance measurements allow the study of the influence of sliding on the elementary processes involved in the corrosion mechanism. By performing a systematic analysis of the changes in the impedance diagrams with the sliding parameters like normal force and sliding speed or contact frequency, a model can be developed incorporating the sliding effects in the mechanism.

Impedance measurements recorded on a non-rubbed and a rubbed Fe-31% nickel alloy are shown in Figure 6. The measurements were done in 0.5 M sulfuric acid at a potential of -675 mV / SSE located in the active region (Boutard et al., 1985). The impedance diagrams were recorded in a limited range of measurement frequencies. In particular, under sliding, the impedance was not measured at frequencies below 0.01 Hz, to limit the duration of the measurements and to avoid in this way the influence of the long-term evolution of the electrochemical state of the surface due to wear - induced changes in the "rubbed area/unrubbed area" ratio. By taking these precautions, we can consider that the condition of average electrochemical steady-state required for impedance measurements is fulfilled.

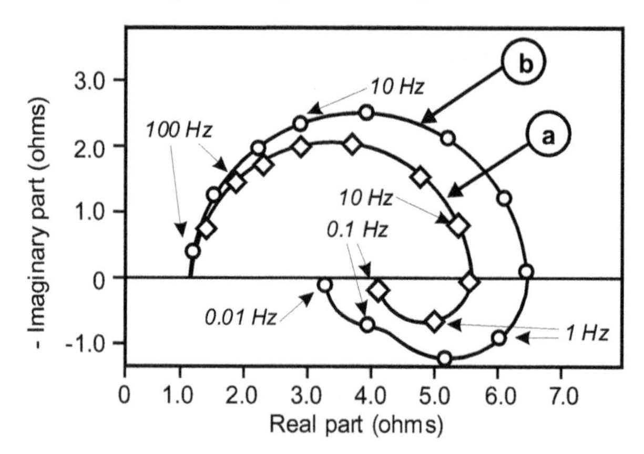

Fig. 6. Electrochemical impedance plots recorded on Fe-31% Ni immersed in 0.5 M sulfuric acid polarized in the active region (-0,675 V/SSE): (a) in absence of any sliding, and (b) under continuous sliding against alumina at a mean contact pressure of 2,6 MPa and a sliding speed of 3,4 cm s^{-1} (Ponthiaux et al., 1999).
SSE: 'mercury/mercurous sulfate/saturated potassium sulfate' reference electrode (E_{SSE} = +0.65 V/NHE).

In absence of friction, the plot consists of two successive semi-circles, a capacitive one reflecting the dielectric properties of the electrochemical double layer, and an inductive one reflecting the contribution of an elementary adsorption step in the reaction mechanism. Under sliding, the shape and size of the plot are modified as follows:

- at high frequencies, the capacitive arc increases reflecting a lower dissolution rate. This might be correlated to a strain hardening effect mentioned above, and highlighted in the polarization curves, and
- at low frequencies, an additional inductive arc appears which highlights the presence of a second adsorbate in the dissolution mechanism.

This can be interpreted by considering that in a non-rubbed condition the semi-circle linked to a possible second adsorbate does not appear because its concentration at the surface of the sample does not vary substantially with the potential. Under sliding conditions, the kinetics of some dissolution steps evolves, in particular the kinetics of steps in which the adsorbate interacts. The variation of the concentration at the surface becomes then detectable.

The in-depth interpretation of these impedance diagrams encounters however some difficulties as those already reported for open circuit potential and polarization measurements which are related to the heterogeneous state of the tested surfaces.

A first difficulty results from the fact that the impedance data reflect the overall state of the tested surface integrating the contributions of non-rubbed and rubbed surfaces. Such data must thus be de-convoluted in order to obtain the specific impedances of these two types of surface states. A first approach to this problem might be to use models similar to those describing the impedance of a sample undergoing a localized corrosion (Oltra & Keddam, 1990). In that specific case, the overall impedance can be considered as the result of two impedances in parallel, namely the impedance of the non-rubbed surface and the one of the rubbed surface. A strict interpretation requires further an evaluation of the areas of these surfaces, e.g. by using profilometry and surface observations by light optical or scanning electron microscopy.

The second difficulty results from the non-uniform state of the rubbed surface. Behind the slider the sample surface can be laid bare for some time before some new surface layers are rebuilt. The restored surface increases gradually with the distance behind the slider along the sliding track. Even if the first difficulty was already solved and the overall impedance of the rubbed surface is obtained, it can not be used as such to characterize the non-uniform distribution of the electrochemical states behind the slider. However, it is expected that impedance measurement procedures already developed for analyzing non-uniform distributions of surface states and the electrochemical models developed for the interpretation of such measurements (Zhang et al., 1987) could be transferred to tribocorrrosion test conditions. Such a study could allow a localized characterization of dissolution and passivation kinetics.

If theoretically this approach seems promising, experimental data analyses show that impedance measurements under sliding are often disturbed at low frequencies due to the random fluctuations of potential or current. This "electrochemical noise" limits unfortunately to some extent the application of impedance measurements. Limitations frequently originate from the sliding action itself and more specifically from the localized damages induced in the contact area by the mechanical interaction. Notwithstanding that, the in-depth analysis of the electrochemical noise will surely in the future be fruitful since it will contain useful information on the progress of the process at both spatial and time scales.

2.5 Electrochemical noise analyses

In the case of tribocorrosion as in cases involving stochastic processes of local surface damages like pitting (Uruchurtu & Dawson, 1987), stress corrosion (Cottis & Loto, 1990), and corrosion-abrasion (Oltra et al., 1986), sliding destroys locally to some extent surface layers. The bare areas created by the slider generate fluctuations of potential or current which consist of the superposition of elementary transients. The sudden increase in current occurs at the time that a bare surface is brought in contact with the electrolyte. The subsequent current decrease reveals the restoration of a protective surface film. The analysis of the transient characteristics (shape, amplitude, duration) provides information on the mechanisms and kinetics of the reactions involved (e.g. dissolution, passivation). The frequency of the transients depends in turn on the number of contacting points in the sliding track at a given time and on the sliding rate of the slider. It is therefore a quantity that provides useful information on the nature of the contact.

Sliding conditions affect the amplitude of the electrochemical noise, namely fluctuations of the open circuit potential as shown in Figure 2 (Ponthiaux et al., 1997). It is usually impossible to isolate the elementary transients. However, the spectral analysis of such a noise allows characteristic quantities to be derived such as the mean amplitude or duration and the average frequency at which transients occur. These characteristics are essential for getting a better understanding of the nature of the contact and the dissolution and passivation kinetics on the bare surface.

3. Modeling approaches

Given the complexity of the tribocorrosion phenomena there is currently no universal predictive model of the wear-corrosion process available. Such a problem solving is largely empirical, and designers rely on expert systems fed by experimental feedback to select the material couples for a given tribological system. In parallel to this technical approach, scientists are developing the modeling elements needed to unravel the phenomena. These models are discussed hereafter.

3.1 Formalism of the wear-corrosion synergy

It was shown here above that electrochemical methods allow under certain conditions to measure in real time the current I related to the corrosion reaction in a sliding track. At the end of a tribocorrosion experiment (e.g. sliding or erosion tests), the electrochemical mass loss, C, can be calculated using Faraday's relationship from the total charge consumed by the corrosion process:

$$C = \frac{M}{nF} Q \tag{3}$$

in which M is the atomic mass of the metal, n is the valency of the oxidized metal in the environment studied, F is the Faraday's constant and Q is the total charge related to the corrosion process, namely

$$Q = \int_0^T I\,(t)\,dt$$

with T the total duration of the test. However, the electrochemical mass loss due to corrosion under sliding, C, is generally not equal to the mass loss of the metal, C_0, obtained under similar test conditions but in absence of any sliding. The electrochemical mass loss under sliding C has thus to be expressed as:

$$C = C_0 + C_M \tag{4}$$

with C_M the amount of corrosion induced by wear.

The total mass loss, *W*, can be determined at the end of a sliding test by some *ex situ* technique like surface profilometry or gravimetry. That total mass loss can be compared to the electrochemical mass loss under sliding, C, from the equation:

$$W = C + M \tag{5}$$

in which M is the mechanical mass loss. This mechanical mass loss M can also be compared to the mass loss, M_0, in a non-corrosive environment as:

$$M = M_0 + M_C \tag{6}$$

in which M_C represents the excess mechanical mass loss due to corrosion. This formalism now allows a general definition of the synergy, S, between corrosion and mechanical processes in case of tribocorrosion as:

$$W = C_0 + M_0 + S \tag{7}$$

In which:

$$S = C_M + M_C \tag{8}$$

The term *S* defined as such reflects the fact that the material mass loss in a corrosive environment cannot be predicted simply by the sum of the mass loss due to corrosion in the environment in absence of any mechanical interaction, and the material loss due to wear recorded under similar testing conditions but in an non-corrosive environment. There is a synergy between these two processes. The formalism of tribocorrosion originally proposed (Watson et al., 1995) has surely an educational value since it allows a diagnostic on the "origin of evil" under the given set of experimental conditions. The wear M_0 is by some authors measured in a test under cathodic polarization in either dry air, de-ionized water or in the presence of corrosion inhibitors (Smith, 1985; Stemp et al, 2003; Takadoum, 1996). There is still some controversy about the validity of such procedures. In practice, the results often depend on the method used which limits the overall benefit of such a decoupling. Moreover, the concepts used do not have a physical meaning, and the synergies between wear and corrosion cannot really be simplified to a summing up. Other approaches that were developed over the past decades are briefly reviewed hereafter.

3.2 Models of oxidative wear and application in aqueous environment

3.2.1 Quinn's model of mild oxidative wear

A two-step model of mild oxidational wear for steel in air was developed (Quinn, 1992, 1994). The author observed that at sliding speeds below 5 ms^{-1} the wear debris consists only of iron oxides and that the particle size is quite constant in the range of several micrometers. In a first step oxides grow on surface asperities as a result of local heating at contact points. When the oxide reaches a critical thickness, x_c, the mechanical stresses generated in the material become too large and a detachment of the oxide layer which reaches the critical thickness takes place on the passage of the slider. The worn volume by unit of sliding distance, w, can be written as:

$$w = \frac{x_c}{Vt_c} \cdot A_r \tag{9}$$

with A_r the real contact area, V the sliding speed, and t_c the time necessary to reach the critical oxide thickness. Quinn's model is thus a law that corresponds to the Archard's wear law with:

$$K_{Archard} = \frac{x_c}{V\,t_c} \tag{10}$$

It is possible to connect x_c and t_c through a thermally activated oxidation kinetics in air which can be considered in first instance as a parabolic function with time:

$$x(t) = \alpha\sqrt{k_0 \exp\left(-\frac{Q_{act}}{R\,T_f}\right)t} \tag{11}$$

where:

- $x(t)$ is the oxide thickness at time t,
- Q_{act} is the activation energy of the oxidation process considered,
- k_0 is the Arrhenius constant associated with the process considered,
- T_f is the flash contact temperature at the interface between two asperities,
- R is the gas constant,
- t is time, and
- α is a constant depending on the density of the oxide and its oxygen volume fraction.

The critical thickness of the oxide, x_c, can only be obtained experimentally by characterizing the debris. The model is thus not a predictive one. Moreover as pointed out (Smith, 1985) the oxide growth laws based on mass can hardly be used under the conditions of contact characterized by a low air supply, and a poor knowledge of the real contact temperature. In practice, the use of oxidation constants leads to wear rates that are several orders of magnitude different from the experimentally verified ones.

3.2.2 Model of Lemaire and Le Calvar

In conclusion, the Quinn's model can not be easily adapted to analyse the depassivation - repassivation process determining the wear laws observed in tribocorrosion conditions. However, the main idea of this model, namely that wear proceeds by a succession of growth and delamination of an oxide layer, can be retained. It was at the origin of the tribocorrosion model presented to explain the the wear of a cobalt-based alloy coating applied on the gripper latch arms of the control rods command mechanisms in PWR (Lemaire & Le Calvar, 2000). In this model, the wear law is given by the following expression:

$$W = W_0 \, N \left(\frac{t}{t_0} \right)^{(1-n)} \tag{12}$$

Where W is the total worn volume, W_0 a constant, and N the number of sliding steps applied to the alloy surface inducing removal of the passive film. t is the mean time interval between two successive sliding steps, and t_0 is a characteristic repassivation time constant. n is a positive exponent whose value was found experimentally close to 0.65 for the cobalt-based alloy coating. The authors explain the wear law expressed by equation (12) by the evolution of the repassivation current Ip given by expression (2). This model implies that the growth of the oxide film between two sliding steps is proportional to $t^{(1-n)}$. The Quinn's law appears as a particular case of such a model for n = 0.5. In triborrosion studies, different values of n (between 0.6 and 0.9) where found depending on the metal, the environment and the electrochemical conditions.

3.2.3 Application to the synergy formalism

Studies in corrosive aqueous solutions (Garcia et al., 2001; Jemmely et al., 2000) are suitable to follow *in situ* the growth of passive films by electrochemical methods, and allow thus the development of more sophisticated models. In these studies performed under continuous or reciprocating sliding conditions, a modeling of currents measured at an applied potential is done. It is then assumed that a unit area of depassivated material repassivates according to a simple repassivation transient, J_a (t), which is not affected by the electrochemical conditions on the areas surrounding the rubbed area. The measured total current is then the sum of the contributions of the different surface areas. A freshly depassivated area produces a large current while a area depassivated some time before produces lower currents. In the case of a reciprocating tribometer operated at a sliding frequency, f, the steady state current I can be expressed as follows assuming that each contact event depassivates the surface:

$$I = A_a \cdot f \int_0^{1/f} j_a(t)dt + (A_t - A_a) f \int_0^{1/f} j_p(t)dt \tag{13}$$

with A_t the total area in contact with the solution, A_a the depassivated area on the sample during one cycle, f the frequency at which the surface is depassivated, j_p the passive current density at the applied potential, and j_a (t) the transient repassivation current density of a unit area at the applied potential.

Taking into consideration the synergy formalism developed above, the components of mass loss per cycle, C_M and C_0 in Equation (4) can now be written as follows:

$$C_M = \frac{M}{nF} \cdot A_a \left(\int_0^{1/f} j_a(t)dt - \int_0^{1/f} j_p(t)dt \right) \tag{14}$$

$$C_0 = \frac{M}{nF} A_t \int_0^{1/f} j_p(t)dt \tag{15}$$

When the component j_p related to passive zones can be neglected, C_0 becomes zero and equation (4) expressing the mass loss by corrosion under sliding becomes then:

$$C = C_M = \frac{M}{nF} A_a \int_0^{1/f} j_a(t)dt \tag{16}$$

The depassivated area during one cycle, A_a, can hardly be assessed. Some authors assume in first instance that it is equal to the apparent area of the sliding track. However, it is well known that the contact takes place only on a fraction of that area. An evaluation of the depassivated area from currents resulting from an electrochemical depassivation achieved by a potential jump was proposed (Garcia et al., 2001). This method also allowed them to evaluate the oxide thickness formed in between two successive depassivation events. They obtained oxide layer thicknesses in the range of a few nanometers.

Another approach was developed (Jemmely et al., 2000). The authors proposed to express the depassivated area in terms of a depassivation ratio per unit of time, R_{dep} :

$$R_{dep} = f.A_a \tag{17}$$

with f the contact frequency. The currents can then be expressed as:

$$I = R_{dep}.Q_{rep} \tag{18}$$

with Q_{rep} the charge density for repassivation. The rate at which an active area is generated per unit of time depends on the morphology and hardness of the surfaces in contact. A derivation of R_{dep} from the scratching of a ductile material by a hard abrasive one was proposed (Adler & Walters, 1996). That approach was taken over (Jemmely et al., 2000) and extended in more general terms (Mischler et al., 1998) in the following expression:

$$R_{dep} = K.V\frac{F_N^{\beta}}{H} \tag{19}$$

with K an empirical constant, V the sliding velocity, F_N the applied load, H the hardness of the tested material, $\beta = 0.5$ in the case of a contact between two counterparts with a similar roughness, β = 0.5 in the case of a rough and hard body against a smooth and ductile counterpart, β = 1 in the case of a hard and smooth body against a rough and ductile counterpart, and β between 0.5 and 1 in the general case.

One empirical constant K remains in this model which approximates Archard's constant and which is related to the probability that a given contact becomes depassivated. The mass loss by corrosion under sliding, C, can thus finally be written as:

$$C = \frac{M}{nF} KV \frac{F_N^{\beta}}{H} \cdot Q_{rep}(f) \tag{20}$$

Current measurements performed at different loads and sliding speeds for materials with different hardness, allow the validation of the general form of this law.

4. Tribocorrosion testing

4.1 Specificity of laboratory and industrial tribocorrosion tests

Similarly as in classical mechanical testing, tribocorrosion tests can be classified into two categories based on their different but complementary purposes, namely fundamental and technological tests.

Fundamental tests are implemented in research laboratories and their objective is to clearly identify and to understand under well defined testing conditions, the basic mechanisms and their synergy that govern the phenomena of tribocorrosion. These tests require the development of experimental methodologies for both the test themselves and the techniques to be used for analyzing and measuring data and other experimental outcomes. Concerning friction in particular, two types of tests can be considered:

- tests at low displacement amplitude referred to as "**fretting tests**". These tests provide information on the response of materials with respect to the solicitation (displacement amplitude, load, frequency, and environment). One can differentiate stick-slip, partial slip or gross slip contact conditions. The information collected is on the nature of the degradation, its location in the contact, the kinetics of crack initiation and crack growth, and the size and shape of the degradation products that may appear in the contact during testing, and
- tests at large displacement amplitude referred to as either "**reciprocating sliding tests**" or "**continuous sliding tests**", provide information on the nature and kinetics of the wear process in connection with the synergies resulting from the mechanical, chemical or electrochemical coupling taking place on contacting surfaces in relative motion.

These tests allow the following analyses based on *in situ* and *ex situ* measurements, like the determination of the mean and local coefficient of friction, the identification of and study of the interactions between surfaces and environment, the nature of the mechanical-chemical coupling, the electrochemical or galvanic coupling due to a heterogeneous structure, the shape and location of rubbed and non-rubbed areas, or the establishment of local wear laws and their spatial distribution on the surface in view of a modelling of wear aiming at a future predictive approach.

Technological tests are designed to reproduce at lab scale mechanical loading and/or environmental conditions corresponding to actual operating conditions, or to mimic particular conditions intending to accelerate material degradation processes. These tests are widely used to predict precisely the behaviour of mechanical devices in actual conditions of service and to improve their reliability and durability. In that respect, they are very useful tools.

The full investigation of the tribocorrosion tests requires generally the use of *in situ* tools like open circuit measurements, polarization measurements, current transients, impedance spectroscopy, and noise measurements, and *ex situ* tools like elemental surface analysis techniques, optical or electron microscopy, micro-topography, micro and nanohardness measurements texture and internal stress analyses.

4.2 Testing protocol: A multiscale analysis of tribocorrosion phenomena

A promising approach of synergy in tribocorrosion has been proposed (Diomidis et al., 2009) based on the fact that the surface state of a wear track evolves with time in a cyclic manner. That evolution is due to the repeated removal and subsequent re-growth of a passive surface film when a mechanical loading is applied. During the latency time, t_{lat}, defined as the time between two successive contacts at a given point in the sliding track, the passivation reaction tends to restore the passive film. The fraction of the sliding track surface covered by this re-grown passive film increases with t_{lat}. By controlling the frequency of such depassivation-repassivation events with respect to the time necessary for film growth, it is possible to measure the properties of the surface at different stages of activity and repassivation. For performing tests at different latency times, t_{lat}, two approaches are possible, each approach having own advantages and limitations as detailed hereafter:

- first approach in the case of **continuous sliding** tests: the latency time t_{lat} can be modified by changing the rotation period t_{rot}. Another way to obtain the same t_{lat} is to keep the sliding speed constant and increase the radius of the track. However, for practical reasons, it is not realistic to consider to multiply the radius of the track by a large factor.
- second approach in the case that the latency time is changed by performing **intermittent sliding** tests: during such tests the counterbody slides for one cycle, and then stays immobile for a certain period of time to allow a part of the passive film to re-grow. Thus, an off-time, t_{off}, is imposed at the end of each cycle. As a result, the latency time between two subsequent contact events, t_{lat}, differs from the rotation period, t_{rot}:

$$t_{lat} = t_{rot} + t_{off} \tag{21}$$

It is clear that in continuous sliding tests, Equation (21) is still valid but t_{off} is zero. Such an approach can thus result in a protocol that provides information on the evolution of the surface with testing time, and the identification of the resulting mechanisms of material loss and surface degradation (Pourbaix, 1974).

4.2.1 Selection of test conditions

In order to characterize the sensitivity of the one or more material systems to tribocorrosion, a careful selection of the test conditions has to be done prior to any testing, so as to obtain discriminating results. The following steps are of large importance in that approach:

- Selection of environmental conditions: the electrolyte should be selected in view of its known oxidative or reducing power. The test temperature can be ambient temperature or any temperature relevant for the field application it should reproduce. A decisive parameter in the selection of the electrolyte is its pH. The selection of the pH can be based on pH-potential diagrams (Pourbaix, 1974). In the case of a metallic alloy, a pH range should be selected by preference where at least one of the constituents passivates. An electrolyte that may cause localized corrosion, like pitting corrosion in particular, should be avoided,
- Parameters linked to the sliding tests need to be selected by considering the following recommendations. The normal force, F_N: the normal force should be selected so as to

avoid plastic deformation of the tested materials. The maximum Hertzian contact pressure on the test material before starting sliding should be smaller than the yield strength. Concerning the track radius, R_{tr}, it should be selected in such a way that edge effects are avoided. E.g. in the case of a disc, the track radius should be by preference about half the test sample radius. Finally the number of cycles, n, depends on the type of material tested and the test conditions. It should be selected so that the wear volume is large enough to be measured accurately, while avoiding too long test durations for practical reasons. A preliminary sliding test might be necessary to determine n. In some particular cases, the selection of the number of cycles can also be done so as to reflect the behaviour of the test material in a real life application.

4.2.2 First step in the testing protocol: Electrochemical tests on passive material without any sliding

After selecting the set of appropriate test conditions, measurements are done to collect information on the electrochemical behaviour of a material fully covered by a passive film. This is done by electrochemical tests in absence of any sliding. After immersion in the electrolyte, the open circuit potential, E_{oc}, is measured versus a reference electrode. In general, a stable value of E_{oc} is obtained after some time of immersion. From an electrochemical point of view, a stable E_{oc} is obtained when the long-term fluctuations of E_{oc} are below 1 mV min^{-1} during a minimum of 1 hour. The time necessary to reach such a stationary open circuit potential in the test electrolyte is an important characteristic of a passivating process, and is called in this protocol as the reaction time characteristic, t_{reac}. The evolution of E_{oc} from immersion time on provides useful information on the electrochemical reactivity of the tested material in the test electrolyte (see Figure 7).

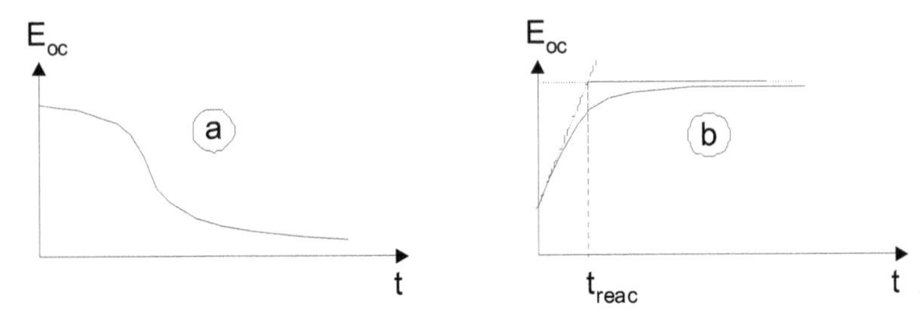

Fig. 7. Schematic representation of the evolution of E_{oc} with immersion time in the case of (a) corrosion, and (b) passivation. The graphical determination of t_{reac} in the case of passivation is shown in (b).

From this figure one can derive that when E_{oc} decreases with time, general corrosion may be suspected, and that when E_{oc} increases with time, passivation or adsorption is probably taking place. In this latter case, t_{reac} can be estimated from the evolution of E_{oc} with immersion time by drawing the tangent to the curve at the point where the slope is maximum, together with a straight line tangential to the data in the part of the curve where E_{oc} is stable. After achieving a long-term stable open circuit potential indicative of passivation, the polarization resistance of the passive material, R_p, is measured by electrochemical impedance spectroscopy. Based on R_p, the specific polarization resistance of passive material, r_{pass}, can be calculated for a test sample with a surface area, A_o, as:

$$r_{pass} = R_p.A_o \tag{22}$$

Specific polarization resistance values for metallic materials of 10^3 Ωcm^2 or lower indicate the presence of an active sample surface, while values around 100×10^3 Ωcm^2 or higher indicate a passive sample surface. The corrosion current density of the material covered by a passive surface film, i_{pass}, is then calculated as follows:

$$i_{pass} = \frac{B}{r_{pass}} \tag{23}$$

with B a constant. For metallic materials, B normally varies between 13 and 35 mV, depending on the nature of the material and the environment. In the protocol worked out hereafter as an example, a value of 24 mV is assumed. This passive current density, i_{pass}, is considered to correspond to the dissolution current of the material through the passive film at stationary state.

4.2.3 Second step in the testing protocol: Electrochemical tests on a fully active sliding track during sliding

The next step is the determination of the corrosion rate of the depassivated material. In order to keep continuously a part of the immersed sample surface in an active state, the passive film has to be removed by mechanical contacts. It is thus necessary to select a rotation period, t_{rot}, which is small compared to t_{reac} so that the passive film has no time to regrow in between two successive contact events. It was proposed (Diomidis et al., 2009, 2010) to take the rotation period t_{rot} equal to:

$$t_{rot} = \frac{t_{reac}}{10000} \tag{24}$$

A first value of t_{lat1} is then taken equal to t_{rot}. It is thus assumed that during such sliding tests the whole wear track area is in an active state, so that:

$$A_{tr} = A_{act} \tag{25}$$

Despite the fact that the width of the sliding track increases progressively due to wear during sliding tests performed against a curved counter-body, a mean sliding track area, A_{tr}, is taken for simplicity, and defined as:

$$A_{tr} = \frac{1}{2}\left(A_{tr\max} + A_{tr\min}\right) \tag{26}$$

with $A_{tr\ max}$ the maximum value measured at the end of the test, and $A_{tr\ min}$ the minimum value at the end of the first cycle. $A_{tr\ min}$ is calculated by multiplying the length of the wear track, L, by the diameter of the Hertzian static contact area, e:

$$e = 2\left(\frac{3\,F_N\,R}{4\,E}\right)^{1/3} \tag{27}$$

with F_N the applied normal load, R the radius of the tip of the curved counter-body, and E the equivalent elastic modulus given by:

$$\frac{1}{E} = \frac{1-\nu_1^2}{E_1} + \frac{1-\nu_2^2}{E_2} \tag{28}$$

with ν_1 and ν_2 the Poisson's ratios, and E_1 and E_2 the elastic moduli of the test sample and counterbody respectively.

Sliding is initiated at the time a stable E_{oc} is achieved. That E_{oc} recorded during sliding is a mixed potential resulting from the galvanic coupling of material inside (A_{tr}) and outside (A_o - A_{tr}) the sliding track. It is assumed that the kinetics of the redox reactions taking place on each of these areas, do not vary with the real potential of the sliding track. In other words, the ohmic drop effect is considered to be negligible in the galvanic coupling between the sliding track and the surrounding area. Electrochemical impedance spectroscopy measurements are performed during sliding to obtain the polarization resistance, R_{ps}, of the sample surface. Similarly to E_{oc}, R_{ps} may be considered as the combination of two polarization resistances, namely R_{act} related to the active area A_{act} which is equal in this case to the wear track, and R_{pass} which corresponds to the surrounding unworn area, (A_o - A_{tr}):

$$\frac{1}{R_{ps}} = \frac{1}{R_{act}} + \frac{1}{R_{pass}} \tag{29}$$

where

$$R_{act} = \frac{r_{act}}{A_{tr}} \tag{30}$$

and

$$R_{pass} = \frac{r_{pass}}{A_0 - A_{tr}} \tag{31}$$

It is then possible to calculate the specific polarization resistance of the active surface, r_{act}, as:

$$r_{act} = \frac{A_{tr}\, R_{ps}\, r_{pass}}{r_{pass} - R_{ps}\left(A_0 - A_{tr}\right)} \tag{32}$$

The corrosion current density of the active material, i_{act}, can now be obtained by substituting r_{pass} with r_{act} in Equation (23):

$$i_{act} = \frac{B}{r_{act}} \tag{33}$$

It can be noted that, the specific resistance of the active bare material in the sliding track, r_{act}, is generally several orders of magnitude lower than the specific resistance of the material covered with a passive film, r_{pass} , outside of the sliding track. Therefore, if A_{tr} is not too small a fraction of the total area A_0 of the sample, the resistance R_{act} related to the sliding track is small compared to the resistance R_{pass} of the area of the sample remained in passive state, outside of the sliding track: if $R_{act} << R_{pass}$, according to equation (29), the measured resistance R_{ps} gives then straight an approximate value of R_{act}.

4.2.4 Third step in the testing protocol: Electrochemical tests on a partially active sliding track during sliding

In the preceding steps, two extreme cases were characterized, namely on the passive material and on the active one repectively. Under tribocorrosion conditions at high latency time, the surface of the material undergoes sequential events of depassivation and repassivation in-between successive contacts. This means that a part of the surface at any given time repassivates progressively. The latency time is then selected so that the regrowth of a surface film between two successive contact events is not anymore negligible as it was the case under sliding at low latency time. To achieve partially active sliding tracks, the latency times can be selected as $t_{lat2} = t_{reac}/1000$ and $t_{lat3} = t_{reac}/100$ (Diomidis et al., 2009). As a result of the increase of the latency time in this step, the wear track can now be assumed to consist of two distinct zones (Diomidis et al., 2010), namely:

- a fraction of the sliding track from which the initial passive film has been removed during sliding. In this area, the test material may be either bare, or covered by a reaction layer different from the initial passive film. This area is named the active area, A_{act}., and
- the remaining sliding track area covered by a surface film that is in the same state as the surface before sliding. This film is either not removed by the counterbody during sliding, or it had the time to restore in its initial state. This area is referred to as the repassivated area, A_{repass} with:

$$A_{tr} = A_{act} + A_{repass} \quad (34)$$

It must be stressed that under continuous sliding, these active and repassivated areas remain constant because of stationary electrochemical state conditions. Under intermittent sliding, these active and repassivated areas on the sliding track evolve with time between two successive contact events since a gradual increase of the coverage of the repassivated area takes place within the "off period". By hypothesis, in both cases, the fraction of the sliding track surface covered by the passive film, A_{repass}/A_{tr}, is assumed to be constant and given by the ratio t_{lat}/t_{reac}:

$$\frac{A_{repass}}{A_{tr}} = \frac{t_{lat}}{t_{reac}} \quad (35)$$

and:

$$\frac{A_{act}}{A_{tr}} = 1 - \frac{t_{lat}}{t_{reac}} \quad (36)$$

At the latency times $t_{lat2} = 0.001\ t_{reac}$ and $t_{lat3} = 0.01\ t_{reac}$, the relationship between repassivated and total wear track area are respectively $A_{repass\ 2} = 0.001 A_{tr}$ and $A_{repass\ 3} = 0.01\ A_{tr}$, and thus the active wear track areas are $A_{act\ 2} = 0.999\ A_{tr}$ and $A_{act\ 3} = 0.99\ A_{tr}$.

4.3 Analysis and interpretation of sliding test results

The total wear W_{tr} in a wear track of area A_{tr} can be expressed as a sum of components related to both types of areas present on the track, the area A_{act} in active state, and the area A_{repass} in re-passivated state:

$$W_{tr} = W^{c}_{act} + W^{m}_{act} + W^{c}_{repass} + W^{m}_{repass} \quad (37)$$

with :

- W_{tr} material loss in wear track,
- W^{c}_{act} material loss due to corrosion of active material in wear track,
- W^{m}_{act} material loss due to mechanical wear of active material in wear track,
- W^{c}_{repass} material loss by corrosion of repassivated material in wear track, and
- W^{m}_{repass} material loss due to mechanical wear of repassivated material in the wear track.

In order to assess the values of these different components and to compare them to determine the characteristics of the wear mechanism, the following analyses have to be performed:

- experimental determination of the mass losses due to corrosion and mechanical wear of active material during sliding tests performed at low latency times,
- experimental determination of the mass losses due to corrosion and mechanical wear of active material during sliding tests performed at large latency times. In this case one must differentiate between continuous and intermittent sliding tests,
- experimental determination of the mass losses due to corrosion and mechanical wear of repassivated material during sliding tests performed at large latency times since they allow the study of the periodic removal and re-growth of a passive surface film,

Detailed information on the analysis and related interpretation of sliding tests can be found in a Handbook on Tribocorrosion (Celis & Ponthiaux, 2011).

5. Conclusions

Tribocorrosion is the degradation of material surfaces by the combined action of corrosion, electrochemical passivation, and external mechanical interactions. It is essentially a surface related process, but some events like hydrogen evolution and absorption by the material, can modify the mechanical properties of the sub-surfaces on materials. Under conditions where tribocorrosion is active, the material loss depends in a complex way on many parameters like the tribological conditions in the contact, the composition of the environment, the temperature, the flow rate, the pH and eventually the applied potential. By analogy it is possible to extend this concept when, in the process previously described, a chemical or physical adsorption of inhibiting species strengthens or replaces the electrochemical passivation process. The successive repetition of some of these processes can lead to a possible synergism between mechanical stress and the effect of the environment what results in a damage of surfaces and systems through an accelerated loss of functionality.

Electrochemical methods used for corrosion studies are of interest in tribocorrosion since they allow the *in situ* monitoring and analysis of the interactions between surfaces and their environment as well as changes induced by a mechanical action like sliding or impact. In combination with conventional tribological measurements, they allow to understand the evolution of the surface state in a time space. The choice of methods to be used depends on the type of mechanical action. The data interpretations must be adapted to the heterogeneous state of surfaces undergoing a mechanical interaction, and also to the contact conditions.

Taking into account the surface state heterogeneity that results from successive mechanical and corrosive interactions, the development of an analysis of local phenomena on surfaces is needed to predict the impact of varying environmental conditions, tribological, and even geometrical ones on the operating behavior of a tribological system. Besides the methods already mentioned above, *ex situ* techniques to characterize surfaces, like the determination of residual stresses, micro-and nano-hardness, topography by 3D profilometry, chemical and structural micro-analysis of surfaces, microscopy at different length scales, must be implemented to acquire the spatial response of materials subjected to combined corrosion and mechanical loadings.

6. References

Adler T.A. & Walters R.P. (1996). Corrosion and wear of 304 stainless steel using a scratch test. *Corrosion science,* Vol 33, No. 12, pp. (1855-1876), ISSN 0010-938X.

Bom Soon Lee, Han Sub Chung, Ki-Tae Kim, Ford F.P. & Andersen P.L. (1999). Remaining life prediction methods using operating data and knowledge on mechanisms. *Nuclear Engineering and Design,* Vol. 191, No. 2, pp. (157-165), ISSN 0029-5493.

Boutard D., Wenger F., Ponthiaux P., & Galland J. (1985). Incidence de méthodes expérimentales diverses sur les diagrammes d'impédance électrochimique d'un alliage fer-31% nickel en cours de corrosion. *Proceedings of 8th European Corrosion Congress,* ISBN 2-88074-228-5, Nice (France), November 1985.

Carton J. F., Vannes A. B. & Vincent L. (1995). Basis of a coating choice methodology in fretting. *Wear,* Vol 185, No.5, 1995, pp. (47-57), ISSN 0043-1648.

Celis J.P. & Ponthiaux P. editors. (2011), *Testing tribocorrosion of passivating materials supporting research and industrial innovation: Handbook", EFC Green Book N° 62,* Maney, ISBN 978 1 907975 20 2, Leeds (UK).

Cottis R. A. & Loto C. A. (1990). Electrochemical noise generation during SCC of a high-strength carbon steel. *Corrosion,* Vol. 46, No.1, pp. (12-19), ISSN 0010-9312.

Déforge D., Huet F., Nogueira R.P., Ponthiaux P., and Wenger F. (2006). Electrochemical noise analysis of tribocorrosion processes under steady-state friction regime. *Corrosion,* Vol. 62, No 6, pp. (514-521), ISSN 0010-9312.

Diomidis N., Celis J.-P., Ponthiaux P. & Wenger F. (2009). A methodology for the assessment of the tribocorrosion of passivating metallic materials. *Lubrication Science,* Vol. 21, No 2, pp. (53-67), ISSN 0954-0075.

Diomidis N., Celis J.P., Ponthiaux P. & Wenger F. (2010). Tribocorrosion of stainless steel in sulfuric acid: Identification of corrosion–wear components and effect of contact area. *Wear,* Vol. 269, No 1-2, pp. (93-103), ISSN 0043-1648.

Garcia, I., Drees, D. & Celis, J.-P. (2001). Corrosion-wear of passivating materials in sliding contacts based on a concept of active wear track area. *Wear,* Vol. 249, No 5-6, pp. (452-460), ISSN 0043-1648.

Godet M., Berthier Y., Lancaster J. & Vincent L. (1991). Wear modelling : using fundamental understanding or practical experience ?, *Wear*, Vol. 149, No 1-2, pp. (325-340), ISSN 0043-1648.

Jemmely, P., Mischler, S. & Landolt, D. (2000). Electrochemical modelling of passivation phenomena in tribocorrosion. *Wear*, Vol. 237, No. 1, pp. (63-76), ISSN 0043-1648.

Law C. G., & Newman J. (1979). A model for the anodic dissolution of iron in sulfuric acid. *Journal of the Electrochemical. Society.*, Vol. 126, No.12, pp. (2150-2155), ISSN 0043-1648.

Lemaire E. & Le Calvar M. (2000). Evidence of tribocorrosion wear in pressurized water reactors. *Wear*, Vol. 249, No. 5-6, pp. (1-7), ISSN 0043-1648.

Lillard R. S., Kruger J., Tait W. S. & Moran P. J. (1995). Using local electrochemical impedance spectroscopy to examine coating failure. *Corrosion*, Vol. 51, No. 4, pp. 251-259 ISSN 0010-9312.

Lim, S.C. & Ashby, M.F. (1987). Wear-mechanism maps. *Acta Metallurgica*, Vol. 35, No. 1, pp. (1-24), ISSN 1359-6454.

Mischler S., Ayrault S., Debaud S., Jemmely Ph., Rosset E. & Landolt D. (1997). Aspects physico-chimiques de la tribocorrosion. *Matériaux et Techniques*, Vol. HS, No. July, pp. (5-10), ISSN 0032-6895.

Mischler, S., Debaud, S. & Landolt, D. (1998). Wear-accelerated corrosion of passive metals in tribocorrosion systems. *Journal of the Electrochemical. Society.*, Vol. 145, No. 3, pp. (750-758), ISSN 0043-1648.

Oltra R. & Keddam M. (1990). Application of E.I.S. to localized corrosion. *Electrochimica Acta*, Vol. 35, No. 10, pp. (1619-1629), ISSN 0013-4686.

Oltra R., Gabrielli G., Huet F. & Keddam M. (1986). Electrochemical investigation of locally depassivated iron. A comparison of various techniques. *Electrochimica Acta*, Vol. 31, No. 12, pp. (1501-1511), ISSN 0013-4686.

Ponthiaux P., Wenger F. & Galland J. (1995). Study of the anodic current-voltage curve of an iron-nickel alloy in normal sulfuric acid. *Journal of the Electrochemical. Society.*, Vol. 142, No. 7, pp. (2204-2210), ISSN 0043-1648.

Ponthiaux P., Wenger F., Galland J., Lederer G. & Celati N. (1997). Utilisation du bruit électrochimique pour déterminer la surface dépassivée par frottement. Cas d'un acier Z2 CND 17-13 en milieu chloruré (NaCl 3%), *Matériaux et Techniques*, Vol. HS, No. July, pp. (43-46), ISSN 0032-6895.

Ponthiaux P., Wenger F., Galland J., Kubecka P. & Hyspecka L. (1999). Effets combinés du frottement et de la corrosion dans le cas d'un alliage fer-nickel en milieu sulfurique. *Matériaux et Techniques*, Vol. HS, No. December, pp. (11-15), ISSN 0032-6895.

Pourbaix M. (1974), *Atlas of Electrochemical Equilibria in Aqueous Solutions*, National Association of Corrosion Engineers, ISBN 0915567989, Houston (USA).

Quinn, T.F.J. (1992). Oxidational wear modelling: part I. *Wear*, Vol. 153, No. 1, pp. (179-200), ISSN 0043-1648.

Quinn, T.F.J. (1994). Oxidational wear modelling: part II. The general theory of oxidational wear. *Wear*, Vol. 175, No. 1-2, pp. (199-208), ISSN 0043-1648.

Rosset E. (1999). Tribologie Systémique. *Oberflächen-Polysurface*, Vol. 1, pp. (7-9), ISSN 1422-3511.

Smith A.F. (1985). Sliding wear of AISI 316 stainless steel in air, 20 - 500°C. *Tribology International*, Vol. 18, No. 1, pp. (35–43), ISSN 0301-679X.

Stemp M., Mischler S. & Landolt D. (2003). The effect of mechanical and electrochemical parameters on the tribocorrosion rate of stainless steel in sulfuric acid. *Wear*, Vol. 255, No.1-6, pp. (466–475), ISSN 0043-1648.

Takadoum J. (1996). The influence of potential on the tribocorrosion of nickel and iron in sulfuric acid solution. *Corrosion Science*, Vol. 38, No. 4, pp. (643–654), ISSN 0010-938X.

Tiedemann W. H., Newman J. & Bennion D. N. (1973). The errors in measurements of electrode kinetics caused by nonuniform ohmic-potential drop to a disk electrode. *Journal of the Electrochemical. Society.*, Vol. 120, No. 2, , pp. (256-258), ISSN 0043-1648.

Uruchurtu J. C. & Dawson J. L. (1987). Noise analysis of pure aluminium under different pitting conditions. *Corrosion*, Vol. 43, No. 1, pp. (19-25), ISSN 0010-9312.

Watson, S.W., Friedersdorf, F.J., Madsen, B.W. & Cramer, S.D. (1995). Methods of measuring wear-corrosion synergism. *Wear*, Vol. 181-183, No. 2, pp. 476-484, ISSN 0043-1648.

Zhang J., Wenger F. & Galland J. (1987). Contrôle de l'état local de corrosion de structures métalliques de grandes dimensions par les mesures d'impédance électrochimiques. *Comptes Rendus de l'Académie des Sciences, Paris,* Série 2, Vol. 304, No. 14, pp. (797-800), ISSN 12518069.

6

Electrochemical Passive Properties of $Al_xCoCrFeNi$ (x = 0, 0.25, 0.50, 1.00) High-Entropy Alloys in Sulfuric Acids

Swe-Kai Chen

Center for Nanotechnology, Materials Science, and Microsystems (CNMM), National Tsing Hua University, Hsinchu, Taiwan

1. Introduction

1.1 Pseudo-unitary lattice with a characteristic parameter as a description of multi-principal alloys – The high-entropy alloys (HEAs)

In the summer of 1995, J.W. Yeh and the author (SKC) started the study of multi-principal-element alloys which was called, then, alloys with high randomness and now the high-entropy alloys (HEAs). SKC checked the first 10 equal-molar alloys, which was designed by Yeh that contained from 6 to 9 elements in the alloys out of one of Al, Cu, and Mo, together with Ti, V, Fe, Ni, Zr, Co, Cr, Pd, and B, with a home-made vacuum-arc remelter, and the author observed that the alloy series containing Mo can be made most easily, while the ones containing 3 at% B are the ones most difficult in melting, and 6 out of 10 can be formed in the water-cooled copper mold of the remelter, i.e., the existence of the HEAs was demonstrated by experiments. The alloys were aimed at that time to design as another kind of bulk glass alloys, and based on the high configurational entropy of R ln(n), n between 5 and 13, similar to the mixing of different gases [1]. No conclusions were drawn with XRD patterns of these alloys that were found two years later to be composed with peaks from a single simple lattice cell like FCC A1 or BCC A2, although some evidence of existence of amorphous phase was observed from TEM diffraction patterns and high resolution images [2,3]. The simple crystalline phases instead of amorphous ones were continuously found in alloys like in AlCoCrCuFeNi during research of HEAs in these 10 to 20 years, and identified with a so-called extended FCC or BCC unit cell that SKC called it a pseudo-unitary lattice in 2010 [4].

As multiple principal element alloys, high-entropy alloys (HEAs) comprise at least five elements whose concentration for each one ranges between 5 at % and 35 at % [5]. Attributes of forming a simple solid solution and nano-particle precipitation, as well as achieving a high hardness and strength, and excellent high-temperature oxidation resistance make HEAs highly promising for application and research and development of these alloys [6-9]. Properties of $Al_xCoCrFeNi$ ($0 \le x \le 1$) HEAs vary significantly with x [10]. For instance, the alloy structure changes from FCC to BCC for increased Al content x. Besides, the coefficient of thermal expansion decreases with x. Both properties are closely related to the bond strength of alloys. Moreover, electrical resistivity of $Al_xCoCrFeNi$ alloys is large, i.e., approximately up to 200 $\mu\Omega$ cm [11].

1.2 Corrosion resistance for HEAs and conventional alloys

Corrosion properties of $AlCoCrCu_{0.5}FeNiSi$ [12,13], $Al_xCrFe_{1.5}MnNi_{0.5}$ [14,15], and $Al_{0.5}CoCrCuFeNiB_x$ [16] HEAs have been extensively studied in recent years. Among these HEAs, $AlCoCrCu_{0.5}FeNiSi$ alloy (HEA 1) displays, at room temperature, a better general corrosion resistance than SS 304 in 1 N H_2SO_4; however, it exhibits a worse pitting corrosion resistance than SS 304 in 1 N H_2SO_4 and in 1 M NaCl, respectively. The general corrosion resistance of each of HEA 1 and SS 304 decreases when exceeding room temperature. The effect of temperature on corrosion resistance of HEA1 is less severe in 1 M NaCl than in 1 N H_2SO_4 [13]. $Al_xCrFe_{1.5}MnNi_{0.5}$ alloys (HEA 2) reveal that in each of the 0.5 M H_2SO_4 and 1 M NaCl solutions, corrosion resistance increases with a decreasing x; in addition, the susceptibility to general and pitting corrosion of HEA 2 increases with an increasing x [14]. $Al_xCrFe_{1.5}MnNi_{0.5}$ alloys (called hereinafter as HEA 2a and 2b for x = 0 and 0.3, respectively) in 0.1 M HCl exhibit different corrosion behaviours for different x values. Although HEA 2a is susceptible to localized corrosion, HEA 2b has a stable passive film on the surface. In 0.1 M HCl, anodized treatment of HEA 2a and 2b alloys in 15 % H_2SO_4 gives higher corrosion resistance than the untreated [15]. In deaerated 1 N H_2SO_4, $Al_{0.5}CoCrCuFeNiB_x$ alloys are more resistant to general corrosion than SS 304, and are not susceptible to localized corrosion. Additionally, the corrosion resistance of $Al_{0.5}CoCrCuFeNiB_{0.6}$ alloy is inferior to $Al_{0.5}CoCrCuFeNi$ alloy [16]. Above HEAs show an extremely close compositional dependence of corrosion behaviour in various solutions.

1.3 Aim of this study

Although many interesting topics have been explored for $Al_xCoCrFeNi$ alloys [10,11], investigation on their corrosion property is still lacking. Therefore, this study elucidates how Al affects their corrosion behaviour. The electrochemical properties of the alloys in sulfuric acids are investigated using the potentiodynamic polarization curve and a weight loss measurement method. Additionally, these alloys are compared with SS 304, especially with respect to the effect of temperature. Moreover, based on use of electrochemical impedance spectroscopy (EIS), the effect of Al on corrosion behaviour is analyzed. Furthermore, the relationship of stability of oxide film with Al content is examined by varying the chloride concentration in a sulfuric solution. Additionally, this study, which extends [17], also attempts to investigate the mechanism of the passive layers influenced by Al content x at various temperatures in detail.

2. Experimental details

2.1 Test materials and conditions for electrochemical and weight loss tests

2.1.1 Test specimens for electrochemical tests and weight loss measurements

As-cast $Al_xCoCrFeNi$ alloys were prepared according to molar ratios of x = 0, 0.25, 0.50, and 1.00 (called C-0, C-0.25, C-0.50, and C-1.00, respectively) in a vacuum arc remelter. Table 1 lists the composition of the alloys. Test specimens were cut in 0.8 cm x 0.8 cm x 0.3 cm and cold-mounted in epoxy with the outside surface from a surface of 0.8 cm x 0.8 cm of specimens. The specimens were subsequently ground and polished with grit #1000 silicon carbide paper, rinsed and dried in preparation for electrochemical tests and weight loss measurements. During determination of the weight loss, six sets of samples were dipped in sulfuric acid for 1, 3, 5, 8, 11, and 15 days, respectively. All tests, except the weight loss test, were performed at least three times to confirm the data reproducibility. Finally, weight loss tests were performed twice and the reproducibility was given in an error bar.

Alloys	Al	Co	Cr	Fe	Ni
C-0	0	27.12	23.74	23.99	25.14
C-0.25	3.05	25.14	22.48	24.15	25.18
C-0.50	5.59	25.25	22.13	22.80	24.22
C-1.00	10.02	23.84	21.11	21.99	23.03
SS 304	0	0	19.40	72.68	7.92

Table 1. Composition (wt %) for alloys C-x and SS 304.

2.1.2 Test solutions and temperatures

The base solution for all tests was 0.5 M of sulfuric acid. Test temperatures were ambient temperature (~25°C). Test solutions bearing chloride ions were with 0.25, 0.50, and 1.00 M sodium chloride in the base solution. To avoid the dissolved oxygen (aeration) affecting the test solutions, deaeration was simultaneously made by a nitrogen gas flow of 120 ml/min in the test solution. The effect of temperature on polarization was examined under thermostatic control at an interval of 15°C in the temperature range of 20°C - 65°C.

2.2 Potentiodynamic polarization curve measurements and electrochemical impedance spectroscopy (EIS)

A three-electrode cell was used for the electrochemical test. The reference electrode was a commercial Ag/AgCl electrode saturated in 3 M KCl electrode (−0.205 V_{SHE} or -0.205 V to standard hydrogen electrode). The auxiliary electrode was made of Pt, and the working electrode was the specimen. Potentiostat was CH Instrument Model-600A. The specimen was cathodically polarized at a potential of −0.4 V_{SHE} for 300 s before the test for the purpose of removing surface oxides. The quasi-steady-state time for an open circuit was 900 s. Scan speed was 1 mV/s for scan potential ranging from −0.6 V_{SHE} to 1.4 V_{SHE}. For EIS, the working potential was that of open circuit at 900 s from the start of immersion with scan amplitude 10 mV and a frequency ranging from 100 kHz to 10 mHz.

2.3 Immersion tests and ICP-AES and XPS analyses

Samples were dipped in sulfuric acid for 15 d to determine the weight-loss rate. Auger electron spectroscopy (AES) and X-ray photoelectron spectroscopy (XPS) analysis were performed with samples after a 0.8 V_{SHE} pretreatment plus a 1-h immersion. Inductively coupled plasma atomic emission spectroscopy (ICP-AES) was performed on the electrolyte after an 8-d immersion of the samples. The effect of temperature on polarization was examined under thermostatic control at an interval of 15°C in the temperature range of 20°C–65°C.

2.4 Scanning electron microscopy (SEM) metallographic examination and energy dispersed X-ray spectroscopy (EDS) analysis

Samples were fine polished, up to 0.05 µm Al_2O_3 powder and, then, examined with SEM (JEOL JSM-840A) equipped with an Oxford EDS for topography and elemental compositions. Finally, samples were examined before and after 3 days immersion of 0.5 M H_2SO_4.

3. Results and discussion

3.1 Potentiodynamic polarization curve and weight loss at 25 °C

Fig. 1 shows the anodic dissolution behaviour of alloys in 0.5 M H_2SO_4, while Table 2 summarizes relevant data. This figure reveals a well-defined passive region of 0 V_{SHE} to 1.2 V_{SHE} in all curves. All curves, except for the one at x = 0.25 (C-0.25), show a secondary passive region at 0.15 V_{SHE}. This passivation is attributed mainly to the further oxidation or hydroxidation of the passive oxide film, thus altering the valence of Cr [18,19]. Fig. 1 also indicates that the secondary passive regions of C-0.50 and C-1.00 are more prominent than those of C-0 and C-0.25. This observation is due to the selective dissolution in the duplex FCC-BCC structure for C-0.50 and in the BCC-ordered BCC structure for C-1.00, as compared with C-0 and C-0.25 which are single FCC phase. In the active-passive transition region, different compositions at different secondary passivation potentials reveal different dissolution rates owing to a selective dissolution. This observation resembles that observed in duplex phase stainless steel [20,21].

Alloys C-x & SS304	E_{corr} (V_{SHE})	I_{corr} ($\mu A/cm^2$)	E_{pp} (V_{SHE})	I_{crit} ($\mu A/cm^2$)	I_{pass} ($\mu A/cm^2$)
C-0	-0.081	15.8	0.002	42.8	4.5
C-0.25	-0.095	16.7	0.008	87.4	7.1
C-0.50	-0.084	13.4	0.017	117.2	6.4
C-1.00	-0.094	13.1	0.010	198.0	13.9
SS 304	-0.185	45.3	-0.071	603.0	19.1

Table 2. Potentiodynamic polarization curve diagram parameters of alloys C-x and SS 304 at 25°C.

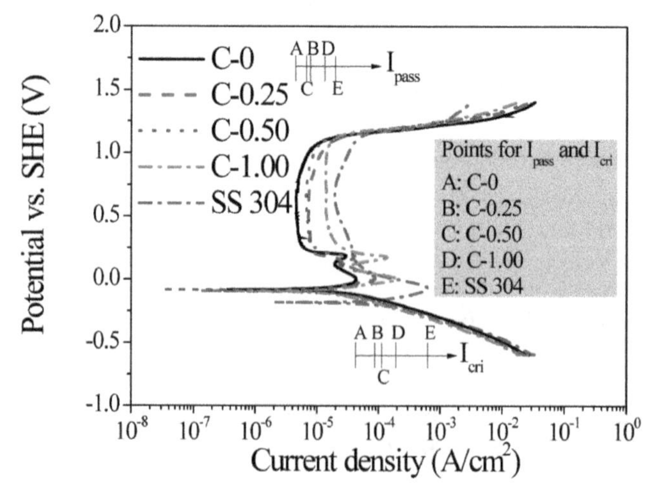

Fig. 1. Potentiodynamic polarization curve diagrams for alloys C-x and SS 304 at 25°C.

Table 2 reveals that the corrosion potential (E_{corr}) and the corrosion current density (I_{corr}) for all of the alloys differ only slightly, and no obvious trends occur for E_{corr} and I_{corr} vs. x variation. The above phenomenon can be attributed to the spontaneous passivation of pure Al in H_2SO_4 [22]. Al metal spontaneously passivates in H_2SO_4, explaining why its corrosion potential is ready in the passive region, i.e., this passivation explains why the polarization curve of Al does not display an apparent active-passive transition region. However, elements such as Cr and Fe exhibit a large critical current density (I_{cri}) for passivation, explaining why Cr and Fe dissolve more than Al before the alloy reaches its passive state. Thus, the variation of Al affects the active region of the polarization curves slightly. Furthermore, in H_2SO_4, all Al, Co, Cr, Fe, and Ni metals show passivity. Among them, Al has a relatively high passive current density (I_{pass}) [22,23] because only Al oxide can easily form a porous film on the metal surface [24]. Therefore, protection by oxide layer on the alloys with higher Al content is inferior to that with lower Al content. Fig. 1 thus reveals that I_{pass} increases with x.

The results of potentiodynamic polarization were compared via performing 15-day-dipping and weight loss experiments. In the 15-day-dipping and weight loss experiments, the corrosion rates for C-0.50 and C-1.00 were markedly higher than those of C-0 and C-0.25 (Fig. 2). This observation differs substantially from the values of I_{corr} obtained from polarization experiment (Fig. 1), in which the two groups only differ slightly, despite the fact that the trend is the same. A previous study found a similar deviation in corrosion current densities obtained from weight loss test and potentiodynamic polarization method [25].

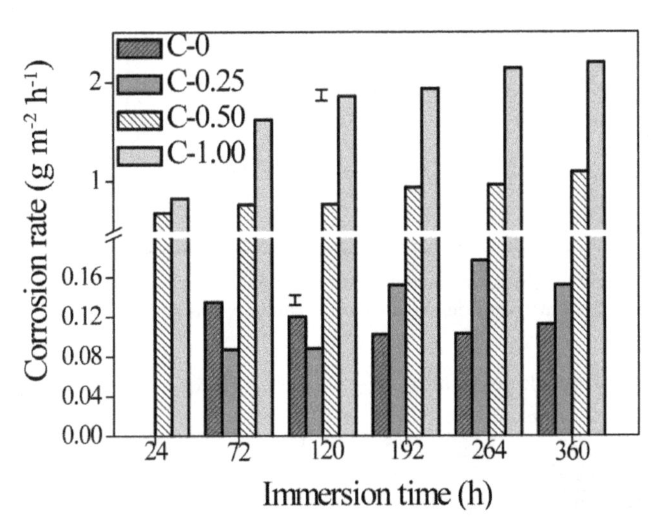

Fig. 2. Diagram showing change in corrosion rate (g m^{-2} h^{-1}) in the 15-day-dipping and weight loss measurement for alloys C-x.

Fig. 1 shows potentiodynamic polarization diagrams for the $Al_xCoCrFeNi$ alloys and SS 304. The alloys have better overall general corrosion behaviour, with a larger E_{corr} and smaller I_{corr}, I_{cri}, and I_{pass} than SS 304.

3.2 Effect of temperature on potentiodynamic polarization

Fig. 3 shows polarization diagrams of $Al_xCoCrFeNi$ alloys at various temperatures. A rising temperature decreased the Tafel slopes of anode (Table 3), increased I_{corr}, and increased E_{corr} and E_t (the transpassive potential) slightly. The corrosion rate is directly related to I_{corr}, according to Arrhenius equation, $I_{corr} = A \exp(-E_a/RT)$ [13,26], where the pre-exponential factor A is generally independent of temperature and is a constant of alloys, where R denotes the gas constant, T denotes temperature, and E_a denotes activation energy for corrosion. In the case of small experimental temperature range, E_a is assumed to be independent of T. Consequently, E_a can be obtained from $\ln(I_{corr})$ vs. 1/T plot. Fig. 4 shows such plots for the alloys and SS 304, indicating that E_a increases with x. This finding suggests that the corrosion rate is more sensitive to temperature for a larger Al content than for a smaller Al content. The $\ln(I_{corr})$ vs. 1/T curves intersect with each other in a range of 23°C - 27°C. Beyond this temperature range, I_{corr} increases with x. The situation is reversed at temperatures lower than 23°C, which is inconsistent with a situation in which E_as for all alloys increase with x from 20°C to 65°C. Hence, E_a, i.e., an intrinsic property of metal, and A, i.e., a surface property of metal, are determinative factors of I_{corr}. While E_a depends only on x, A depends on both x and temperature (Table 4). Therefore, although E_a increases with x, A also increases with x. Combining the effects of E_a and A explains the different corrosion behaviours of the alloys with an increasing x at temperatures exceeding 27°C and lower than 23°C. Thus, the performance of passive films, when Al is added, at higher temperatures becomes inferior to that without addition of Al. In determining I_{corr}, A is more important than E_a at temperatures exceeding 27°C, while E_a is more important than A at temperatures lower than 23°C.

Alloys	20 °C		35 °C		50 °C		65 °C	
C-x	β_a[a]	β_c[b]	β_a	β_c	β_a	β_c	β_a	β_c
C-0	158	218	128	158	134	162	89	158
C-0.25	158	178	103	167	89	168	92	149
C-0.50	94	158	113	178	138	159	89	198
C-1.00	104	148	93	173	98	242	100	220

[a] Anodic Tafel slope β_a in mV/decade, the measured Tafel regions are with 40~50 mV of overvoltage.
[b] Cathodic Tafel slope β_c in mV/decade, the measured Tafel regions are with 150~170 mV of overvoltage.

Table 3. Fit data for Tafel slopes of alloys C-x in 20 °C - 65 °C.

Alloys	A(x, T), A/cm²				E_a, kJ/mol
	20 °C (293 K)	35 °C (303 K)	50 °C (323 K)	65 °C (338 K)	
C-0	1.16×10^{-4}	1.16×10^{-4}	1.16×10^{-4}	1.07×10^{-4}	3.96
C-0.25	1.90×10^{-4}	2.03×10^{-4}	1.77×10^{-4}	1.90×10^{-4}	5.35
C-0.50	7.17	5.64	3.70	8.41	31.24
C-1.00	1.78×10^{9}	7.46×10^{8}	1.89×10^{9}	1.31×10^{9}	78.61
SS 304	1.18×10^{-4}	1.28×10^{-4}	1.70×10^{-4}	2.07×10^{-4}	9.87

Table 4. The fits for A(x, T) and $E_a(x)$ in $I_{corr}(x, T) = A(x, T) \exp(-E_a(x)/RT)$.

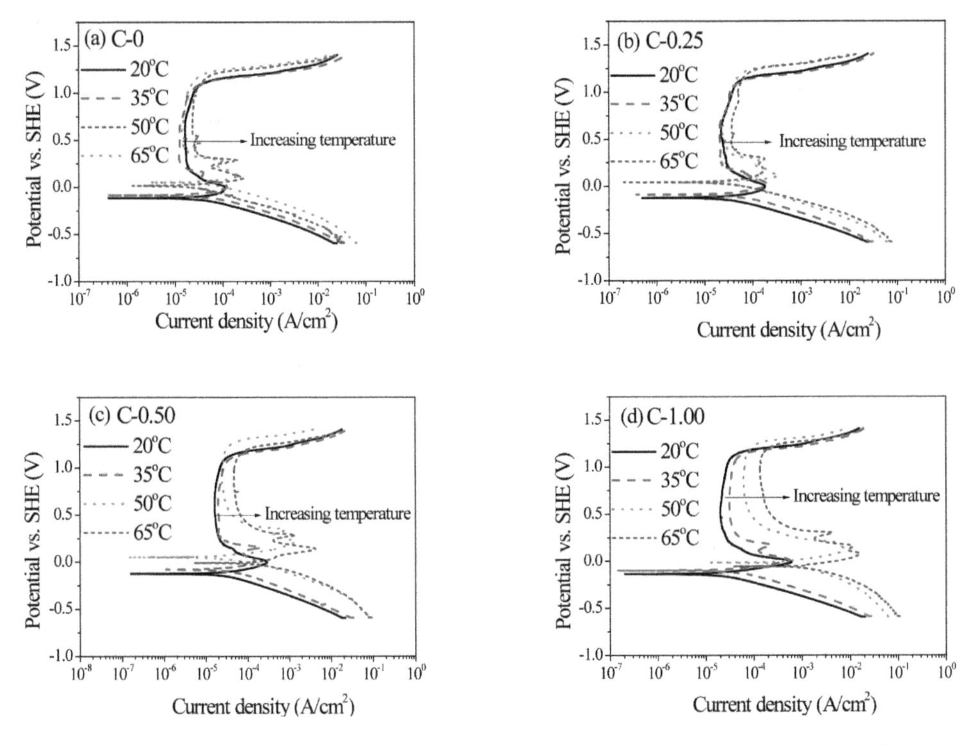

Fig. 3. Potentiodynamic Polarization Curve diagrams for (a) C-0, (b) C-0.25, (c) C-0.50, and (d) C-1.00 at various temperatures.

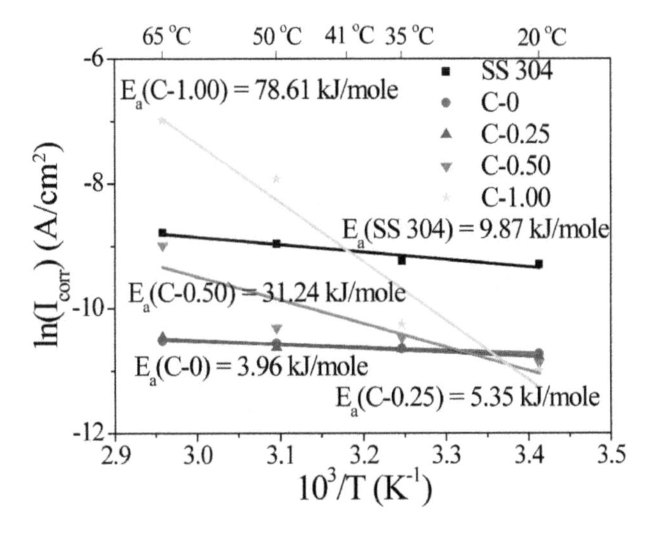

Fig. 4. The Arrhenius plots for alloys C-x and SS 304 at 20°C - 65°C.

3.3 EIS test at 25°C

Figs. 5 and 6 summarize the EIS results of alloys in a sulfuric solution and their schematic equivalent circuit diagrams, respectively. Table 3 lists related parameters of the equivalent circuit diagrams, where R_s, R_f, and R_{ct} denote impedances of the sulfuric solution, oxide layer, and adsorption layer, Q_f and Q_{ad} denote capacitances of constant phase element (CPE) for oxide layer and adsorption layer, respectively. Next, the oxide layer thickness is evaluated by using the Helmholtz model [27] and expressing the layer thickness of the oxide layer, d, as $d = \varepsilon\varepsilon_o S/Q_f$, where ε_o denotes the permittivity of free space (8.85×10^{-14} F/cm), ε denotes the dielectric constant of the medium, and S denotes the surface area of the electrode. Assuming that ε and S for all oxide layers of alloys are the same allows us to compare relative values of d for all samples by $1/Q_f$. Fig. 7 reveals that $1/Q_f$ values are proportional to x, implying that d increases with Al content x. However, according to this figure, the impedance of oxide layer R_f decreases with x and, in Fig. 8, the impedance of the oxide layer is inversely proportional to I_{pass}. Restated, a thinner oxide layer implies a larger value of impedance. To explain this phenomenon, besides the thickness of oxide layer, the density of oxide layer is also considered. As mentioned in Section 3.1, Al oxide easily forms a porous film on the metal surface [24]. Therefore, it is easily understood that in addition to causing a thicker oxide layer, Al element promotes the dispersive oxide layer. Combining these two effects obviously reveals that R_f decreases with x.

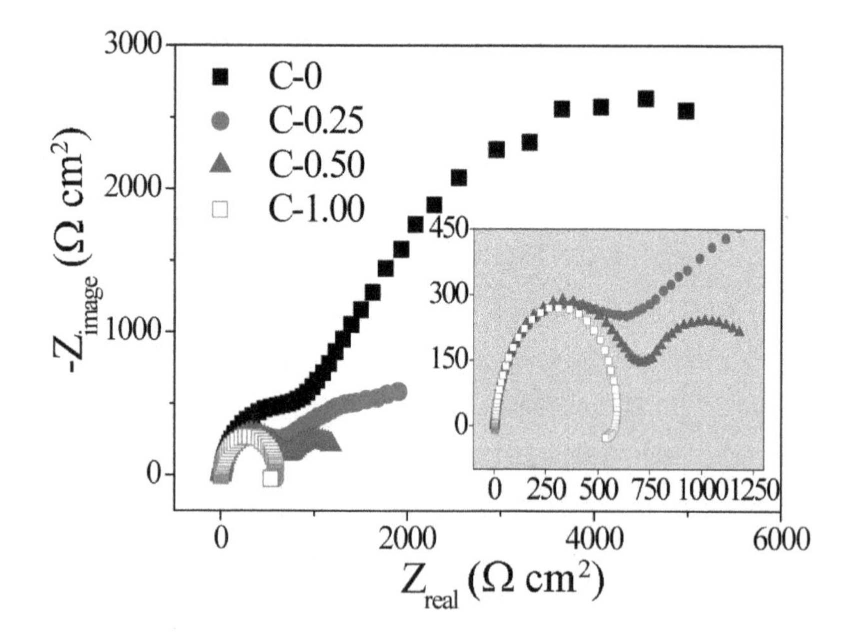

(a)

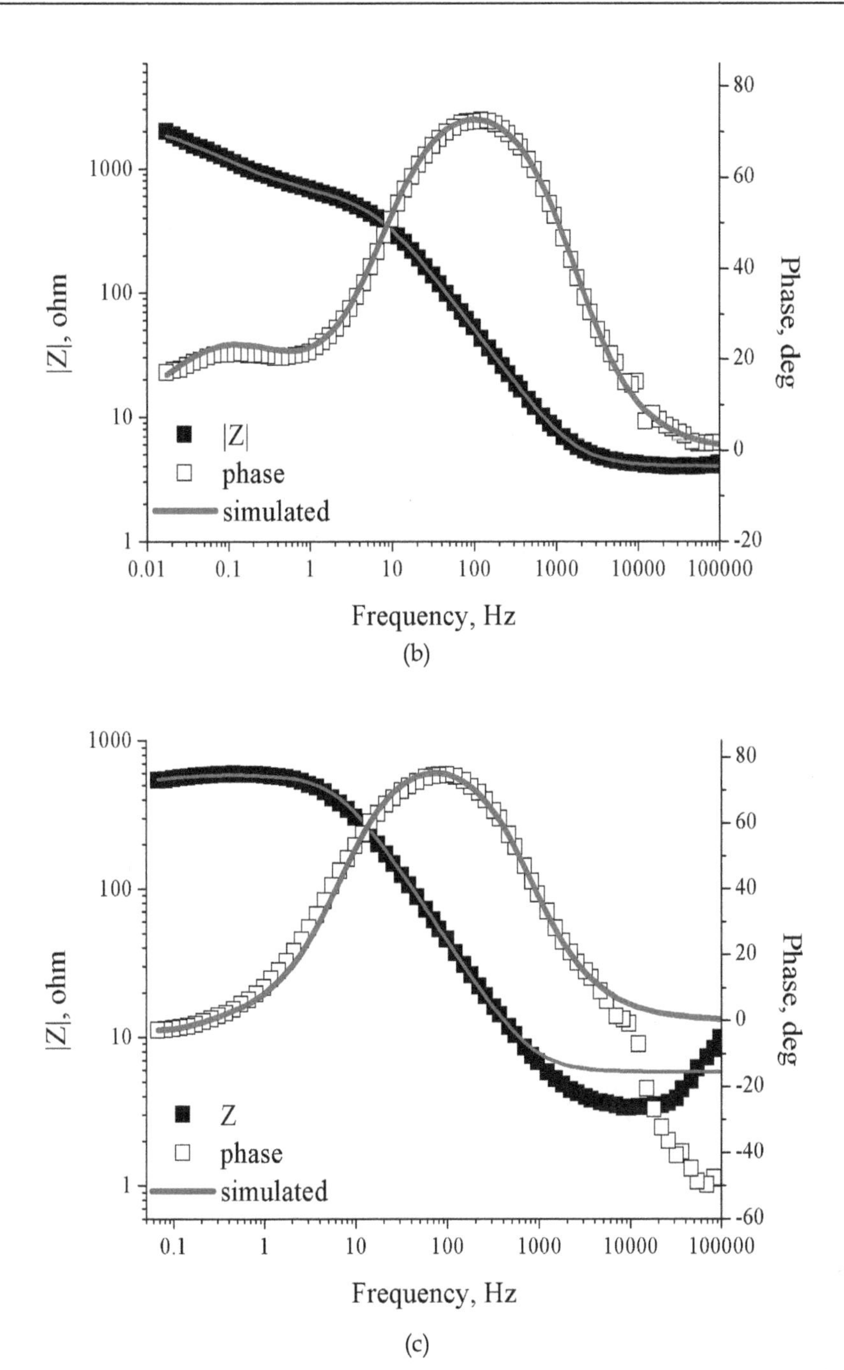

Fig. 5. (a). The Nyquist plots for alloys C-x at 25°C. (b). The Bode plot for C-0.25. (c). The Bode plot for C-1.00.

According to Fig. 6, only C-1.00 reveals a component of inductance in the equivalent circuit (See Fig. 5(c)). In previous studies [28-31], alloys with Al and Ni readily react with $(OH)^-$ and $(SO_4)^{2-}$ in a sulfuric solution and adsorbed on the surface of the alloys, which increases the amount of the ions in the adsorption layer. Therefore, Q_{ad} increases with x, as listed in Table 5. As x value increases to 1.00, the inductance appears in the equivalent circuit in Fig. 6(b). This effect normally occurs in the case of a severe corrosive condition [32]. Origin of the inductance can generally be influenced by some adsorbed intermediates or can be attributed to a space at the interfaces [33]. In C-1.00, a microstructure with an Al and Ni-rich phase which is seen as a reactive phase from metallograph, not only causes adsorption in these Al and Ni-rich areas in corrosion process, but also decreases the impedance in the low frequency area owing to their continuous dissolution. The fact that R_{ct} decreases with x demonstrates a higher dissolution rate for alloys with a higher Al content.

Alloys C-x	R_s (Ω cm^2)	Q_f (μF/cm^2)	n_f	R_f (Ω cm^2)	Q_{ad} (μF/cm^2)	n_{ad}	R_{ct} (Ω cm^2)
C-0	3.271	54.57	0.9094	992.2	636.7	0.7444	7691
C-0.25	3.758	56.61	0.9081	610.5	1525	0.6347	1932
C-0.50	2.994	46.55	0.9223	642.8	3221	0.6454	819.1
C-1.00	3.462	47.16	0.9614	518.1	L_{ad}	-	66.81

* Lad = 122.4 Henry

Table 5. EIS equivalent circuit parameters for alloys C-x.

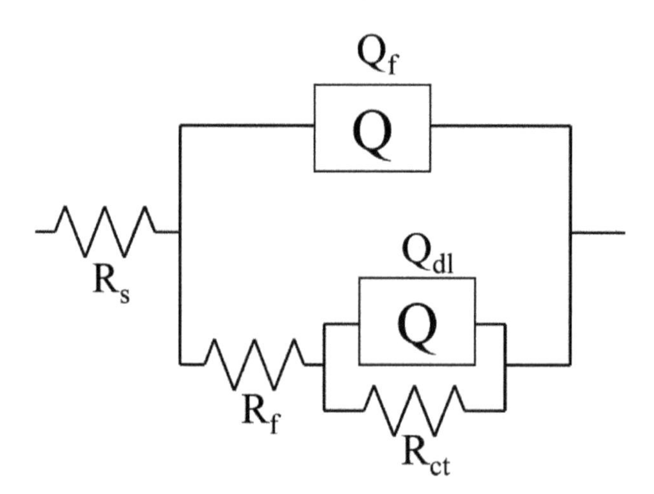

Fig. 6. EIS equivalent circuits for alloys C-0, C-0.25, C-0.50, and C-1.00.

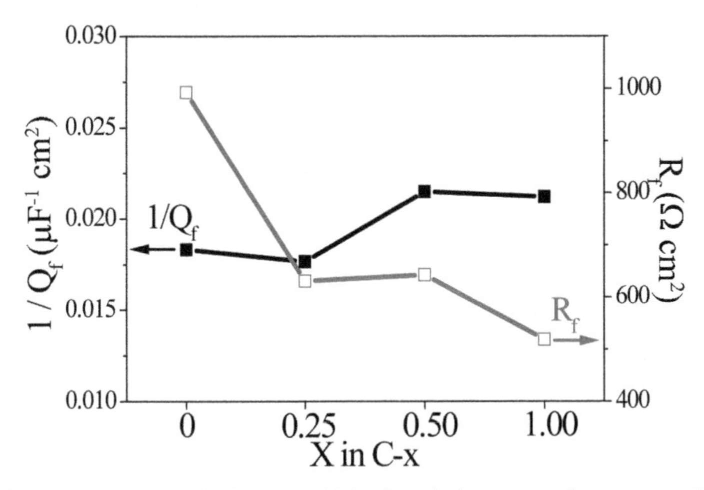

Fig. 7. Impedance and relative thickness ($1/Q_f$) of oxide layer vs. Al content x plots.

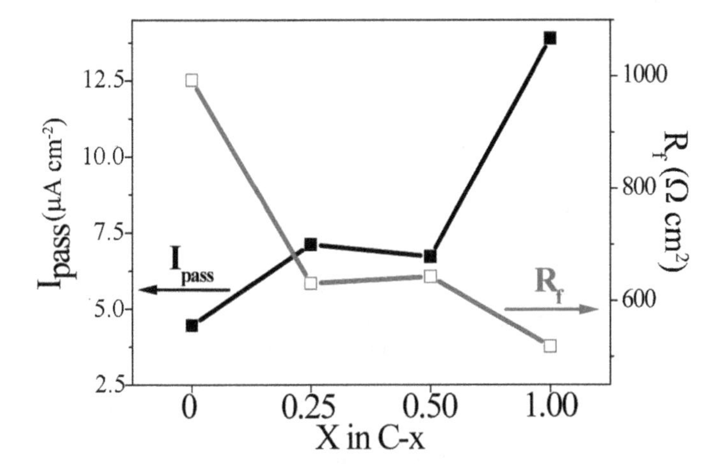

Fig. 8. Impedance and I_{pass} of oxide layer vs. Al content x plots.

3.4 Polarization behaviour for alloys in a chloride-containing H_2SO_4 solution

Fig. 9 shows potentiodynamic polarization curve diagrams for the alloys in 0.5 M H_2SO_4 solution containing various concentrations of chloride ions, as well as in simple 0.5 M H_2SO_4 solution as a comparison. According to Fig. 9(a), oscillation occurs in a passive region for C-0 in 0.5 M H_2SO_4 containing 0.5 M and 1 M of chloride. This phenomenon has been attributed to the cycling process for small pitting and re-passivation with the duration of several seconds for each cycle [34]. Oscillation in the passive region in potentiodynamic polarization curve is a metastable state [35]. This metastable state generally reflects the difficulty of pitting, i.e., alloy C-0 has good anti-pitting ability, while those containing aluminum (C-0.25, 0.50, and 1.00) with no metastable state show an inferior anti-pitting ability (Figs. 9 (b) to (d)).

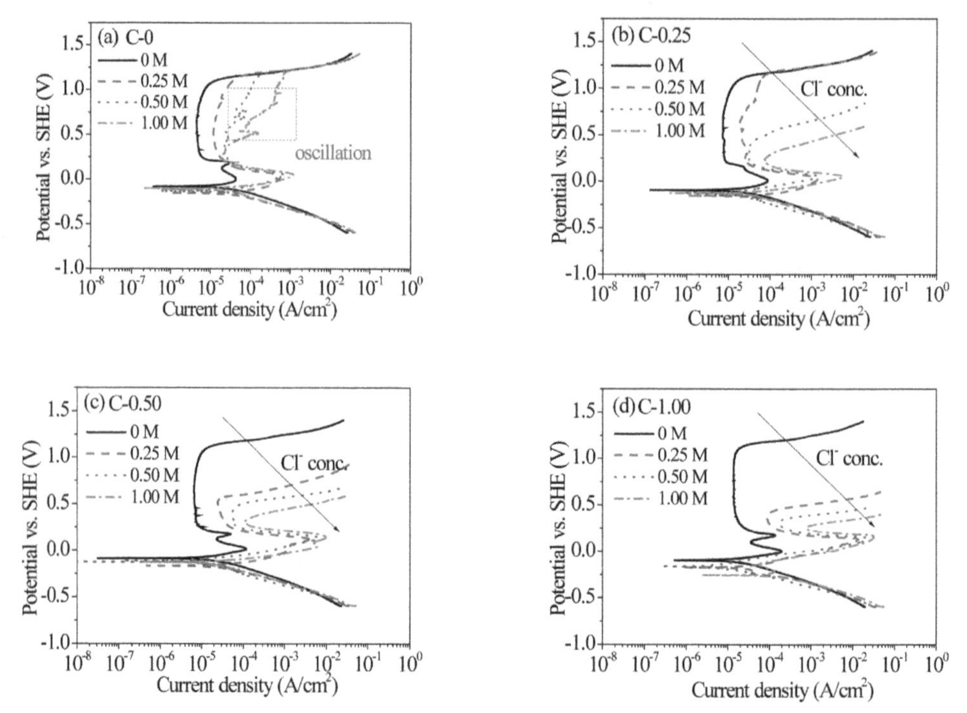

Fig. 9. Potentiodynamic Polarization Curve diagrams for (a) C-0, (b) C-0.25, (c) C-0.50, and (d) C-1.00 at 25 °C in chloride-containing sulfuric acid solution at various Cl^- molarity (M) values.

From an adsorption viewpoint, adsorption competition always prevails on the alloy surface between chloride ions and dissolved oxygen atoms. Notably, no oxide layer forms once chloride ions adsorb on the alloy surface, in which the metal ions readily dissolve. Therefore, the adsorption of chloride ions increases the reacting current density (as indicated by a comparison of Figs. 1 and 9), subsequently increasing the rate of metal dissolution.

Rapid dissolution of alloys in chloride-containing solution is discussed next. When chloride ions are adsorbed on the interface of passive layer and a sulfuric solution, metastable ion complexes gradually form from the anions of a passive layer. These metastable ion complexes enable the anions to dissolve. Once the ion complexes that are on the passive layer/solution interface dissolve into the sulfuric solution, the inner ion complexes of the passive layer move to the passive layer/solution interface in order to correlate with the applied potential. The inability of the anions to form oxide implies the continuous formation of metastable ion complexes and dissolution of ions. Since Al easily forms $[Al(SO_4)]^+$ with $(SO_4)^{2-}$, and $Al(OH)SO_4$ with $(SO_4)^{2-}$ and $(OH)^-$, respectively [36], these metastable ion complexes combine with Cl^- and dissolve afterwards. Therefore, pitting easily occurs on the surface of aluminum alloys. Next, the aluminiferous passive layer and non-aluminiferous passive layer are compared. Fig. 10 shows the pitting potential (E_{pit}) of the alloys and SS 304 in different solutions. The value of E_{pit} for C-0 is almost independent of chloride concentration. The value of E_{pit} for C-0.25 decreases abruptly for a chloride concentration exceeding 0.50 M. This value is close to that of SS 304. The values of E_{pit}, for C-0.25, C-0.50, and C-1.00, decrease to 0.2-0.5 V_{SHE} at a chloride concentration of 0.25 M (Fig. 10). A higher Al concentration in the alloys implies a lower value of E_{pit}. For C-0, deterioration of the passive layer is attributed to the evolution of oxygen. Meanwhile, for C-0.25, C-0.50, and C-1.00, the deterioration of passive layer is attributed to the pitting process. An increasing chloride ion concentration causes the chloride ions to cluster at the defect sites of the passive layer and severely attack the passive layer. Consequently, E_{pit} shifts to a more active region.

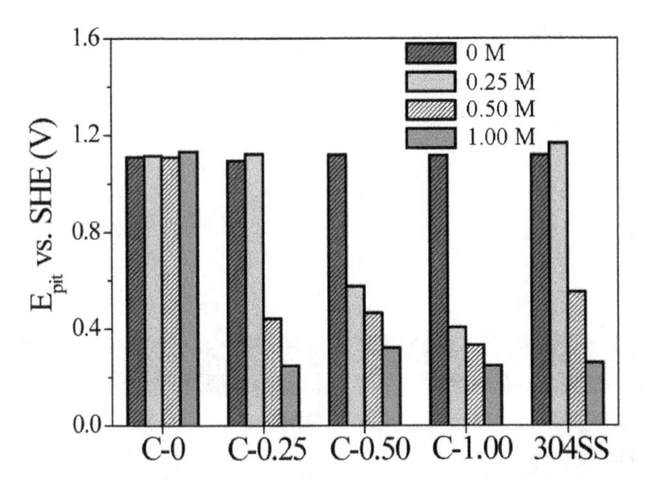

Fig. 10. Histogram of E_{pit} for alloys C-x and SS 304 in solution of different Cl^- ion molarity (M).

3.5 Metallographic examination and EDS analysis

Microstructures for not H_2SO_4-immersed alloys C-0, C-0.25, C-0.50, and C-1.00 are with single FCC, single FCC, duplex FCC-BCC, and BCC-ordered BCC phases, respectively [6]. Table 6 lists the EDS composition for each phase in different alloys.

Alloys	Phases and states	Al	Co	Cr	Fe	Ni
C-0	Overall, not immersed	0	25.93	25.73	24.21	24.13
	Overall, immersed	0	24.45	26.39	24.83	24.32
C-0.25	Overall, not immersed	6.16	23.27	23.58	23.59	23.40
	Overall, immersed	6.18	23.65	24.41	23.04	22.71
C-0.50	Overall, not immersed	11.01	22.77	22.61	21.70	21.92
	FCC matrix, not immersed	8.36	24.74	23.48	22.77	20.65
	BCC, not immersed	**13.94**	21.11	20.48	20.53	**23.94**
	Overall, immersed	8.35	22.75	27.19	23.60	18.11
	FCC matrix, immersed	9.96	22.57	23.16	23.59	20.72
	Wall-shaped BCC, immersed	**3.82**	22.50	**36.33**	25.58	**11.77**
C-1.00	Overall, not immersed	18.88	20.55	20.63	20.01	19.93
	Overall, immersed	12.45	19.80	29.87	23.42	14.45
	BCC, immersed	17.14	20.67	21.96	20.89	19.34
	Ordered BCC, immersed	**3.04**	17.34	**47.53**	27.96	**4.14**

Table 6. EDS analyses (at %) for alloys C-0, C-0.25, C-0.50, and C-1.00.

Figs. 11(a)-(b) show the microstructure of C-0 before and after 3-d immersion in 0.5 M H_2SO_4, respectively. Figs. 11(c)-(d) show the microstructure of C-0.25 before and after 3-d immersion in 0.5 M H_2SO_4, respectively. General corrosion occurs for both C-0 and C-0.25, as revealed by EDS analyses (Table 6).

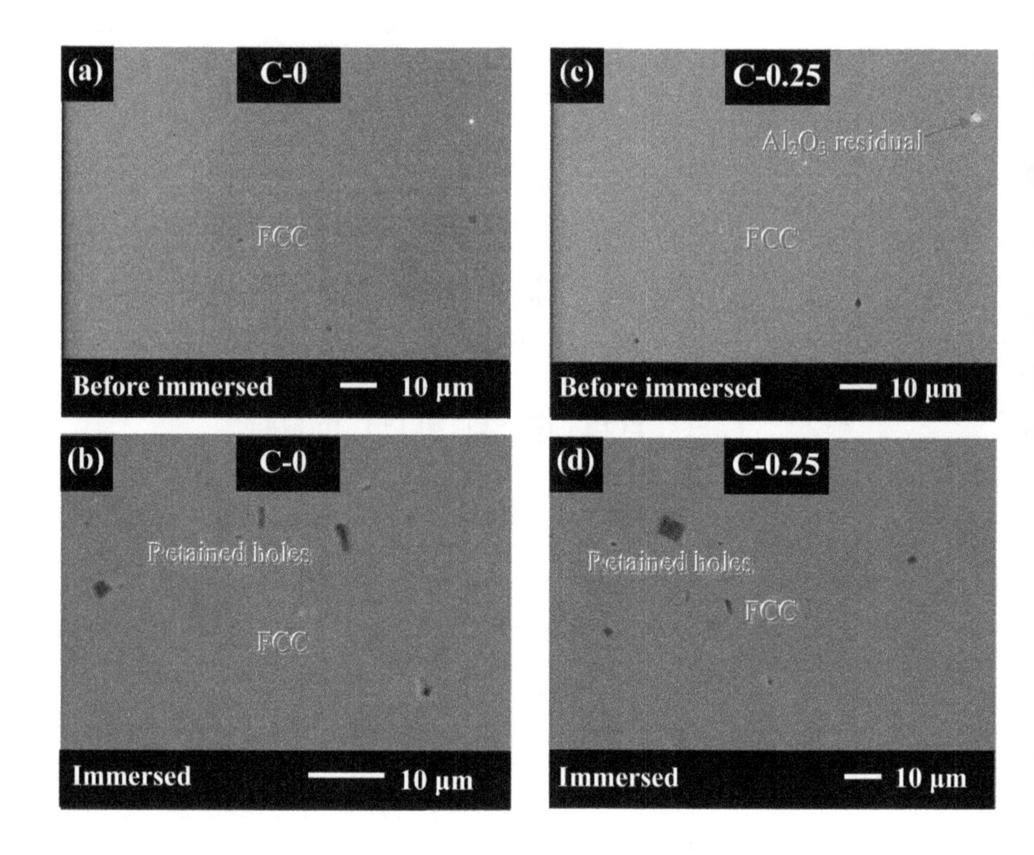

Fig. 11. Metallograph of alloys C-0 ((a) & (b)) and C-0.25 ((c) & (d)). (a) & (c), before; and (b) & (d), after immersion. Retained holes were from cast procedure, and Al_2O_3 residuals were from polishing procedure.

Figs. 12(a)-(b) show the microstructure of C-0.50 before and after 3-d immersion in 0.5 M H_2SO_4, respectively. According to these figures, after immersion the FCC phase remains smooth while the BCC phase shows a rough morphology. Fig. 12(c) shows a line-scanned area across the FCC and BCC phases for an immersed sample. Fig. 12(d) summarizes the line-scanned results, indicating that the BCC phase of C-0.50 before immersion is rich in Al and Ni. However, after immersion, it is poor in Al and Ni and rich in Cr.

Fig. 13 shows the microstructure and line-scan analysis of C-1.00 before and after immersion. Before immersion, BCC and ordered BCC phases cannot be resolved from the microstructure. The composition of BCC phase after alloy immersion is close to the overall alloy composition before immersion, indicating that the BCC phase is a corrosion-resistant phase. Moreover, the change in overall composition after immersion is attributed to the selective dissolution of Al and Ni in the ordered BCC phase of this alloy (Table 6).

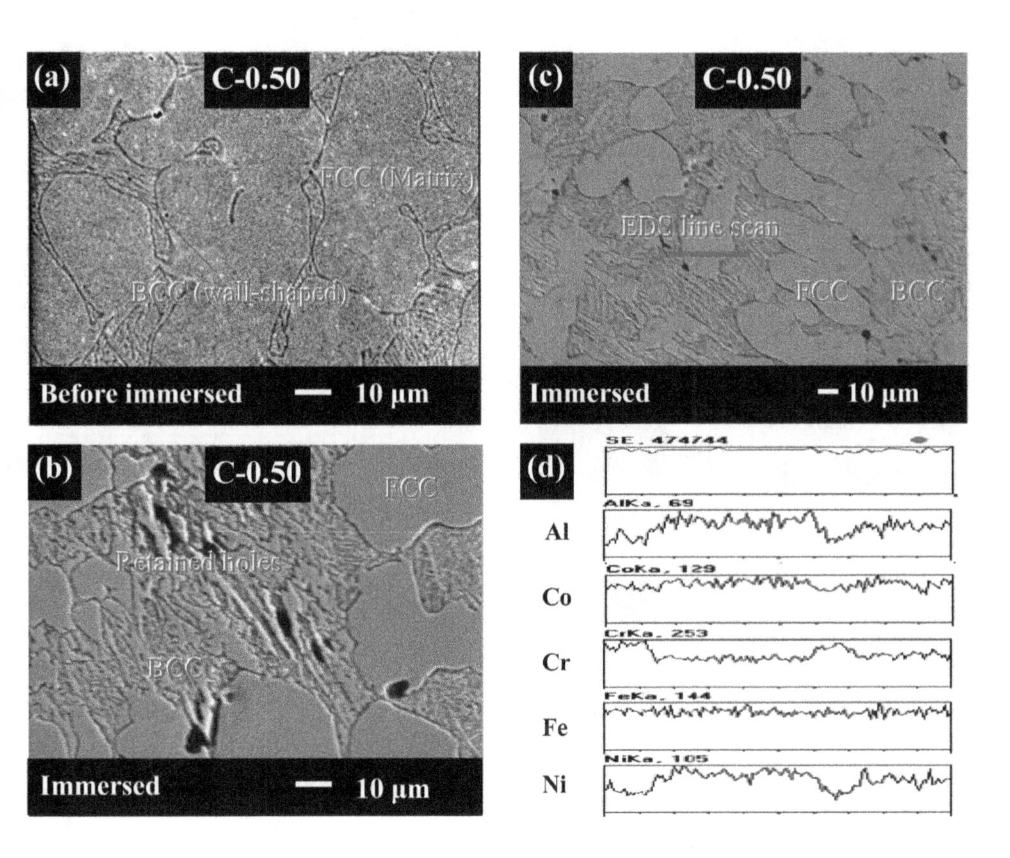

Fig. 12. Metallograph of alloy C-0.50. (a), before; (b) & (c), after immersion; and (d), EDS line-scan results of the location indicated in (c). Retained holes were from cast procedure.

This selective corrosion in Al and Ni-rich phase in C-0.50 and C-1.00, which results in the corrosion attack on Al and Ni, is due to the large bonding in Al and Ni [37]. Alloys containing this bonding readily react with $(OH)^-$ and $(SO_4)^{2-}$ to form Al and Ni complexes and dissolve in a sulfuric solution. Accordingly, after immersion, the remaining compound in the less corrosive-resistant Al and Ni-rich phase is an oxide, rich in Cr, in the residual passive film.

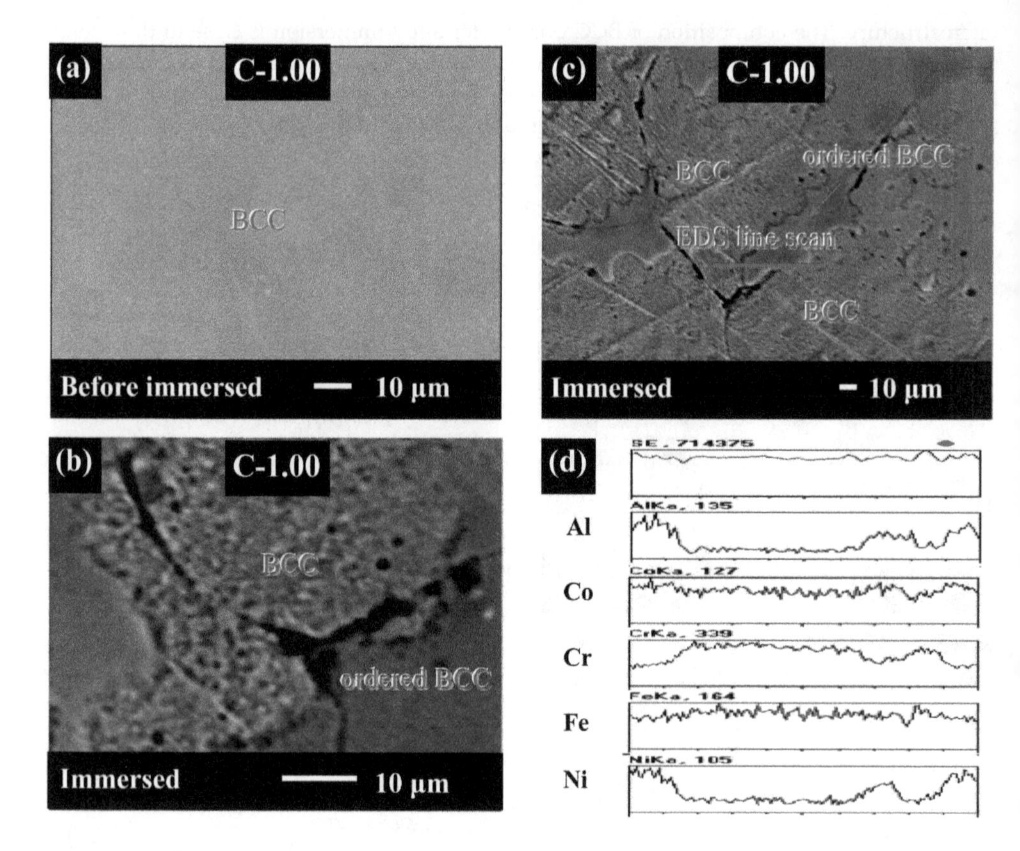

Fig. 13. Metallograph of alloy C-1.00. (a), before; (b) & (c), after immersion; and (d), EDS line-scan results of the location indicated in (c).

3.6 Comparison among potentiodynamic polarization, electrochemical impedance spectroscopy, and weight-loss immersion tests

As discussed above, the corrosion current density (I_{corr}), the critical current density (I_{cri}), and the passive current density (I_{pass}) were obtained from potentiodynamic polarization. The capacitance (Q_f) and the resistance (R_f) of oxide layer were obtained from electrochemical impedance spectroscopy (EIS) equivalent circuits. And the weight-loss rate (W_{loss}) was obtained from weight-loss immersion test. All these data were taken from experiments at ambient temperature (25°C) in 0.5 M H_2SO_4.

Figs. 14(a)-(b), whose data were listed in Table 2, show I_{cri} and I_{pass} vs. Al content x plots, respectively. One can easily see that both I_{cri} and I_{pass} increase with x. This implies that the passive corrosion property of $Al_xCoCrFeNi$ decreases with Al content x. Fig. 14(c) shows W_{loss} vs. x plot. Like I_{cri} and I_{pass}, W_{loss} also increases with x. Notice that, unlike potentiodynamic polarization, immersion weight-loss test is a natural electrochemical reaction, i.e., without applying any voltage on the test sample. On the other hand, I_{cri} and

I_{pass} locate at the passive region of polarization curve. The same tendency for I_{cri}, I_{pass}, and W_{loss} here indicates that the spontaneous passivation occurs for $Al_xCoCrFeNi$, i.e., the open circuit potential (OCP) is readily in the passive region of polarization curve. The above phenomenon can be attributed to the spontaneous passivation of pure Al in H_2SO_4 [24]. EIS equivalent circuits reveal that the passive layers of $Al_xCoCrFeNi$ consist of an oxide layer and an adsorption layer mentioned in Section 3.3. Here, only parameters associated with the oxide layer, i.e., Q_f and R_f, are discussed. The oxide layer thickness is evaluated by using the Helmholtz model mentioned above and denoted by d, as $d = \varepsilon\varepsilon_0 S/Q_f$, where ε_0 denotes the permittivity of free space (8.85×10^{-14} F/cm), ε denotes the dielectric constant of the medium, and S denotes the surface area of the electrode. Assuming that ε and S for oxide layers of alloys are the same allows us to compare relative values of d for all samples by $1/Q_f$. Figs. 14(d)-(e), whose data were listed in Table 5, show the Q_f and R_f vs. x plot, respectively. Both Q_f and R_f decreases with x. This represents that d increases with Al content x, and a thicker oxide layer implies a smaller value of impedance. Therefore, one can explain this phenomenon by considering both the thickness and the density of oxide layer. Related study reported Al oxide easily forms a porous structure [25]. Hence, it is easily understood that in addition to causing a thicker oxide layer, Al promotes the dispersive and porous oxide layer. In summary, Al has a negative effect to the passive parameters, including I_{cri}, I_{pass}, W_{loss}, and R_f, for $Al_xCoCrFeNi$ in H_2SO_4.

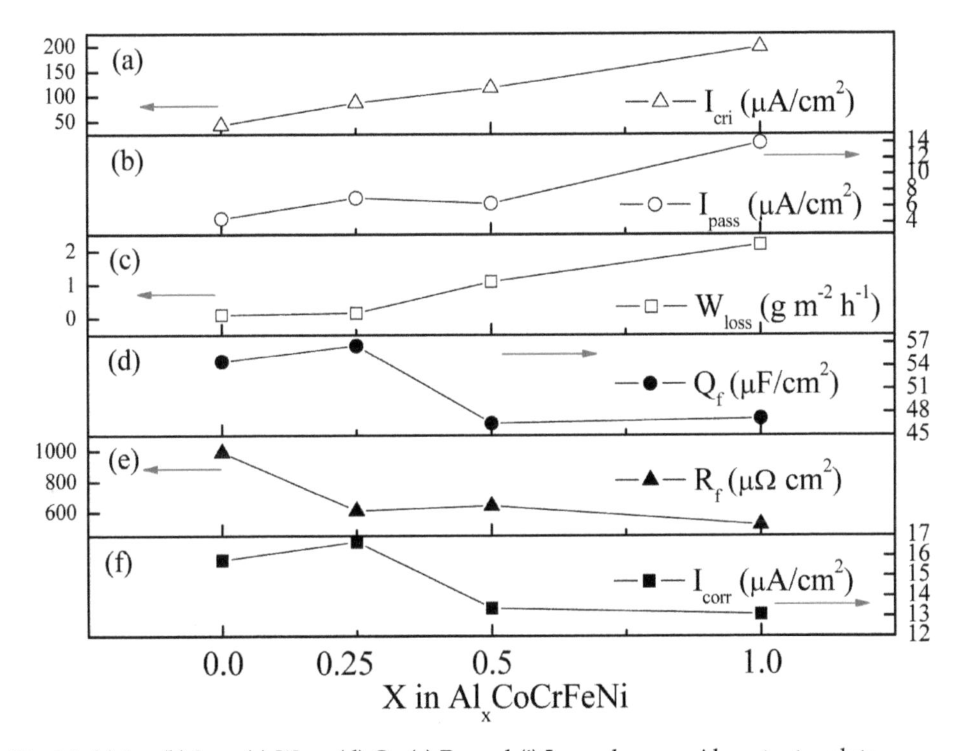

Fig. 14. (a) I_{cri}, (b) I_{pass}, (c) W_{loss}, (d) Q_f, (e) R_f, and (i) I_{corr} values vs. Al content x plots.

Interestingly, Al makes a different effect on general corrosion. Fig. 14(f) shows I_{corr} vs. x plot. One can see that I_{corr} decreases with x. This implies that Al promotes the general corrosion resistance, but degrades the passive one.

3.7 AES, XPS, and ICP-AES analyses of oxide layers

Figs. 15(a)-(d) show the AES results for C-0, C-0.25, C-0.50, and C-1.00, respectively. Owing to the slight difference of atomic number, the signals of Fe, Co, and Ni overlap in AES analysis. Hence, one can see the signals of Co are higher than that of Fe or Ni even for the equal-mole nominal chemical composition of Fe, Co, and Ni. What mentioned above, only the longitudinal composition profiles of O are discussed. A negative and a near-zero slopes are revealed in the relative concentration vs. sputter time profiles in Figs. 15(a)-(d).

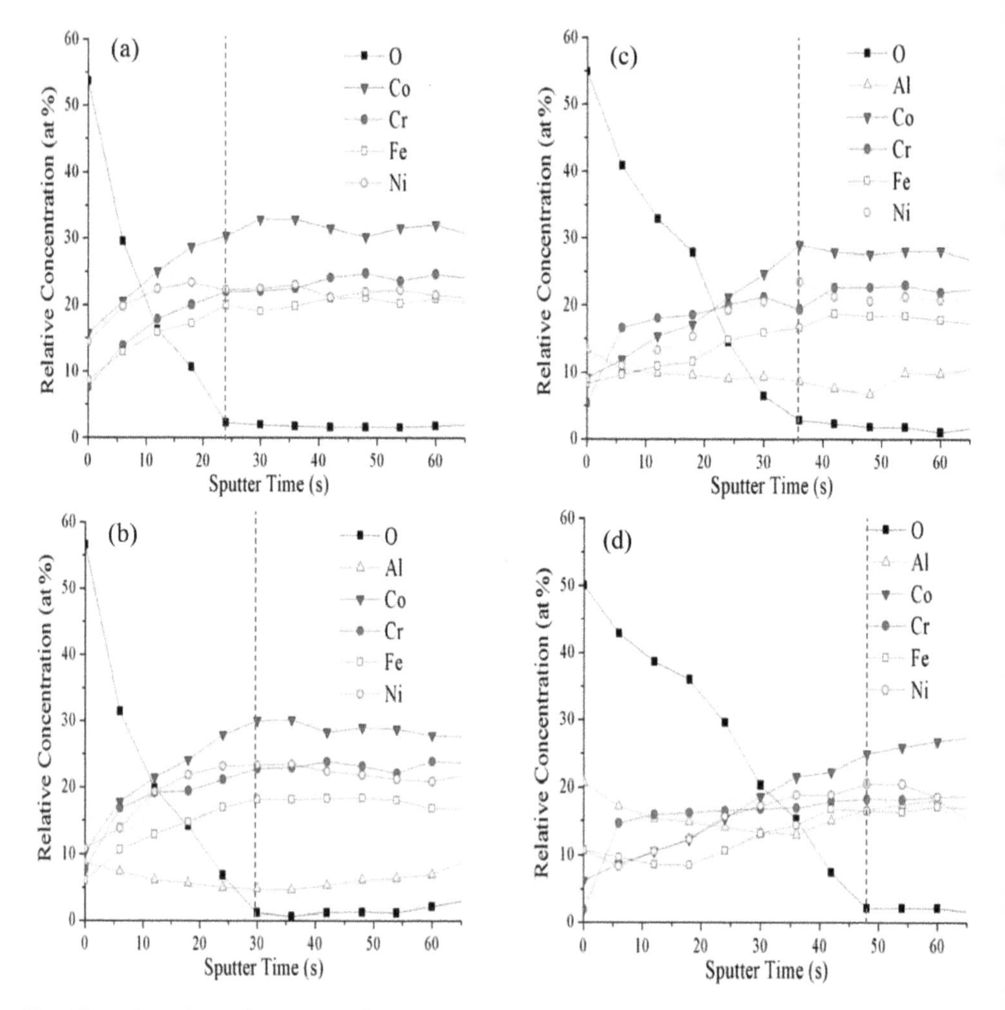

Fig. 15. AES analyses for (a) C-0, (b) C-0.25, (c) C-0.50, and (d) C-1.00.

A value for the identified sputter rate for SiO_2 in this AES device is 7.5 nm per min. Because the immersion for the samples is in the same 1-h period of time, the oxide layers of C-0, C-0.25, C-0.50, and C-1.00 can be distinguished from the terminals of the negative slope shown in each of the profiles. Look at the vertical red-dashed line, i.e., the end-terminal of each profile. It represents the interface between the oxide layer and the intrinsic metal. One can see that the thickness of the oxide layer increases with Al content x that is in accordance with the results of EIS.

XPS analyses attempt to investigate the binding energy profile of 2psub<3/2>sub for Al, Co, Cr, Fe, and Ni. Compared to Co, Cr, Fe, and Ni, Al reveals relatively low atomic sensitivity factor [38]. The signals of Al for alloys C-0 to C-0.50 are too small to identify. Hence, only C-1.00 was used for XPS analysis. Figs. 16(a)-(b) show the Al(2psub<3/2>sub) spectra of C-1.00 after the sputter times of 20 and 35 s, respectively. The raw profile revealing two main peaks represents the exhibition of the selective dissolution. The oxides consists of Al_2O_3, $Al(OH)_3$, and $Al_{25}Ni_{75}O_x$. Al tends to form oxides in H_2SO_4 [24] can explain the formation of Al_2O_3 and $Al(OH)_3$. The existence of $Al_{25}Ni_{75}O_x$ results from the relatively negative enthalpy of Al and Ni. Corresponding to the ICP-AES analysis in the next section, Al-Ni selective dissolution undoubtedly exists for C-1.00. Figs. 16(c)-(d) show the Co(2psub<3/2>sub) spectra of C-1.00 after the sputter times of 20 and 35 s, respectively. The binding types of Co^{2+} and Co^{3+} can be seen. Compare Fig. 16(c) with Fig. 16(d), one can see that the peak intensity of Co_2O_3 is very small in the deep region of the oxide layer. Figs. 16(e)-(f) show the Cr(2psub<3/2>sub) spectra of C-1.00 after the sputter times of 20 and 35 s, respectively. Three kinds of oxide, including Cr_2O_3, $Cr(OH)_3$, and CrO_3, exist for Cr [39]. However, CrO_3 merely forms at high temperatures. Hence, only Cr_2O_3 and $Cr(OH)_3$ are revealed in the profile. One can see that the deep region of oxide layer remains in relatively small amount Cr_2O_3. Figs. 16(g)-(h) show the Fe(2psub<3/2>sub) spectra of C-1.00 after the sputter times of 20 and 35 s, respectively. Similar to references [23,40] Fe_3O_4 and Fe_2O_3 oxides can be found. Figs. 16(i)-(j) show the Ni(2psub<3/2>sub) spectra of C-1.00 after the sputter times of 20 and 35 s, respectively. In resemblance with reference [41], NiO and $Ni(OH)_2$ can be observed. However, a very small amount of $Ni(OH)_2$ appears in our case.

Table 7 lists the results of ICP-AES of immersion solutions for C-x. To trace the ions resulting from the intrinsic metal, one can study the selective dissolution of the alloy elements. Compared with C-0 and C-0.25, C-0.50 and C-1.00 reveal relatively greater Al-Ni selective dissolution. This event is consistent with the results of the XPS analysis.

Alloys		Al	Co	Cr	Fe	Ni	Remarks
C-0	alloy	0	25.93	25.73	24.21	24.13	
	solution	0	24.91	24.89	25.17	25.01	*
C-0.25	alloy	6.16	23.27	23.58	23.59	23.40	
	solution	7.90	23.01	23.07	23.20	22.80	*
C-0.50	alloy	11.01	22.77	22.61	21.70	21.92	
	solution	14.92	21.90	16.51	19.96	26.70	**
C-1.00	alloy	18.88	20.55	20.63	20.01	19.93	
	solution	31.52	20.98	4.86	14.92	27.71	**

*General corrosion, **Selective dissolution in Al and Ni

Table 7. ICP-AES composition (at%) of immersion solution for alloys C-0, C-0.25, C-0.50, and C-1.00.

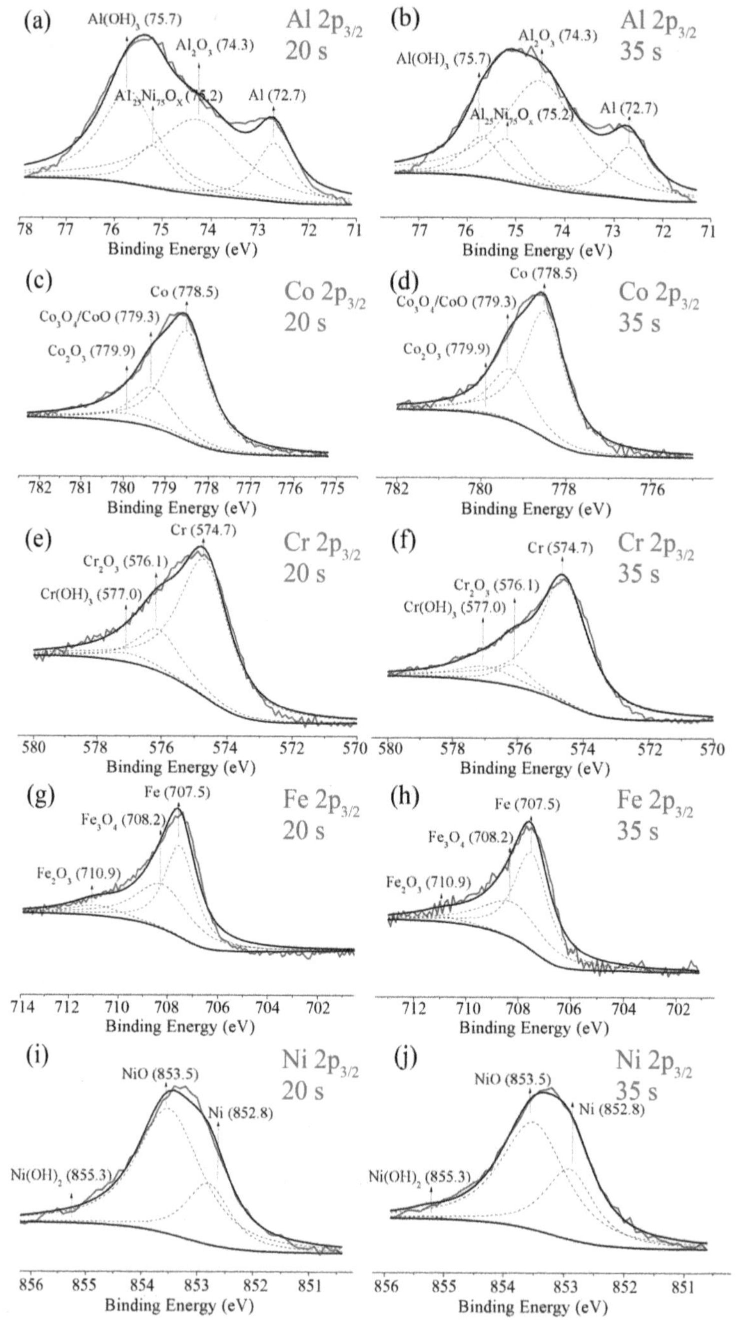

Fig. 16. XPS analyses after pre-sputtering for (a) Al-20 s, (b) Al-35 s, (c) Co-20 s, (d) Co-35 s, (e) Cr-20 s, (f) Cr-35 s, (g) Fe-20 s, (h) Fe-35 s, (i) Ni-20 s, and (j) Ni-35 s.

3.8 Corrosion current density (I_{corr}) at various temperatures

As mentioned in Section 3.1, I_{corr} decreases with Al content at 25°C. However, this differs from temperatures to temperatures. Fig. 17 shows the I_{corr} values of C-x at various temperatures. One can see that I_{corr} decreases with Al content x at low temperatures (< 27°C), and, conversely, at high temperatures (> 27°C). The EIS results (Section 3.1) indicate that more Al content x makes the oxide layers thicker and more dispersive. At low temperatures, the thicker oxide is the dominator for I_{corr}; whereas, at high temperatures, the dispersive oxide dominates. Therefore, this special phenomenon occurs.

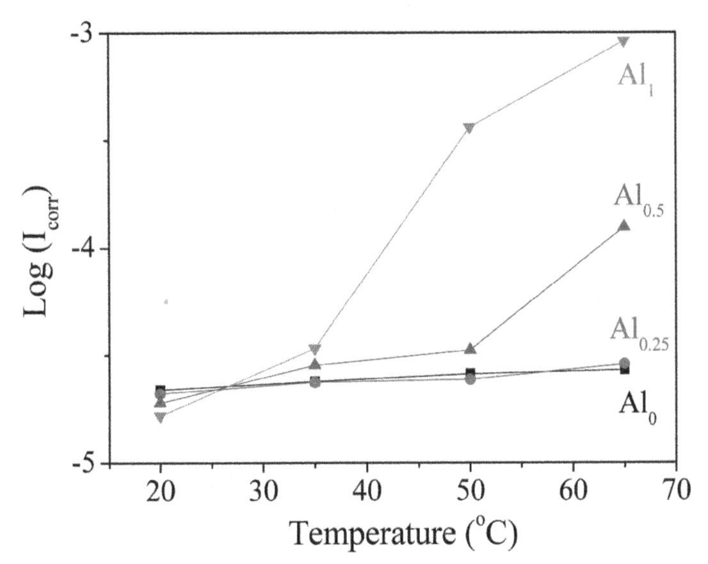

Fig. 17. I_{corr} values for alloys C-x (Al_x) at various temperatures.

4. Conclusions

Owing to the spontaneous passivation of Al element in H_2SO_4, the variation of Al reveals a more apparent effect in a passive region rather than in an active one. Therefore, in contrast with I_{pass}, which increases with x, no obvious trends occur for E_{corr} and I_{corr} vs. x variation. In particular, the weight loss experiment indicates that I_{pass} is a proper index to evaluate the weight loss of samples since $Al_xCoCrFeNi$ alloys are found to have passive behaviour in long-term dipping.

EIS results indicate that the passive films of $Al_xCoCrFeNi$ alloys become increasingly thicker and more dispersive with an increasing x. Therefore, I_{pass} increases with x. As x value increases to 1.00, the inductance effect appears in the equivalent circuit for severe dissolution of Al and Ni-rich phase. As for the effect of chloride on the anti-corrosion property, chloride eases the passive layer to form metastable ion complexes and further dissolve into H_2SO_4. With an increasing chloride concentration and Al content, the metastable ion complexes easily form, allowing E_{pit} to shift to a more active region. Additionally, the microstructure of both C-0 and C-0.25 is single FCC phase, while those of C-0.50 and C-1.00 are duplex FCC-BCC and complex BCC-ordered BCC phase, respectively.

In C-0.50 and C-1.00, the secondary passivation phenomenon in polarization curve results from selective dissolution of the Al and Ni-rich phase.

Moreover, I_{corr} increases with x at higher temperatures (> 27°C), while I_{corr} decreases with x at lower ones (< 23°C). That more closely examining Arrhenius plots of I_{corr} reveals that both pre-exponential factor A and activation energy E_a increase with Al content. However, A affects I_{corr} more significantly than it does so for E_a at higher temperatures (> 27°C) and, conversely, at lower temperatures (< 23°C).

Al is an inferior factor to the passive corrosive resistance but helpful for the general corrosive resistance for $Al_xCoCrFeNi$ in H_2SO_4. The thickness and the density of oxide layers promoted by the addition of Al compete with each other at various temperatures. At ambient temperature, the thick oxide layer dominates I_{corr} value; at temperatures higher than 27°C, the loss oxide layer does. Intuitionally, one may improve the corrosion performance for $Al_xCoCrFeNi$ by adjusting Al content.

5. Acknowledgements

The author would like to thank the financial support of this research from the National Science Council of the Republic of China, Taiwan (NSC96-2221-E007-066-MY3). Mr. Yih-Farn Kao is grateful for his help in compilation of this manuscript. This work is mainly from the 2009 master thesis of the Department of Materials Science Engineering of the National Tsing Hua University by Mr. Tsung-Dar Lee, who was guided by the author.

6. References

[1] See for example, D.R. Gaskell, Introduction to the Thermodynamics of Materials, 3rd ed., Taylor & Francis, Washington D.C., 1995, p. 204.

[2] K.H. Huang, Multicomponent alloy systems containing equal-mole elements, M.S. thesis, Department of Materials Science and Engineering, NTHU, Taiwan, 1996.

[3] K.T. Lai., Microstructure and properties of multicomponent alloy system with equal-mole elements, M.S. thesis, Department of Materials Science and Engineering, NTHU, Taiwan, 1998.

[4] Y.F. Kao, S.K. Chen, J.H. Sheu, J.T. Lin, W.E. Lin, J.W. Yeh, S.J. Lin, T.H. Liou, C.W. Wang, Hydrogen storage properties of multi-principal-component $CoFeMnTi_xV_yZr_z$ alloys, Int. J. Hydrogen Energy 35 (2010) 9046-9059.

[5] J.W. Yeh, S.K. Chen, S.J. Lin, J.Y. Gan, T.S. Chin, T.T. Shun, C.H. Tsau, S.Y. Chang, Nanostructured high-entropy alloys with multiple principal elements: Novel alloy design concepts and outcomes, Adv. Eng. Mater. 6 (2004) 299-303.

[6] P.K. Huang, J.W. Yeh, T.T. Shun, S.K. Chen, Multi-principal-element alloys with improved oxidation and wear resistance for thermal spray coating, Adv. Eng. Mater. 6 (2004) 74-78.

[7] C.Y. Hsu, J.W. Yeh, S.K. Chen, T.T. Shun, Wear resistance and high-temperature compression strength of FCC $CuCoNiCrAl_{0.5}Fe$ alloy with boron addition, Metall. Mater. Trans. A 35A (2004) 1465-1469.

[8] J. Tong, S.K. Chen, J.W. Yeh, T.T. Shun, C.H. Tsau, S.J. Lin, S.Y. Chang, Mechanical performance of the $Al_xCoCrCuFeNi$ high-entropy alloy system with multiprincipal elements, Metall. Mater. Trans. A 36A (2005) 1263-1271.
[9] J.W. Yeh, S.K. Chen, J.Y. Gan, S.J. Lin, T.S. Chin, T.T. Shun, C.H. Tsau, S.Y. Chang, Formation of simple crystal structures in Cu-Co-Ni-Cr-Al-Fe-Ti-V alloys with multiprincipal metallic elements, Metall. Mater. Trans. A 35A (2004) 2533-2536.
[10] Y.F. Kao, T.J. Chen, S.K. Chen, J.W. Yeh, Microstructure and mechanical property of as-cast, -homogenized, -deformed $Al_xCoCrFeNi$ ($0 \leq x \leq 2$) high-entropy alloys, J. Alloys & Comp. 488 (2009) 57-64.
[11] H.P. Chou, Y.S. Chang, S.K. Chen, J.W. Yeh, Microstructure, thermophysical and electrical properties in $Al_xCoCrFeNi$ ($0 \leq x \leq 2$) high-entropy alloys, Mater. Sci. Eng. B 163 (2009) 184-189.
[12] Y.Y. Chen, T. Duval, U.D. Hung, J.W. Yeh, H.C. Shih, Microstructure and electrochemical properties of high entropy alloys—a comparison with type-304 stainless steel, Corros. Sci. 47 (2005) 2257-2279.
[13] Y.Y. Chen, U.T. Hong, H.C. Shih, J.W. Yeh, T. Duval, Electrochemical kinetics of the high entropy alloys in aqueous environments—a comparison with type 304 stainless steel, Corros. Sci. 47 (2005) 2679-2699.
[14] C.P. Lee, C.C. Chang, Y.Y. Chen, J.W. Yeh, H.C. Shih, Effect of the aluminium content of $Al_xCrFe_{1.5}MnNi_{0.5}$ high-entropy alloys on the corrosion behaviour in aqueous environments, Corros. Sci. 50 (2008) 2053-2060.
[15] C.P. Lee, Y.Y. Chen, C.Y. Hsu, J.W. Yeh, H.C. Shih, Enhancing pitting corrosion resistance of $Al_xCrFe_{1.5}MnNi_{0.5}$ high-entropy alloys by anodic treatment in sulfuric acid, Thin Solid Films 517 (2008) 1301-1305.
[16] C. P. Lee, Y. Y. Chen, C. Y. Hsu, J. W. Yeh, and H. C. Shih, The Effect of Boron on the Corrosion Resistance of the High-Entropy Alloys $Al_{0.5}CoCrCuFeNiB_x$, J. Electrochem. Soc., 154 (2007) C424-C430.
[17] Y.F. Kao, T.D. Lee, S.K. Chen, Y.S. Chang, Electrochemical passive properties of $Al_xCoCrFeNi$ (x = 0, 0.25, 0.50, 1.00) alloys in sulfuric acids, Corros. Sci. 52 (2010) 1026-1034.
[18] V. Ashworth, P.J. Boden, Potential-pH diagrams at elevated temperatures, Corros. Sci. 10 (1970) 709-718.
[19] Y.Y. Chen, L.B. Chou, L.H. Wang, J.C. Oung, H.C. Shih, Electrochemical polarization and stress corrosion cracking of alloy 690 in 5-M chloride solutions at 25°C, Corros. 61 (2005) 3-11.
[20] M. Femenia, J. Pan, C. Laygraf, In situ local dissolution of duplex stainless steels in 1 M H_2SO_4 + 1 M NaCl by electrochemical scanning tunneling microscopy, J. Electrochem. Soc. 149 (2002) B187-B197.
[21] I.H. Lo, W.T. Tsai, Effect of selective dissolution on fatigue crack initiation in 2205 duplex stainless steel, Corros. Sci. 49 (2007) 1847-1861.
[22] F. D. Bogar, M. H. Peterson, A comparison of actual and estimated long-term corrosion rate of mild steel in seawater, Laboratory Corrosion Test and Standards, ASTM STP 866 (1985) 197-206.
[23] P. Marcus, Corrosion Mechanisms in Theory and Practice, 2nd ed., Marcel Dekker, New York, 2002.

[24] V. Shankar Rao, V.S. Raja, Anodic polarization and surface composition of Fe-16Al-0.14C alloy in 0.25 M sulfuric acid, Corros. 59 (2003) 575-583.
[25] L. Young, Anodic Oxide Films, 1st ed., Academic Press, London, 1961.
[26] G.K. Gomma, Corrosion of low-carbon steel in sulphuric acid solution in presence of pyrazole-halides mixture, Mater. Chem. & Phys. 55 (1998) 241-246.
[27] C.F. Zinola, A.M. Castro Luna, The inhibition of Ni corrosion in H_2SO_4 solutions containing simple non-saturated substances, Corro. Sci. 37 (1995) 1919-1929.
[28] M.R.F. Hurtado, P.T.A. Sumodjo, A.V. Benedetti, Electrochemical studies with a Cu-5 wt.% Ni alloy in 0.5 M H_2SO_4, Electrochimica Acta 48 (2003) 2791-2798.
[29] F.M Reis, H.G. de Melo, I. Costa, EIS investigation on Al 5052 alloy surface preparation for self-assembling monolayer, Electrochimica Acta 51 (2006) 1780-1788.
[30] T.M. Yue, L.J. Yan, C.P. Chan, C.F. Dong, H.C. Man, G.K.H. Pang, Excimer laser surface treatment of aluminum alloy AA7075 to improve corrosion resistance, Surface and Coating Technology 179 (2004) 158-164.
[31] I. Epelboin, C. Gabrielle, M. Keddam, H. Takenouti, Achievements and tasks of electrochemical engineering, Electrochimica Acta 22 (1975) 913-920.
[32] M. Metikoš-Huković, R. Babić, S. Brinić, EIS-in situ characterization of anodic films on antimony and lead-antimony alloys, J. Power Sources 157 (2006) 563-570.
[33] A.R. Trueman, Determining the probability of stable pit initiation on aluminium alloys using potentiostatic electrochemical measurements, Corros. Sci. 47 (2005) 2240-2256.
[34] Y.M. Tang, Y. Zuo, X.H. Zhao, The metastable pitting behaviours of mild steel in bicarbonate and nitrite solutions containing Cl^-, Corros. Sci. 50 (2008) 989-994.
[35] R.T. Foley, T.H. Nguyen, The chemical nature of aluminum corrosion, J. Electrochem. Soc. 129 (1982) 464-467.
[36] S. Van Gils, C.A. Melendres, H. Terryn, E. Stijns, Use of in-situ spectroscopic ellipsometry to study aluminium/oxide surface modifications in chloride and sulfuric solutions, Thin Solid Films 455 (2004) 742-746.
[37] H. Nakazawa, H. Sato, Bacterial leaching of cobalt-rich ferromanganese crusts, International Journal of Mineral Processing 43 (1995) 255-265.
[38] C.D. Wagner, W.M. Riggs, L.E. Davis, J.F. Moulder, G.E. Muilenberg, Handbook of X-Ray Photoelectron Spectroscopy, 1st ed., Perkin-Elmer Corporation, Minnesota, 1979.
[39] A.A. Hermas, M. Nakayama, K. Ogura, Formation of stable passive film on stainless steel by electrochemical deposition of polypyrrole, Electrochimica Acta 50 (2005) 3640-3647.
[40] K. Varga, P. Baradlai, W.O. Barnard, G. Myburg, P. Halmos, J.H. Potgieter, Comparative study of surface properties of austenitic stainless steels in sulfuric and hydrochloric acid solutions, Electrochimica Acta 42 (1997) 25-35.
[41] R. Wang, An AFM and XPS study of corrosion caused by micro-liquid of dilute sulfuric acid on stainless steel, Appl. Surf. Sci. 227 (2004) 399-409.

7

Reinforcement Fibers in Zinc-Rich Nano Lithium Silicate Anticorrosive Coatings

Carlos Alberto Giudice
UTN (Universidad Tecnológica Nacional),
CIDEPINT (Centro de Investigación y Desarrollo en Tecnología de Pinturas),
La Plata
Argentina

1. Introduction

Well-known the electrochemical nature of most processes of corrosion, the technology of anticorrosive coatings is oriented in the direction of making products that control the development of electrode reactions and that generate the isolating of metal surface by applying films with very low permeability and high adhesion (Sorensen et al., 2011).

The zinc-rich coatings and those modified with extenders and/or metal corrosion inhibitors display higher efficiency than other coatings. A problem that presents this type of primers is the extremely reactive characteristic of metallic zinc; consequently, the manufacturers formulate these coatings in two packages, which imply that the zinc must be incorporated to the vehicle in previous form to coating application (Jianjun et al., 2008 & Lei-lei & De-liang, 2010).

Considering the concept of sacrificial anode (cathodic protection), coatings that consist of high purity zinc dust dispersed in organic and inorganic vehicles have been designed; in these materials, when applied in film form, there are close contacts of the particles among themselves and with the base or metallic substrate to be protected.

The anodic reaction corresponds to the oxidation of zinc particles (loss of electrons) while the cathodic one usually involves oxygen reduction (gain of electrons) on the surface of iron or steel; the "pressure" of electrons released by zinc prevents or controls the oxidation of the metal substrate. Theoretically, the protective mechanism is similar to a continuous layer of zinc applied by galvanizing with some differences because the coating film initially presents in general a considerable porosity (Jegannathan et al., 2006).

In immersion conditions, the time of protection depends on the zinc content in the film and on its dissolution rate. The mechanism is different for films exposed to the atmosphere, because after the cathodic protection in the first stage, the action is restricted substantially to a barrier effect (inhibition resistance) generated by the soluble zinc salts from corrosion by sealing the pores controlling access to water, water vapor and various pollutants. Due to the above, it is necessary to find the appropriate formulation for each type of exposure in service (Hammouda et al., 2011).

With regard to zinc corrosion products, they are basic compounds whose composition varies according to environmental conditions (Wenrong et al., 2009); they are generally soluble in water and can present amorphous or crystalline structure. In atmospheric exposure, zinc-based coatings that provide amorphous corrosion products are more efficient since these seal better the pores and therefore give a higher barrier effect (lower permeability). Fortunately, zinc-rich coatings of satisfactory efficiency in outdoor exposure display in the most cases amorphous corrosion products.

The durability and protective ability depends, in addition to environmental factors, on the relationship between the permeability of the film during the first stage of exposure and the cathodic protection that takes place (Xiyan et al., 2010). The protection of iron and steel continues with available zinc in the film and effective electrical contact; therefore, particularly in outdoor exposure, the time of satisfactory inhibitory action may be more prolonged due to the polarizing effect of the corrosion products of zinc (Thorslund Pedersen et. al., 2009).

A cut or scratch of the film applied on polarized panel allows again the flow of protective electrical current: metallic zinc is oxidized and the film is sealed again. A substantial difference with other types of coatings is that the corrosive phenomenon does not occur under the film adjacent to the cut (undercutting).

With respect to spherical zinc, the transport of current between two adjacent particles is in tangential form and consequently the contact is limited. With the purpose of assuring dense packing and a minimum encapsulation of particles, the pigment volume concentration (PVC) must be as minimum in the order of the critical pigment volume concentration (CPVC).

The problems previously mentioned led to study other shapes and sizes of zinc particles. The physical and chemical properties as well as the behaviour against the corrosion of these primers are remarkably affected by quoted variables and in addition, by the PVC; thus, for example, it is possible to mention the laminar zinc, which was intensely studied by the authors in other manuscripts (Giudice et al., 2009 & Pereyra et al., 2007).

The objective of this paper was study the influence of the content and of the nature of reinforcement fibers as well as the type of inorganic film-forming material, the average diameter of spherical zinc dust and the pigment volume concentration on performance of environmentally friendly, inorganic coatings suitable for the protection of metal substrates. The formulation variables included: (i) two binders, one of them based on a laboratory-prepared nano solution lithium silicate of 7.5/1.0 silica/alkali molar and the other one a pure tetraethyl silicate conformed by 99% w/w monomer with an appropriate hydrolysis degree; (ii) two pigments based on spherical microzinc (D 50/50 4 and 8 μm); (iii) three types of reinforcement fibers used to improve the electric contact between two adjacent spherical zinc particles (graphite and silicon nitride that behave like semiconductor, and quartz that is a non-conductor as reference); (iv) three levels of reinforcement fibers (1.0, 1.5 and 2.0% w/w on coating solids) and finally, (v) six values of pigment volume concentration (from 57.5 to 70.0%).

2. Materials and methods

2.1 Characterization of main components

2.1.1 Film-forming materials

Water-based nano lithium silicate of 7.5/1.0 silica/alkali molar ratio. Previous experiences with these solutions on glass as substrate allowed infer that as silicon dioxide content in the composition increases the film curing velocity also increases and that in addition the dissolution rate decreases.

For this study, a commercial colloidal lithium silicate (3.5/1.0 silica/alkali molar ratio in solution at 25% w/w) was selected; with the aim of increasing the silica/alkali ratio, a 30% w/w colloidal alkaline solution of nanosilica was used (sodium oxide content, 0.32%). The aim was to develop a system consisting of an inorganic matrix (alkaline silicate) and a nanometer component (silica) evenly distributed in that matrix with the objective of determining its behaviour as binder for environment friendly, anticorrosive nano coatings.

Solvent-based, partially hydrolyzed tetraethyl orthosilicate. The tetraethyl orthosilicate is synthesized from silicon tetrachloride and anhydrous ethyl alcohol. This product commercializes as condensed ethyl silicates and usually contains approximately 28% w/w of SiO_2 and at least 90% w/w monomer. The additional purification removes waste products of low boiling point (mainly ethanol) and the dimmers, trimmers, etc.; in some cases, this treatment allows obtaining pure tetraethyl silicate conformed by 99% w/w monomer.

Theoretically, the complete hydrolysis of ethyl silicate generates silica and ethyl alcohol. Nevertheless, the real hydrolysis never produces silica in form of SiO_2 (diverse intermediate species of polysilicates are generated). Through a partial hydrolysis under controlled conditions, it is possible to obtain a stable mixture of polysilicate prepolymers. The stoichiometric equation allows calculating the hydrolysis degree X (Giudice et al., 2007 & Hoshyargar et al., 2009).

The pure or condensed ethyl silicate does not display good properties to form a polymeric material of inorganic nature. In this paper, ethyl silicate was prepared with 80% hydrolysis degree in an acid medium since catalysis carried out in advance in alkaline media led to a fast formation of a gel.

The empirical equation of ethyl silicate hydrolyzed with degree X was used to estimate the weight of the ethyl polysilicate and the hydrolysis degree, through the calculation of the necessary amount of water. The weight was obtained replacing the atomic weights in the mentioned empirical formula; the result indicates that it is equal to 208-148 X.

The percentual concentration of the silicon dioxide in the ethyl polysilicate is equal to the relation molecular weight of SiO_2 x 100 / weight of the ethyl polysilicate; consequently, SiO_2, % = 60 x 100 / (208-148 X). On the other hand, to calculate the water amount for a given weight of tetraethyl orthosilicate and with the purpose of preparing a solution of a predetermined hydrolysis degree, the equation weight of water = 36 (100 X) / 208 was used.

Finally, the amount of isopropyl alcohol necessary to reach the defined percentual level the silica content was calculated. It is possible to mention that after finishing the first hydrolysis stage of tetraethyl orthosilicate that leads to the silicic acid formation, the absence of alcohol would generate the polycondensation of mentioned acid with silica precipitation and the null capacity to conform a polymeric silicic acid (Wang et al., 2009 & Yang et al., 2008).

In a first stage, the pure tetraethyl silicate and the isopropyl alcohol were mixed under agitation. Later, the water and the hydrochloric acid solution selected as catalyst were added (the final pH of the solution was slightly acid, 0.01% w/v, expressed as hydrochloric acid); agitation continued until the end of the dissipation of heat (exothermic reaction).

The conclusions from the experiences indicate that: an excessive amount of water (higher than calculated) generates a rapid gelling in the package, a high pH leads to a fast silica precipitation that reduces the capacity of formation of an inorganic polymer of elevated molecular weight and, in addition, a large quantity of acid retards the condensing reaction due to the repulsion of protonated hydroxyl groups (Giudice et al., 2007).

2.1.2 Pigmentation

In this study, two samples of commercial spherical zinc dust were used; the D 50/50 average particle diameters were 4 μm (fine) and 8 μm (regular), Figure 1. The main features were respectively 98.1 and 98.3% of total zinc and 94.1 and 94.2% of metal zinc, which means 5.0 and 5.1% of zinc oxide; in addition, metal corrosion inhibitors displayed respectively 2282 and 1162 $cm^2.g^{-1}$ values of specific area (BET).

2.1.3 Reinforcement fibers

Nowadays reinforcement fibers are used in many materials to improve their physical and chemical properties (Huang et al., 2009 & Amir et al., 2010). A composite (FRP, fiber-reinforced polymer) is formulated and manufactured with the purpose of obtaining a unique combination of properties; the incorporation of reinforcement fibers to coatings forms a hybrid structure. Fiber is defined as any material that has a minimum ratio of length to average transverse dimension of 10 to 1; in addition, the transverse dimension should not exceed 250 μm.

In anticorrosive coating formulations with hybrid structures, the following reinforcing fibrous materials were used: graphite, silicon nitride and quartz. The levels selected for the experiment were 1.0, 1.5 and 2.0% w/w on coating solids.

Graphite. It is an allotropic form of carbon (hexagonally crystallized). It displays black color with metallic brightness and it is non-magnetic; it has 2.267 $g.cm^{-3}$ density at 25 °C. Usually, it is used as pigment in coatings to give conductive properties to the film. Graphite is formed by flakes or crystalline plates attached to each other, which are easily exfoliated. The electrons that are between layers are those that conduct electricity, and these are what give the quoted brightness (the light is reflected on the electron cloud) (Yoshida et al., 2009 & Abanilla et al., 2005). In perpendicular direction to the layers, it has a low electrical conductivity, which increases with the temperature (it behaves like a semiconductor); on the other hand, throughout the layers, the conductivity is greater and it is increased with the temperature, behaving like a semi-metallic conductor. In this work, graphite in fiber form was used with average values of 1020 μm and 82 μm for length and transverse dimension, respectively (Figure 2.A).

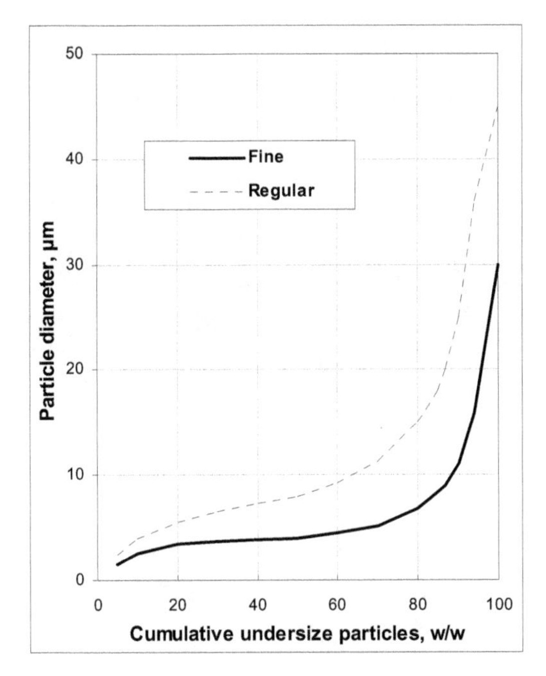

Fig. 1. Commercial spherical zinc dusts. D 50/50 average particle diameters: 4 µm (fine) and 8 µm (regular).

Silicon nitride (Si_3N_4). It displays three different crystal structures (α, β and γ). It is industrially obtained by direct reaction between silicon and nitrogen at temperatures between 1300 and 1400 °C. Silicon nitride is a material frequently used in the manufacture of structural ceramics with high requests of mechanical stress and wear resistance; it displays a moderately high modulus of elasticity and an exceptionally high tensile strength, which makes it attractive for its use in the form of fiber as reinforcement material for coatings films. It behaves like a semiconductor and it has 3.443 $g.cm^{-3}$ density at 25 °C. For this experience, hexagonal β phase in the form of fiber has been selected, with average values of 1205 µm and 102 µm for length and transverse dimension, respectively (Figure 2.B).

Quartz. It is rhombohedral crystalline silica reason why it is not susceptible of exfoliation; chemically it is silicon dioxide (SiO_2). Usually it appears colorless (pure), but it can adopt numerous tonalities if it has impurities; its hardness is such that it can scratch the common steels. It is often used in coatings as extender after being crushed and classified by size (average diameter between 1.5 and 9.0 µm). It is an insulating material from the electrical point of view; it has 2.650 $g.cm^{-3}$ density at 25 °C (Chen et al., 2010 & Lekka et al., 2009). In this experience, quartz fibers were used with average values of 1118 µm and 95 µm for length and transverse dimension, respectively (Figure 2.C).

These reinforcements cannot be classified as nano materials since they no have at least one of the dimensions inferior to 100 nm (Aluru et al., 2003; Radhakrishnan et al., 2009; Behler et al., 2009 & Li & Panigrahi 2006).

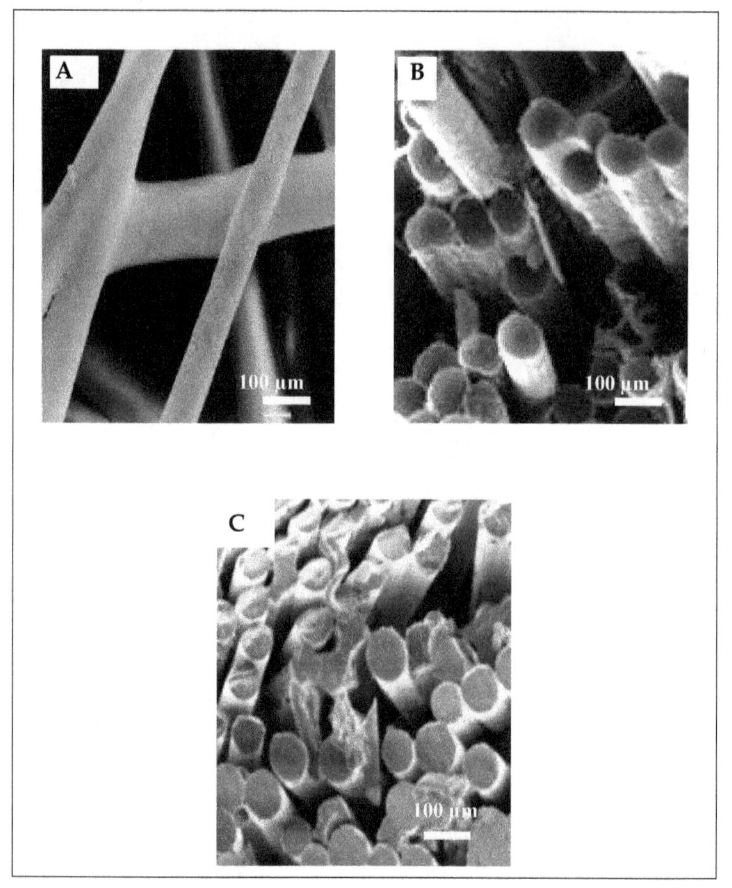

Fig. 2. SEM micrograph of reinforcement fibers: A. Graphite, B. Silicon nitride and C. Quartz.

2.1.4 Rheological agent

Clay modified with amines in gel form was chosen, which was added to the system after finishing the pigment dispersion (1.0% w/w on coating).

2.2 Pigment volume concentration

The physical and mechanical properties of the film and its protective capacity depend on the shape and size of the zinc particles and in addition of the PVC. The critical value (CPVC) is strongly influenced by the ability of binder for wetting the pigment particles.

In this study, PVC values ranged from 57.5 to 70.0%; the variation of two consecutive PVC values was 2.5% in all cases. Preliminary laboratory tests (salt spraying chamber), with values of PVC from 10 to 70% for all formulations, helped to define the range of PVC more convenient to study (Sonawane et al., 2010).

2.3 Manufacture of coatings

In this paper, the manufacture of hybrid coatings was made by ultrasonic dispersion in the vehicle. The efficiency of the fiber dispersion in the polymeric matrix was rheological monitored (viscosity of the system, measured at 10^{-3} sec^{-1}, decreased during the sonification process until reaching a stationary value); the SEM observation corroborated both the efficiency of fiber dispersion and its stability after 3 months in can.

Finally, it should also be mentioned that this type of compositions is provided in two packages with the purpose of avoiding the reaction of metallic zinc with any vestige of moisture in some of the components, which would lead to the formation of gaseous hydrogen. In addition, the system could form a gel because of the reaction between silicic acids and zinc cations. Accordingly, prior to the primer application, the metallic zinc was dispersed for 180 seconds at 1400 rpm in high-speed disperser.

2.4 Preparation of panels

SAE 1010 steel panels were previously degreased with solvent in vapor phase; then they were sandblasted to ASa 2½ grade (SIS Specification 05 59 00/67) obtaining 25 μm maximum roughness Rm. The application was made in a single layer reaching a dry film thickness between 75 and 80 μm. In all cases, and to ensure the film curing before beginning the tests, the specimens were kept in controlled laboratory conditions (25±2 °C and 65±5% relative humidity) for seven days.

The study was statistically treated according to the following factorial design: 2 (type of binder) x 2 (average diameter of spherical microzinc particles) x 3 (type of reinforcement fibers) x 3 (level of reinforcement fibers) x 6 (PVC values), which make 216 combinations. In addition, reference primers (without reinforcement fiber) based on both binders and both spherical microzinc particles for the six mentioned PVC values were also considered, which means in total 24 reference primers. All panels were prepared in duplicate; primer identification is displayed in Table 1.

2.5 Laboratory tests

After curing process, panels of 150 x 80 x 2 mm were immersed in 0.1 M sodium chloride solution for 90 days at 25 °C and pH 7.0. A *visual inspection* was realized throughout the experience; in addition, the *electrode potential* was determined as a function of exposure time; two clear acrylic cylindrical tubes were fixed on each plate (results were averaged).

The tube size was 10 cm long and 5 cm diameter, with the lower edge flattened; the geometric area of the cell was 20 cm^2. A graphite rod axially placed into the tubes and a saturated calomel electrode (SCE) were selected respectively as auxiliary and reference electrodes. The potential was measured with a digital electrometer of high input impedance.

Similar panels were tested in salt spraying (fog) chamber (1500 hours) under the operating conditions specified in ASTM B 117. After finishing the tests, the panels were evaluated according to ASTM D 1654 (Method A, in X-cut and Method B, in the rest of the surface) to establish the *degree of rusting*.

Film-forming material	(A) Water-based nano lithium silicate of 7.5/1.0 silica/alkali molar ratio
	(B) Solvent-based, partially hydrolyzed tetraethyl orthosilicate
Spherical microzinc	(I) Spherical microzinc (fine), D 50/50 4 μm
	(II) Spherical microzinc (regular), D 50/50 8 μm
Reinforcement fibers	Types: (1) Without, (2) Graphite, (3) Silicon nitride (4) Quartz
	Level: (a) 1.0, (b) 1.5 and (c) 2.0% w/w on coating solids
PVC	Values: 57.5, 60.0, 62.5, 65.0, 67.5 and 70.0%

Table 1. Primer identification.

3. Result and discussion

3.1 Visual observation

Immersion test in 0.1 M sodium chloride solution allowed to observe, particularly in those panels with X-cut, that coatings based on nano-structured film-forming material as binder showed greater amount of white products from corrosion of metallic zinc than in those panels protected with primers made with partially hydrolyzed ethyl silicate as binder. This performance would be supported in the less zinc dispersion ability that displays the first binder, which would generate films more porous.

On the other hand, primers that included fine zinc particles (4 μm) also showed a galvanic activity more important than those made from regular zinc particles (8 μm). A similar conclusion was reached with the primers based on microzinc/conducting reinforcement fibers (graphite and silicon nitride) with respect to those based just on spherical microzinc dusts and mixture with insulating reinforcement fiber (quartz). In turn, it was also observed a rise of the galvanic activity of metallic zinc when the amount of conducting fibers was increasing in the film.

For the lower values of PVC studied, the incorporation of conductive reinforcing fibers in increasing levels led to primers with a galvanic activity also increasing (similar amount of white salts than in primers formulated with PVC nearest to CPVC).

3.2 Corrosion potential

Immediately after finishing the immersion of all coated panels in the electrolyte, the potential was inferior to -1.10 V, a value located in the range of protection of the electrode. It is worth mentioning that cathodic protection is considered finished when the corrosion potential of coated panel increased to more positive values (anodic ones) than -0.86 V (referring to SCE) since the characteristic corrosion points of the iron oxides were visually observed.

The electrode potential measurements as a function of immersion time indicates that both types of binders had a significant influence on the electrode potential: in general, more negative values were obtained with nano-structured film-forming materials, which means that the primers based on lithium silicate showed better cathodic protection than those manufactured with ethyl silicate.

On the other hand, slight differences in electrode potential could also be attributed to the average diameter of zinc particles; it was observed greater galvanic activity in samples prepared with 4 μm than with 8 μm (values more negative of electrode potentials for the former than for the latter).

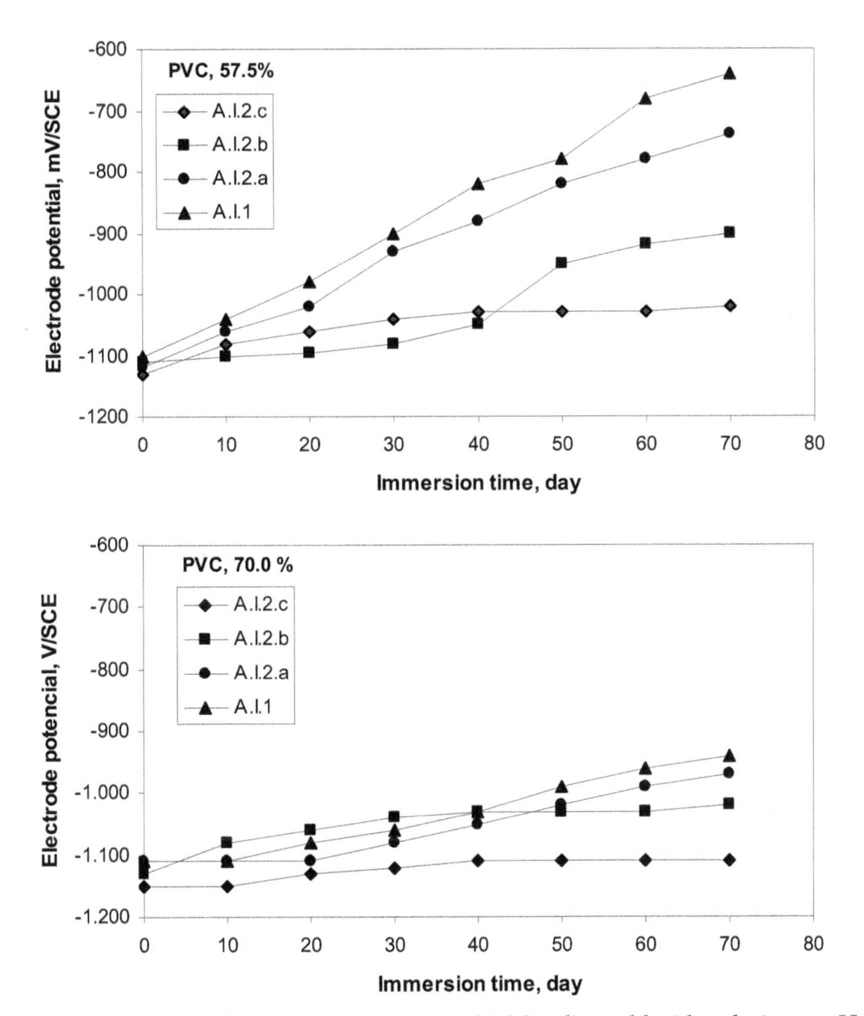

Fig. 3. Electrode potential vs. immersion time in 0.1 M sodium chloride solution at pH 7.0 and 25 °C for primers based on 7.5/1.0 nanosilica/lithium oxide molar ratio, fine microzinc and graphite fibers.

The experimental values indicate a significant shift towards more positive values of potential in those primers with decreasing amounts of conducting reinforcement fibers in their composition (2.0, 1.5 and 1.0% w/w ratio, in that order).

Considering the performance, the worst primers have been formulated both with microzinc dusts alone and mixed with insulation reinforcement fibers.

Finally, there was observed only a slightly decreasing efficiency to the lower values of PVC studied with the incorporation of conductive reinforcing fibers in increasing levels.

Figure 3 includes values of potential versus immersion time in 0.1 M sodium chloride solution at pH 7.0 and 25 °C for primers formulated with 57.5 and 70.0% PVC values and based on 7.5/1.0 nano silica/lithium oxide molar ratio as film-forming material, fine microzinc (D 50/50 4 μm) as pigment inhibiting and graphite as reinforcement fiber in the three levels studied. In addition, this figure displays the corresponding reference primers.

There is a total correlation between conclusions of visual observation and results of the electrode potentials obtained during immersion in 0.1 M sodium chloride solution; therefore, the basis of the quantitative results of electrode potentials are the same that those spelled out in the visual observation.

3.3 Degree of rusting

The performance of panels tested during 1500 hours in salt spraying (fog) chamber (35±1 °C; pH 6.5-7.2; continuous spraying of 5±1% w/w of NaCl solution) are shown in Figure 4 and Figure 5. They include only the average values of the tests performed in the failure in X-cut (Method A) and over the general area of panel (Method B).

The results of Method A were evaluated according to the advance from the cutting area: the value 10 defines a failure of 0 mm while zero corresponds to 16 mm or more. Those results corresponding to Method B were measured taking into account the percentage of area corroded by the environment: the scale ranges from 10 to 0, which means respectively no failure and over 75% of the rusted area.

On the other hand, Figure 6 displays one of the primer with best performance in salt spraying (fog) chamber for 1500 hours: A.I.2.c; applying the Method A, this primer showed a degree of rusting 10 (no failure, which means 0 mm of advance from the cutting area).

To study the variables considered (main effects), a statistical interpretation was carried out. First, the variance was calculated and later the Fisher F test was done.

The results indicated that type of binder, average diameter of microzinc particles, type of reinforcement fibers, level of reinforcement fibers and finally PVC values displayed an important influence on the performance of the protective coatings.

According to results, it was considered desirable for the statistical analysis to take into account all values of PVC studied for allowing a certain margin of safety in the performance since it is possible the generation of heterogeneities in primer composition attributable to poor incorporation of metallic zinc and/or sedimentation in container before applying.

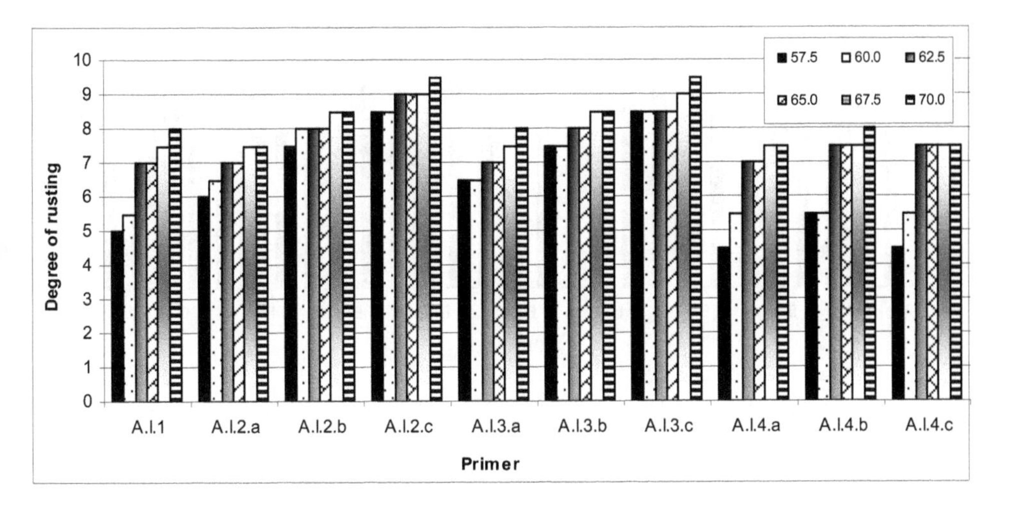

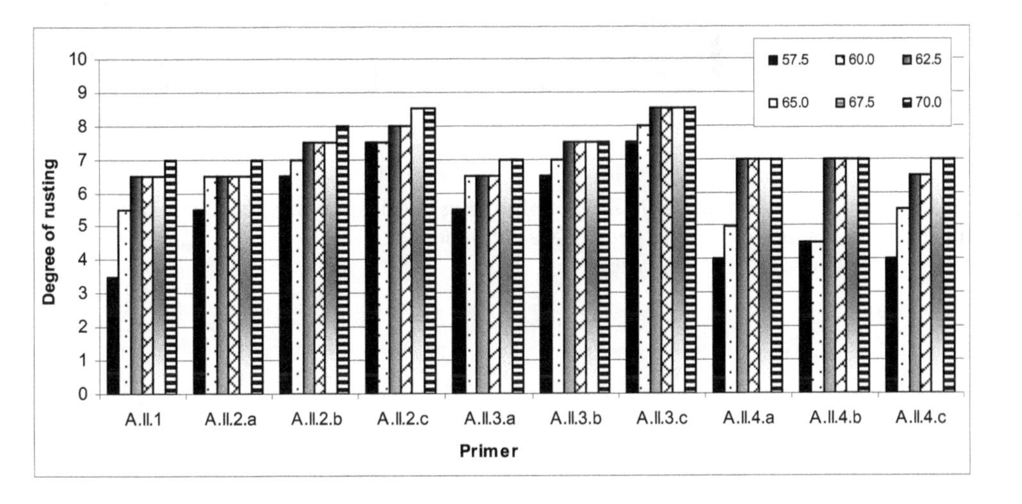

Fig. 4. Coatings based on binder A: Degree of rusting in salt spraying (fog) chamber; average values of failures in X-cut and in general area of panel.

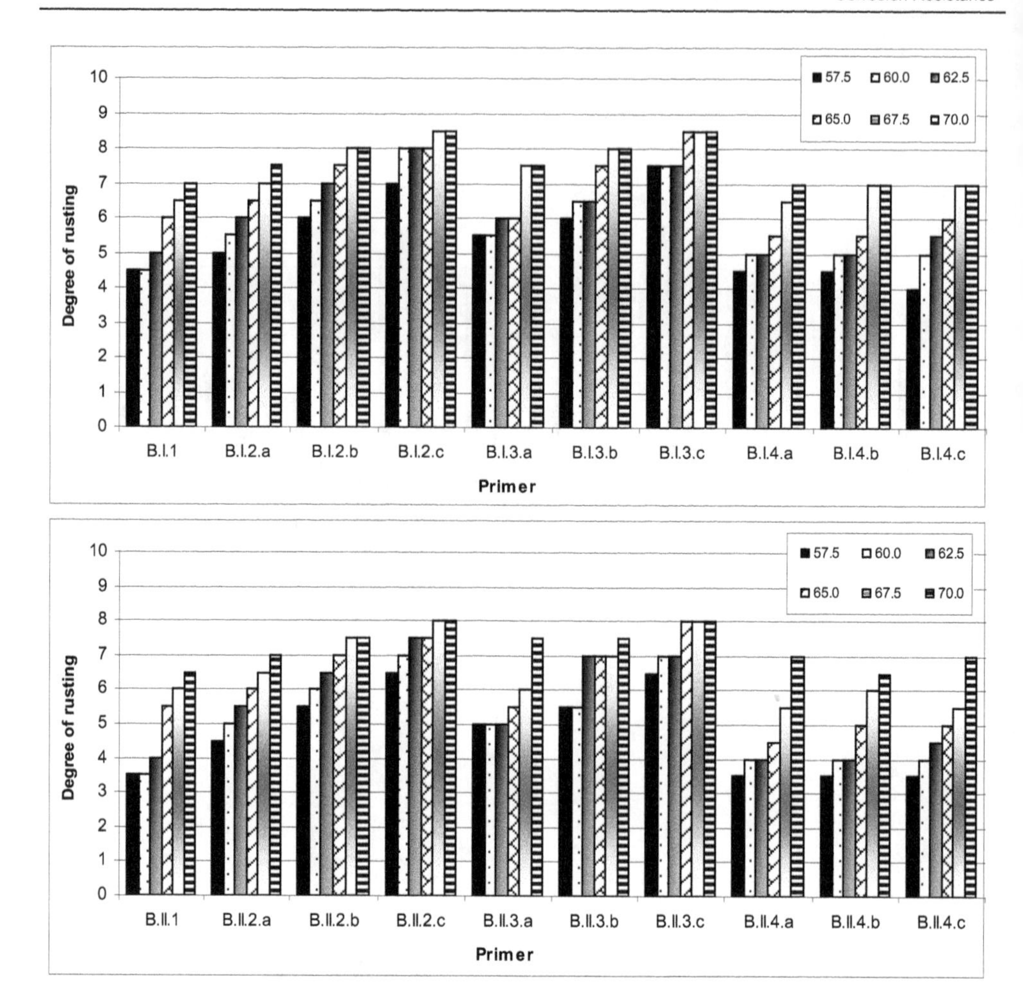

Fig. 5. Coatings based on binder B: Degree of rusting in salt spraying (fog) chamber; average values of failures in X-cut and in general area of panel.

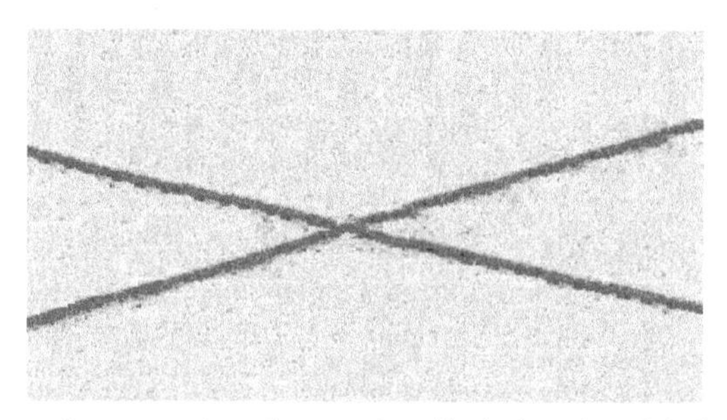

Fig. 6. Primer A.I.2.c, performance in salt spraying (fog) chamber, Method A, degree of rusting: 10.

With the purpose of establishing the efficiency of each protective coating from an anticorrosive point of view, the average value of degree of rusting was calculated for areas with and without cutting. The results of Table 2 confirm the superior performance of water-based nano lithium silicate of 7.5/1.0 silica/alkali molar ratio in relation to solvent-based, partially hydrolyzed tetraethyl orthosilicate, the major efficiency of fine microzinc compared to regular one and the increasing efficiency of primers as the level of conducting reinforcement fibers increased (Ahmed et al., 2010). On this last variable of formulation, it is worth mentioning that primers with 2.0% conducting reinforcement fibers showed the best protective capacity, which would occur due to the improved electric contact between zinc particles and with metallic substrate. On the other hand, the quartz used as reinforcement fiber due to characteristic non-conductive showed a similar performance that the corresponding reference primer (without reinforcement fiber).

Nature of effect	Type of effect	Degree of rusting
		Average values of failures in X-cut and in general area of panel
Type of binder	A	7.1
	B	6.2
Microzinc D 50/50	I	7.0
	II	6.3
Type of reinforcement fibers	1	5.8
	2	7.2
	3	7.2
	4	5.8
Level of reinforcement fibers	a	6.2
	b	6.8
	c	7.3

Table 2. Average values of the simultaneous statistical treatment of all variables.

On the other hand, Table 3 lists the average values and the standard deviations of statistical processing for each primer; in addition, it also displays the general average values for each type of reinforcement fiber taking into account both binders considered. In this table, the highest value also indicates the best performance in terms of the ability to control the metal corrosion. The analysis of the results obtained by using both types of binder displays that the primers based just on two spherical microzinc (reference primers) and non-conducting reinforcement fibers (quartz) in the three considered levels, formulated with reduced values of PVC, showed a sharp decline in corrosion performance. On the other hand, those that included conducting reinforcement fibers (graphite and silicon nitride), despite having been manufactured with a significantly lower level of pigmentation, maintained their efficiency. Corresponding standard deviation values support this conclusion.

These results would be based on the reduced electrical contact between particles of both types of microzinc and the metal substrate, regardless of the corrosion products could not only increase the electrical resistance of the film but also could decrease the amount of available zinc.

The incorporation of conducting reinforcement fibers seems to have favored the conductivity, which leads to reduction of the efficient PVC, according to the abundant amount of zinc corrosion products visually observed, the results of the electrode potentials and those obtained in the salt spraying (fog) chamber. Figures 7 and 8 display the primer films based on binder A (water-based nano lithium silicate of 7.5/1.0 silica/alkali molar ratio), microzinc I (fine, 4 μm) and fiber 2 (graphite) in level c (2.0% w/w) for 57.5% and 70.0% PVC values respectively, after finishing the accelerated aging test. The analysis of the cited figures reveals that despite having larger distance between the particles of microzinc in the case of 57.5% PVC compared with that formulated with 70.0% PVC, the galvanic activity in the two primers is significant in both cases (as evidenced by the amount of white zinc salts). In addition, results of figures show that the conductive reinforcement fibers linked electrically the microzinc particles each other, even in the primer of less PVC (all particles, despite having no direct contact between them, demonstrated activity like sacrificial anodes).

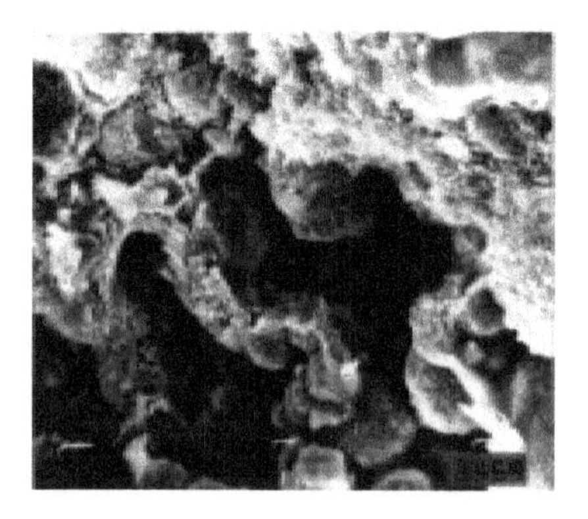

Fig. 7. SEM micrograph of primer A.I.2.c formulated with 57.5% PVC.

Fig. 8. SEM micrograph of primer A.I.2.c formulated with 70.0% PVC.

Primer	Average values	Standard deviation σ_{n-1}	General average values	Primer	Average value	Standard deviation σ_{n-1}	General average value
A.I.1	6.6	1.16	6.6	**B.I.1**	5.6	1.07	5.6
A.I.2.a	6.9	0.58	8.0	**B.I.2.a**	6.2	0.94	7.1
A.I.2.b	8.2	0.41		**B.I.2.b**	7.2	0.82	
A.I.2.c	8.9	0.38		**B.I.2.c**	8.0	0.55	
A.I.3.a	7.0	0.63	7.9	**B.I.3.a**	6.3	0.93	7.1
A.I.3.b	8.0	0.44		**B.I.3.b**	7.1	0.86	
A.I.3.c	8.8	0.42		**B.I.3.c**	8.0	0.55	
A.I.4.a	6.4	1.20	6.5	**B.I.4.a**	5.6	0.97	5.7
A.I.4.b	6.7	1.13		**B.I.4.b**	5.7	1.08	
A.I.4.c	6.4	1.28		**B.I.4.c**	5.8	1.17	
A.II.1	6.0	1.34	6.0	**B.II.1**	4.8	1.33	4.8
A.II.2.a	6.5	0.55	7.2	**B.II.2.a**	5.8	0.93	6.6
A.II.2.b	7.3	0.52		**B.II.2.b**	6.7	0.82	
A.II.2.c	7.8	0.49		**B.II.2.c**	7.4	0.53	
A.II.3.a	6.5	0.55	7.3	**B.II.3.a**	5.7	0.98	6.6
A.II.3.b	7.2	0.42		**B.II.3.b**	6.6	0.86	
A.II.3.c	8.2	0.41		**B.II.3.c**	7.2	0.66	
A.II.4.a	6.1	1.28	6.1	**B.II.4.a**	4.8	1.29	4.8
A.II.4.b	6.1	1.24		**B.II.4.b**	4.8	1.21	
A.II.4.c	6.1	1.16		**B.II.4.c**	4.9	1.24	

Table 3. Average values and standard deviation.

4. Final considerations

To explain the great tendency of zinc particles to corrode at the film surface of water-based nano lithium silicate primers as comparing with those solvent-based, partially hydrolyzed tetraethyl orthosilicate, it is necessary to consider that the first ones are based on binders, as mentioned, with a higher superficial tension. The last one implies inferior wetting, that means lower adhesion, penetration and spreading during metal zinc incorporation previous to application; consequently, they wet with more difficult the zinc particles while the second ones do it in a better way (more reduced interfacial tension).

The above-mentioned characteristic explains the great porosity of zinc-rich nano lithium silicate films and their high cathodic protective activity as comparing with zinc-rich tetraethyl orthosilicate films.

With regard to average diameter of zinc particle, size diminution increases significantly the surface area for a given weight. Since all surfaces have a given level of free energy, the ratio of surface energy to mass in small particles is so great that the particles adhered strongly themselves. For this reason, a lower particle size in a poor dispersion originates a greater flocculates (a high number of unitary particles are associated), which lead to zinc-rich primer films of high porosity and because of good cathodic protective activity. Moreover, a lower particle size could lead to films with a higher electrical contact (better packaging ability); since the current density is inherent to the chemical nature of zinc dust and the operating conditions of the corrosion cell, the increase in specific area elevates not only the current of protection but also generates a better superficial distribution (more efficient primers). During immersion test in 0.1 M sodium chloride solution, visual inspection of plates protective with zinc-rich primers (both types of binders) showed a more localized steel attack when zinc dust of the higher particle size was used.

Concerning incorporation of reinforcement fibers, the conductive or non-conductive characteristic was a very important variable. The first ones improved notably the primer performance since they increased the electrical contact between particles and with the metallic substrate, particularly in the higher levels in the formulations; the performance is correlated with the higher useful zinc in the film. On the other hand, non-conductive reinforcement fibers did not modify the primer efficiency as compared with reference panels (without reinforcement fibers) and for this reason their incorporation is not justified from technical and economical viewpoints.

Referring to PVC values (zinc content in dry film), previous results of laboratory tests demonstrated that a higher amount of microzinc leads to a longer useful life of primers. Nevertheless, it is important to mention that the choice of zinc content must be made by considering the physical characteristic of the primer film required for each particular case. When pigment volume concentration exceeds largely the CPVC, film properties such as adhesion, flexibility, abrasion resistance, etc. are drastically reduced while when the percentual level is slight under the critical value the efficiency is also considerably diminished.

In the case of primers, which have got incorporated conductive reinforcing fibers, results allow concluding that it is possible to reduce appreciably the PVC without affecting significantly the efficiency in service. In addition, it is important to mention that the quoted diminution of zinc content in the film is direct proportional to decrease the primer cost since it is the most expensive component of the composition.

5. References

Abanilla, M.A.; Li, Y. and Karbhari, V.M. (2005). Durability characterization of wet layup graphite/epoxy composites used in external strengthening, *Composites Part B: Engineering*, Vol. 37, No. 2-3, (August 2005), pp. 200-212, DOI: 10.1016 /j.compositesb.2005.05.016

Ahmed Al-Dulaimi, Ahmed (2010). *Evaluation of polyaniline composites and nanostructures as anti-corrosive pigments for carbon steel*. Masters thesis, Universiti Teknologi Malaysia, Faculty of Chemical and Natural Resource Engineering.

Aluru, N. et al., in: Gooddard, Brenner, Lyshevski, Iafrate Ed., (2003). *Nanostructure Studies of the Si-SiO_2 Interface*, Handbook of Nanoscience, Engineering and Technology, CRC Press., Washington D.C., USA, 2003, Chapter 11.2.

Amir, N.; Ahmad, F. and Megat-Yusoff, P. (2010). *Study on Fiber Reinforced Epoxy-based Intumescent Coating Formulations and Their Characteristics*. International Conference on Plant Equipment and Reliability (ICPER 2010), Kuala Lumpur, Malaysia, June 2010.

Behler, D.K.; Stravato, A.; Mochalin, V.; Korneva, G.; Yushin, G. and Gogotsi, Y. (2009). Nanodiamond-Polymer Composite Fibers and Coatings, *ACS Nano*, 2009, Vol. 3, No. 2, pp 363–369, DOI: 10.1021/nn800445z

Chen, Y. H.; Yu Chu, J. and Zhu, Q.J. (2010). Effects of Coating on Interfacial Fatigue of Fiber-Reinforced Composites, *Advanced Materials Research, Manufacturing Science and Engineering* I, Vol. 97, No. 101, (March 2010), pp. 830-833, DOI: 10.4028/www.scientific.net/AMR.97-101.830

Giudice, C.A. and Pereyra, A.M. (2007). Soluble metallic silicates in the anticorrosive inorganic coating formulation with non-flammable properties. *Pitture e Vernici European Coatings,* Vol. 83, No. 7, pp. 48-57, ISSN 0048-4245

Giudice, C.A. and Pereyra, A.M. (2009). *Tecnología de pinturas y recubrimientos. Componentes, formulación, manufactura y control de calidad,* Ed. edUTecNe, Argentina, 2009, pp. 1-444.

Hammouda, N.; Chadli, H.; Guillemot, G. and Belmokre, K. (2011). The Corrosion Protection Behaviour of Zinc Rich Epoxy Paint in 3% NaCl Solution, *ACES,* Vol.1 No.2, (April 2011), pp.51-60, DOI: 10.4236/aces.2011.12009

Hoshyargar, F.; Ali Sherafati, S. and Hashemi, M. (2009). Short communication: A new study on binder performance and formulation modification of anti-corrosive primer based on ethyl silicate resin. *Progress in Organic Coatings,* Vol. 65, No. 3, (July 2009), pp. 410-413, DOI: 10.1016/j.porgcoat.2009.02.006

Huang, K.M.; Weng, C.J.; Lin, S.Y; Yu, Y.H. and Yeh, J.M. (2009). Preparation and anticorrosive properties of hybrid coatings based on epoxy-silica hybrid materials, *Journal of Applied Polymer Science,* Vol. 112, No. 4, (May 2009), pp. 1933–1942, DOI: 10.1002/app.29302

Jegannathan, S.; Sankara Narayanan, T.; Ravichandran, K.; and Rajeswari, S. (2006). Formation of zinc phosphate coating by anodic electrochemical treatment. *Surface and Coatings Technology,* Vol. 200, No. 20-21, (May 2006), pp. 6014-6021, DOI: 10.1016/j.surfcoat.2005.09.017

Jianjun, Y.; Zhenhua, S.; Xiangsheng, M. and Wenlong, L., Deqiang, J. (2008). Corrosion Protective Performance of Coatings. Study on Zinc Ingredient in Zinc Rich Epoxy Primers. *Paint & Coatings Industry,* Vol. 08-2008, (August 2008), DOI: CNKI:SUN:TLGY.0.2008-08-003.

Lei-lei, M. and De-liang, L. (2010). The Classification and Testing Standards of Zinc Rich Coatings, *Shanghai Coatings,* Vol. 05, (May 2010), DOI: CNKI:SUN:SHTL.0.2010-05-015

Lekka, M.; Koumoulis, D.; Kouloumbi, N. and Bonora N.P. (2009). Mechanical and anticorrosive properties of copper matrix micro- and nano-composite coatings. *Electrochimica Acta,* Vol. 54, No. 9, (March 2009), pp. 2540-2546, DOI: 10.1016/j.electacta.2008.04.060

Li, X.; Tabil, L.G. and Panigrahi, S. (2006). Chemical Treatments of Natural Fiber for Use in Natural Fiber-Reinforced Composites: A Review. *Chemistry and Materials Science, Journal of Polymers and the Environment,* Vol. 15, No. 1, (March 2006) pp. 25-33, DOI: 10.1007/s10924-006-0042-3

Pereyra, A.M. and Giudice, C.A. (2007). Shaped for performance: the combination of lamellar zinc and mica improves the efficiency of zinc-rich primers. *European Coatings Journal,* Vol. 9, pp. 40-45, ISSN 0930-3847

Radhakrishnan, S.; Siju, C.R.; Mahanta, D.; Patil, S. and Madras, S. (2009). Conducting polyaniline–nano-TiO_2 composites for smart corrosion resistant coatings, *Electrochimica Acta*, Vol. 54, No. 4, (January 2009), pp. 1249-1254 DOI: 10.1016/j.electacta.2008.08.069

Sonawane, S.; Teo, B.; Brotchie, A.; Grieser, F. and Ashokkumar, M. (2010). Sonochemical Synthesis of ZnO Encapsulated Functional Nanolatex and its Anticorrosive Performance, *Ind. Eng. Chem. Res.*, (January 2010), Vol. 49, No. 5, pp. 2200–2205. DOI: 10.1021/ie9015039

Sorensen, P.A.; Kiil, S.; Dam-Johansen, K. and Weinell C.E. (2011). Anticorrosive coatings: a review, *Chemistry and Materials Science, Journal of Coatings Technology and Research*, Vol. 6, No. 2, pp. 135-176, DOI: 10.1007/s11998-008-9144-2

Thorslund Pedersen, L.; Weinell, C.; Hempel, A. S; Verbiest, P; Van Den Bosch, J. and Umicore J. (2009). *Advancements in high performance zinc epoxy coatings Zinc,* Chemicals Source Corrosion 2009, (March 2009), Copyright NACE International

Wang, D. and Bierwagen, G. (2009). Sol–gel coatings on metals for corrosion protection. *Progress in Organic Coatings*, Vol. 64, No. 4 (March 2009), pp. 327-338. DOI: 10.1016/j.porgcoat.2008.08.010

Wenrong, J. (2009). The Research Progress on Green Coatings, *Guangdong Chemical Industry*, Vol. 5, (May 2009), DOI: CNKI:SUN:GDHG.0.2009-05-035

Xiyan, L.; Jianming, J.; Sibo, K. and Jing, Z. (2010) Preparation of Flexible Epoxy Anticorrosive Coatings, *China Coatings*, Vol. 7, (July 2010), DOI: CNKI:SUN:ZGTU. 0.2010-07-018

Yang, F.; Zhang, X.; Han, J. and Du, S. (2008). Characterization of hot-pressed short carbon fiber reinforced ZrB_2–SiC ultra-high temperature ceramic composites, *Journal of Alloys and Compounds*, Vol. 472, No. 1-2, (20 March 2009), pp. 395-399, DOI: 10.1016/j.jallcom.2008.04.092

Yoshida, K.; Matsukawa, K.; Imai, M. and Yano, T. (2009). Formation of carbon coating on SiC fiber for two-dimensional SiCf/SiC composites by electrophoretic deposition, *Materials Science and Engineering:* B, Vol. 161, No. 1-3, (April 2009), pp. 188-192, DOI: 10.1016/j.mseb.2008.11.032

8

Comparative Study of Porphyrin Systems Used as Corrosion Inhibitors

Adina-Elena Segneanu, Ionel Balcu, Nandina Vlatanescu, Zoltan Urmosi and Corina Amalia Macarie
National Institute of Research and Development for Electrochemistry and Condensed Matter, INCEMC-Timisoara Romania

1. Introduction

Fundamental research in durability of materials and structures have shown great potential for enhancing the functionality, serviceability and increased life span of our civil and mechanical infrastructure systems and as a result, could contribute significantly to the improvement of every nation's productivity, environment and quality of life. The intelligent renewal of aging and deteriorating civil and mechanical infrastructure systems includes efficient and innovative use of high performance composite materials for structural and material systems. [Monteiro et al, 2002]

The word corrosion is as old as the earth, but it has been known by different names.

Corrosion is known commonly as rust, an undesirable phenomena which destroys the luster and beauty of objects and shortens their life. A Roman philosopher, Pliny (AD 23-79) wrote about the destruction of iron in his essay 'Ferrum Corrumpitar'. Corrosion since ancient times has affected not only the quality of daily lives of people, but also their technical progress. There is a historical record of observation of corrosion by several writers, philosophers and scientists, but there was little curiosity regarding the causes and mechanism of corrosion until Robert Boyle wrote his 'Mechanical Origin of Corrosiveness.'

Philosophers, writers and scientists observed corrosion and mentioned it in their writings:

- Pliny the elder (AD 23-79) wrote about spoiled iron.
- Herodotus (fifth century BC) suggested the use of tin for protection of iron.
- Austin (1788) noticed that neutral water becomes alkaline when it acts on iron.
- Thenard (1819) suggested that corrosion is an electrochemical phenomenon.
- Hall (1829) established that iron does not rust in the absence of oxygen.
- Davy (1824) proposed a method for sacrificial protection of iron by zinc.
- De la Rive (1830) suggested the existence of microcells on the surface of zinc.

The most important contributions were later made by Faraday (1791-1867) who established a quantitative relationship between chemical action and electric current. Faraday's first and second laws are the basis for calculation of corrosion rates of metals. Ideas on corrosion control started to be generated at the beginning of nineteenth century. Whitney (1903) provided a scientific basis for corrosion control based on electrochemical observation. As early as in eighteenth century it was observed that iron corrodes rapidly in dilute nitric acid but remains intact in concentrated nitric acid. Schonbein in 1836 showed that iron could be made passive. It was left to U.R. Evans to provide a modern understanding of the causes and control of corrosion based on his classical electrochemical theory in 1923. Corrosion laboratories established in M.I.T., USA and University of Cambridge, UK, contributed significantly to the growth and development of corrosion science and technology as a multi disciplinary subject. In recent years, corrosion science and engineering has become an integral part of engineering education globally. [Ahmad, 2006]

The strong damaging effects of corrosion require establishing and taking some control measures. In accordance with the ways in which corrosion manifest, the supporting material and the specific local conditions, corrosion control can take different forms. Surely that here should also be added the use of materials maximum resistant to corrosive environment in order to limit the corrosion effects.

The practice of corrosion prevention by adding substances which can significantly retard corrosion when added in small amounts is called inhibition. Inhibition is used internally with carbon steel pipes and vessels as an economic control alternative to stainless steels and alloys, and to coatings on non-metallic components. One unique advantage is that adding inhibitor can be implemented without disruption of a process. The addition of an inhibitor (any reagent capable of converting an active corrosion process into a passive process) results in significant suppression of corrosion.

Corrosion inhibitors are substances when added in small amounts in a corrosive environment reduces significantly the corrosion rate for metallic material in contact with the environment.

A typical good corrosion inhibitor will give 95% inhibition at concentration of 80ppm, and 90% at 40ppm. Some of the mechanism of its effect are formation of a passivation layer (a thin film on the surface of the material that stops access of the corrosive substance to the metal), inhibiting either the oxidation or reduction part of the redox corrosion system (anodic and cathodic inhibitors), or scavenging the dissolved oxygen.

Some corrosion inhibitors are hexamine, phenylenediamine, dimethythanolamine, sodium nitrite, cinnamaldehyde, condensation products of aldehydes and amines, chromates, nitrites, phosphates, hydrazine, ascorbic acid, and others.

The corrosion inhibitors are added not only in aqueous solutions, but also in oils and fuels, the liquid cooling etc. can also be organic additives and coatings (varnishes, paints) on metallic surfaces.

The presence of a chemical compound in an environment, even in small concentrations, can lead to significant changes in speed and form of corrosion of a metallic material in contact with the environment. The acceleration or inhibition of corrosion processes are specific methods, dependent of metal-corrosive environment characteristics.

Corrosion inhibitors are selected on the basis of solubility or dispersibility in the fluids which are to be inhibited.[Rahimi,2004]

Porphyrins are well-known for their biological, catalytic, and photochemical properties. Considerable effort has been devoted to confining porphyrin molecules in microporous inorganic matrixes because such a hostguest approach can improve the efficiency of photoinduced charge separation by preventing back electron transfer. Moreover, organic-inorganic composite materials sometimes offer unique properties that are not available in any of the individual parts.[Bose et al,2002].

In this paper we intend to test the corrosion resistance of two other types of organic inhibitors and to study in which conditions they behave similarly to an anticorrosive paint. The corrosion resistance was studied by cyclic voltammetry, in 20% Na_2SO_4 electrolyte solution, Tafel tests and in the salt spray chamber, using diverse exposure conditions.

The two types of organic inhibitors used in this study for comparison are:

- 5,10,15,20 tetrakis(1-methyl-4pyridyl)21H,23H-porphine,tetra-p-fosylate salt
- 4,4',4',4'''(porphine-5,10,15,20-tetrayl)-tetrakis (benzeric sulfonic acid)

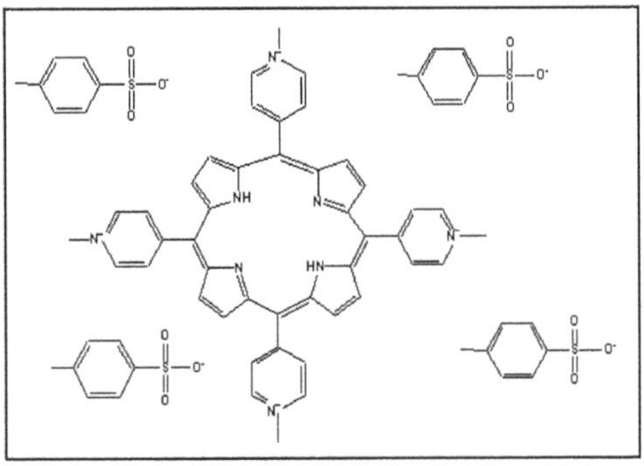

Fig. 1. Structure of 5,10,15,20 tetrakis(1-methyl-4pyridyl)21H,23H-porphine,tetra-p-fosylate salt.

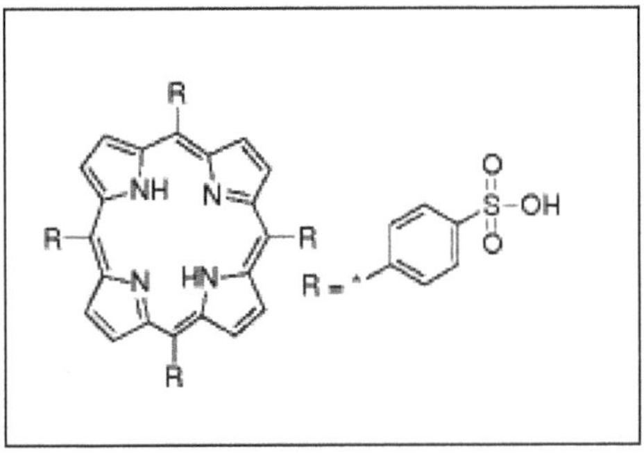

Fig. 2. Structure of 4,4',4',4'''(porphine-5,10,15,20-tetrayl)-tetrakis (benzeric sulfonic acid).

2. Experimental

The initial data consisted of 2 types of porphyrins dissolved KOH, H_2SO_4 and benzonitrile, as presented bellow, as the first set:

a. 0.2 g of Na_4TFP Ac porphyrin ($C_{44}H_{26}N_4Na_4O_{12}S_4 \times H_2O$) dissolved in 40 ml 10% KOH, mentioned as system A.

b. 0.2 g of Na_4TFP Ac porphyrin ($C_{44}H_{26}N_4Na_4O_{12}S_4 \times H_2O$) dissolved in 40 ml 10% H_2SO_4, mentioned as system B.
c. 0.2 g of H_2TPP porphyrin (5,10,15,20 tetrakis 4 phenyl-21H,23H) dissolved in benzonitrile, mentioned as system C.

System C presented the best anticorrosive properties.

The **second set** consists of:

a. 0.2 g of 5,10,15,20 tetrakis(1-methyl-4pyridyl)21H,23H-porphine,tetra-p-fosylate salt dissolved in 40 mL benzonitrile, mentioned from this point forward as **system I**
b. 0.2 g of 4,4′,4′,4‴(porphine-5,10,15,20-tetrayl)-tetrakis (benzeric sulfonic acid) dissolved in the same solvent namely benzonitrile (40mL) mentioned from this point forward as **system II.**

Various apparatuses were used, like the DCTC 600 salt spray chamber or the Dynamic EIS Voltalab. The results are presented as mm/year corrosion speed, thus evaluating the different coating systems.

The electrochemical studies namely cyclic voltammetry and Tafel curves carried out to test the protective layer were conducted using the PGZ 402 Dynamic EIS Voltalab. For the data acquisition the Voltamaster 4, version 7.08, was used. This specialized software can determine, based on references, from the Tafel test's values, the exact corrosion speed, measured in mm/year.

The voltammetry measurements, the Tafel tests, were conducted between -1000 and 1000 mV potentials at a sweep rate of 100 mV/s. Before each experiment, the working electrodes were polished with a series of wet sandpapers of different grit sizes (320, 400, 600, 800, 1000 and 1200). After polishing, the carbon-steel electrode are washed with ultrapure water and dried at room temperature and then the active part was immersed in porphyrin solution.

The working electrode is the carbon-steel electrode, (prepared as mentioned earlier) with 0,28 cm^2 active surface, (coated or uncoated); platinum counter electrode with 0.8 cm^2 active surface and saturated calomel electrode, (SCE), as reference electrode; all of which connected to the PGZ 402 Dynamic EIS Voltalab, from Radiometer Copenhagen.

20% Na_2SO_4 solution was used as base electrolyte.

The thickness loss and weight loss tests were not conducted, due to the relatively small size of the electrodes.

To test the corrosion resistance of the porphyrin systems eighteen electrodes were used.

For a good repeatability and accuracy the eighteen electrodes are pretreated as follows:

Three electrodes are uncoated/untreated, three electrodes are immersed for 5 minutes in **system I,** three electrodes are immersed for 60 minutes in **system I,** three electrodes are immersed for 5 minutes in **system II,** three electrodes are immersed for 60 minutes in **system II,** three electrodes are coated with anticorrosive paint. The porphyrin systems are dissolved and then applied on the electrodes; the electrodes are immersed in the solution for 5, respectively for 60 minutes, thus simulating a shorter and a longer exposure time.

After testing the porphyrin systems by electrochemical studies, the DCTC 600 dry salt spray chamber, (Figure 2) and the ASTM B 117 method (dry salt spray corrosion test), were used. The method establishes the spraying and drying times respectively the spraying and drying frequencies; in our case, the 5% NaCl solution is sprayed for 5 minutes at 35°C, afterwards is dried at 50°C for 55 minutes. The salt solution was prepared using 1 kg of pure NaCl dissolved in 20 liters of distilled water, resulting the 5% NaCl solution, mentioned earlier.

The exposure time in the salt spray chamber of the probes was 336 hours.

3. Results and discussions

Regarding the electrochemical studies of the corrosion resistance of the protective layers formed from the first set of porphyrin it has been demonstrated that the electrodes which have been treated with the system C (0.2 g of H_2TPP porphyrin (5,10,15,20 tetrakis 4 phenyl-21H,23H) disolved in 40 mL benzonitrile) gave the best results. The immersion time was 5 minutes.

Parameters	Electrodes			
	Uncoated	System A	System B	System C
$i^{\rightarrow}_{peak}$ [mA/cm²]	290	190	90	280
$\varepsilon^{\rightarrow}_{pic}$ [mV]	900	750	600	1300
$i^{\leftarrow}_{peak}$ [mA/cm²]	-	60	100	85
$\varepsilon^{\leftarrow}_{peak}$ [mV]	-	50	100	50
ε_{O2} [mV]	1500	1500	1500	1500
ε_{pas}[mV]	1350	900	850	1600
i_{pas} [mA/cm²]	25	8	0	50

Table 1. Results of cyclic voltammograms.

Parameters	Electrodes			
	Uncoated	System A	System B	System C
i_{cor} [mA/cm²]	0.7666	0.9792	0.6506	0.0718
v_{cor} [mm/year]	8.99	11.48	7.63	0.842
Rp	50.91	129.51	59.57	147.67
C	0.9962	0.9996	0.9997	1.000

Table 2. Results of Tafel test.

The notations from the table: $i^{\rightarrow}_{peak}$ – peak current density for anodic polarization; $\varepsilon^{\rightarrow}_{peak}$ – peak potential for anodic polarization; $i^{\leftarrow}_{peak}$ - peak current density for cathodic polarization; $\varepsilon^{\leftarrow}_{peak}$ - peak potential for cathodic polarization; ε_{O2} - oxygen generation potential; ε_{pas} - passivation potential; i_{pas} - passivation current.

Continuing with the tests, after we have demonstrated that porphyrin (5,10,15,20 tetrakis 4 phenyl-21H,23H) had the best results, we have realized a study to compare this porphyrin with other two types of porphyrins, namely: 5,10,15,20 - tetrakis(1-methyl-4pyridyl)21H,23H-porphine,tetra-p-fosylate salt and 4,4',4'',4'''(porphine-5,10,15,20-tetrayl)-tetrakis(benzeric sulfonic acid); also used as organic inhibitors.

From the obtained voltammograms were determined the anodic i_{peak} and ε_{peak} and from the Tafel curves were determined the corrosion current, polarisation resistance (Rp), corrosion rate and the correlation coefficient.

We continued the studies, cyclic voltammetry and Tafel method for carbon steel electrodes treated in different ways.

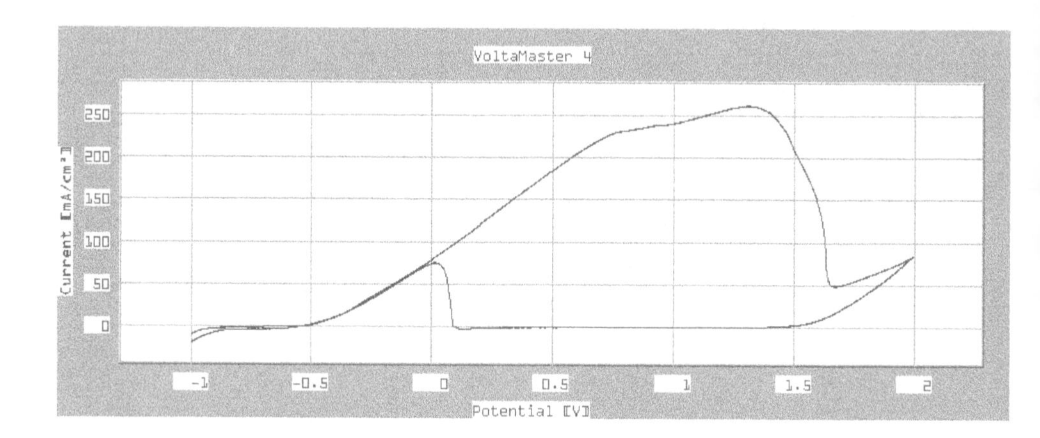

Fig. 3. Cyclic voltammograms of coated electrode with system C.

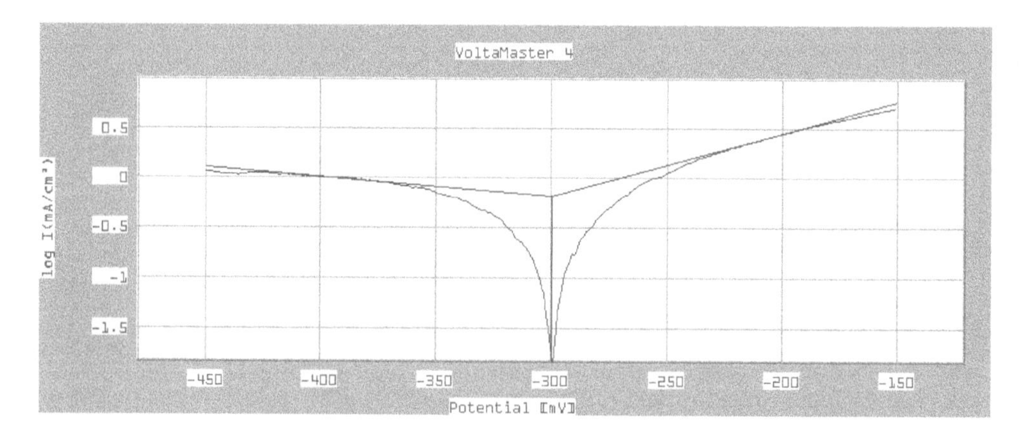

Fig. 4. Tafel tests of coated electrode with system C.

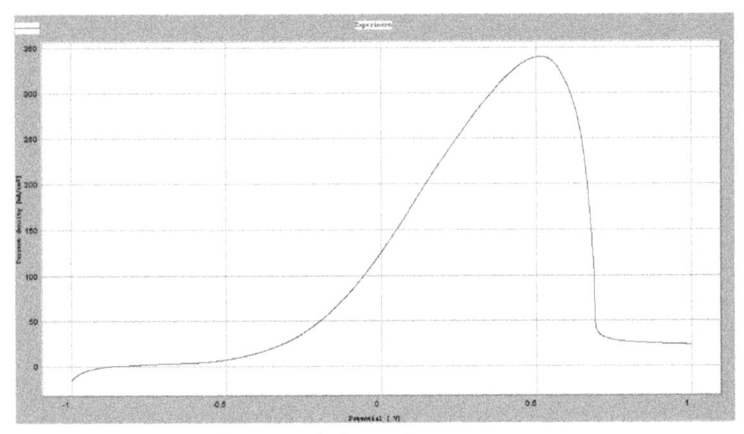

Fig. 5. Cyclic voltammogram of uncoated electrodes.

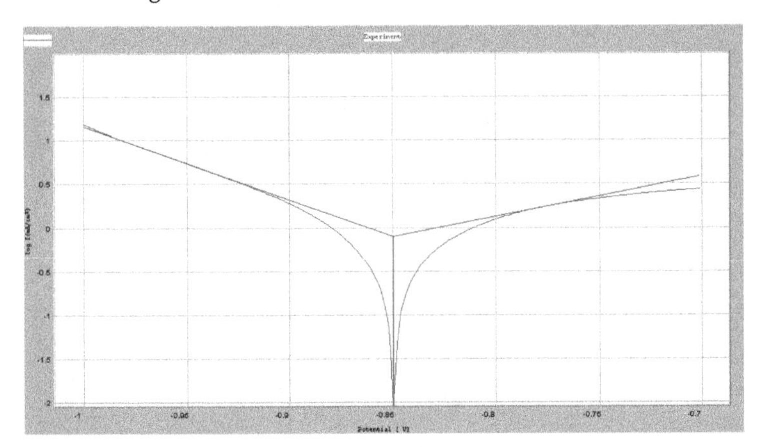

Fig. 6. Tafel tests of uncoated electrodes.

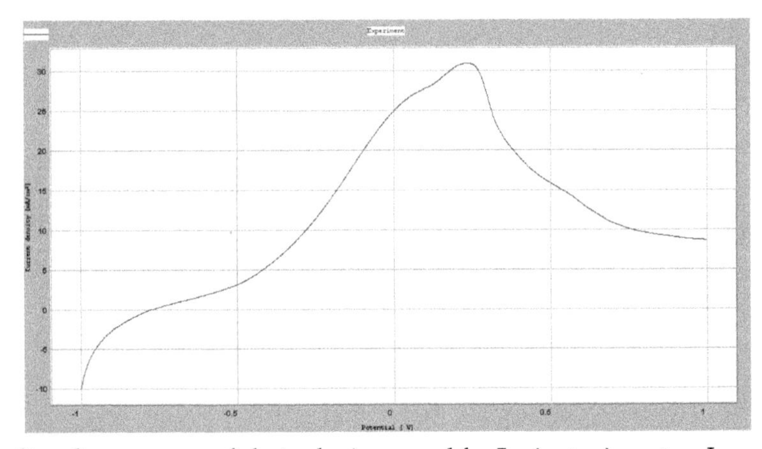

Fig. 7. Cyclic voltammogram of electrodes immersed for 5 minutes in system I.

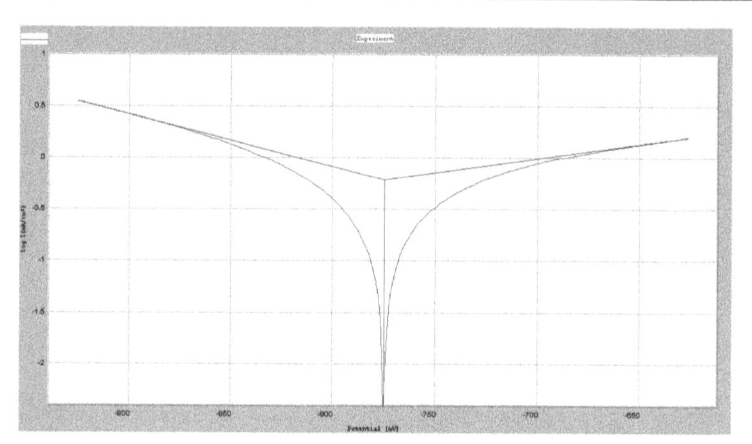

Fig. 8. Tafel tests of electrodes immersed for 5 minutes in system I

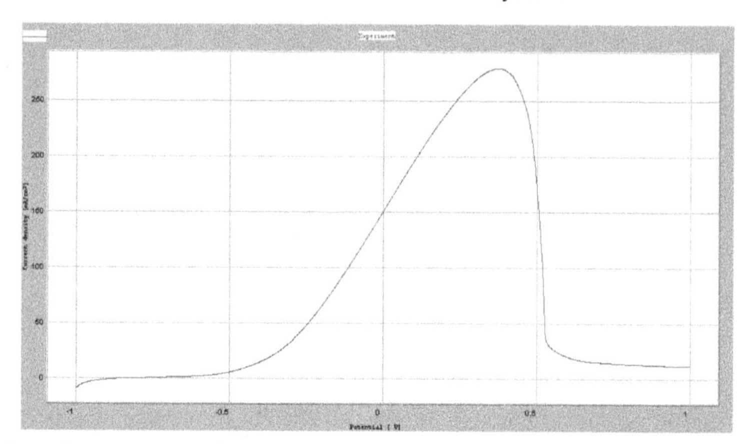

Fig. 9. Cyclic voltammogram of electrodes immersed for 5 minutes in system II

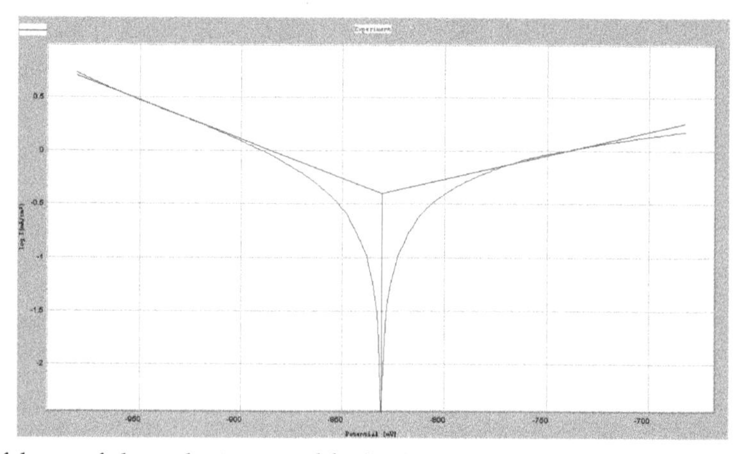

Fig. 10. Tafel tests of electrodes immersed for 5 minutes in system II

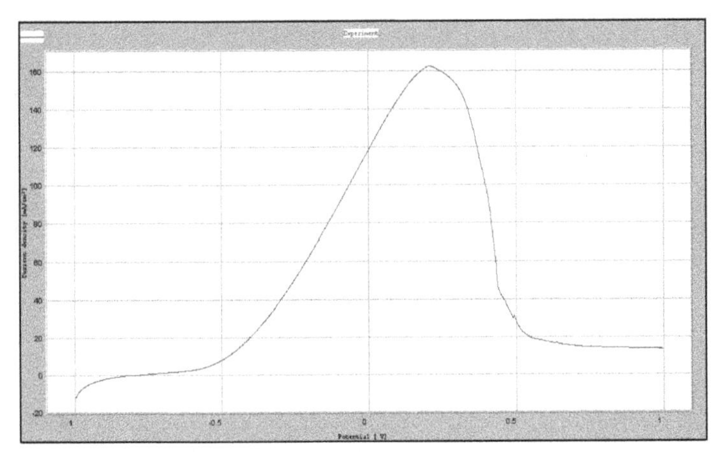

Fig. 11. Cyclic voltammogram of electrodes immersed for 60 minutes in system I.

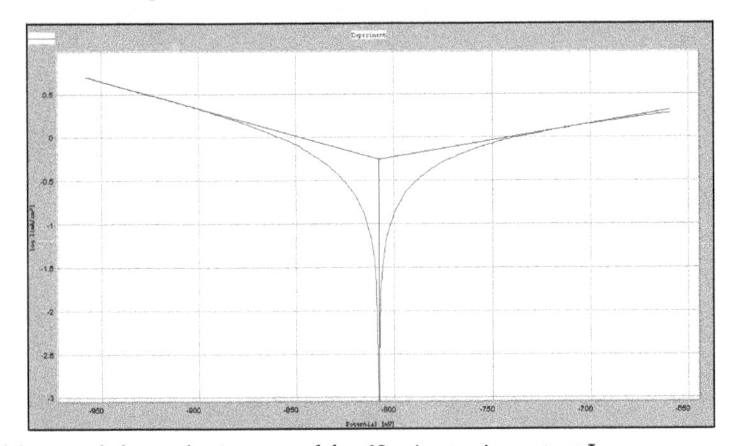

Fig. 12. Tafel tests of electrodes immersed for 60 minutes in system I.

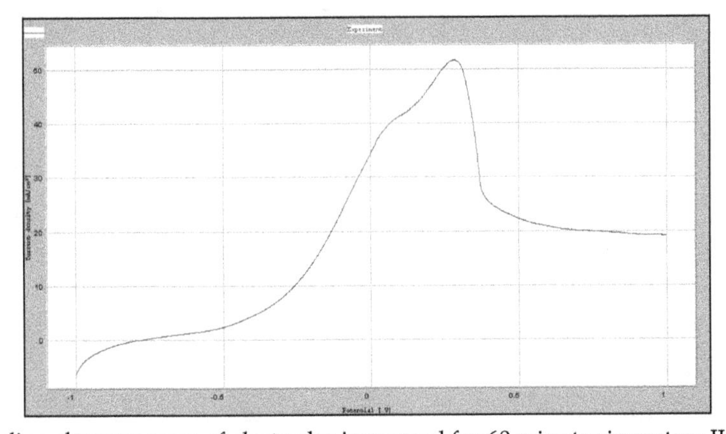

Fig. 13. Cyclic voltammogram of electrodes immersed for 60 minutes in system II.

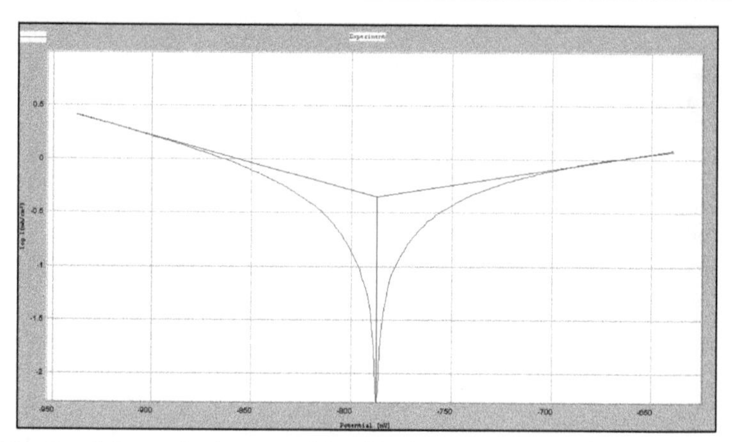

Fig. 14. Tafel tests of electrodes immersed for 60 minutes in system II.

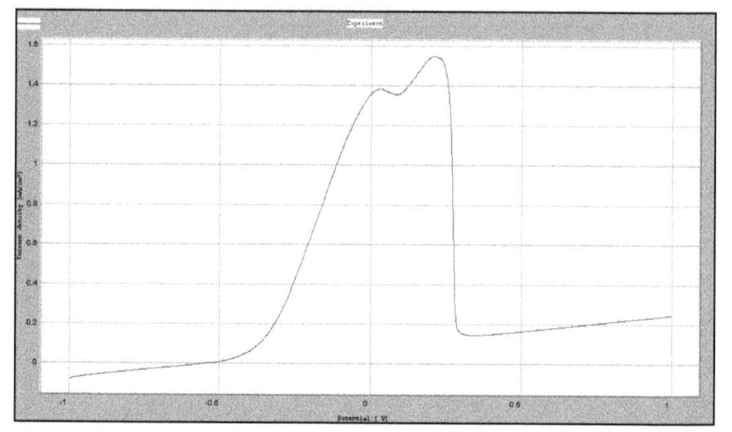

Fig. 15. Cyclic voltammogram of electrodes coated with paint.

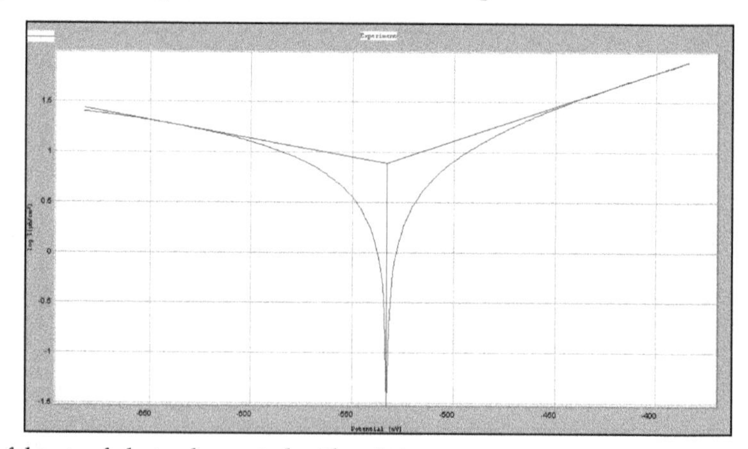

Fig. 16. Tafel tests of electrodes coated with paint.

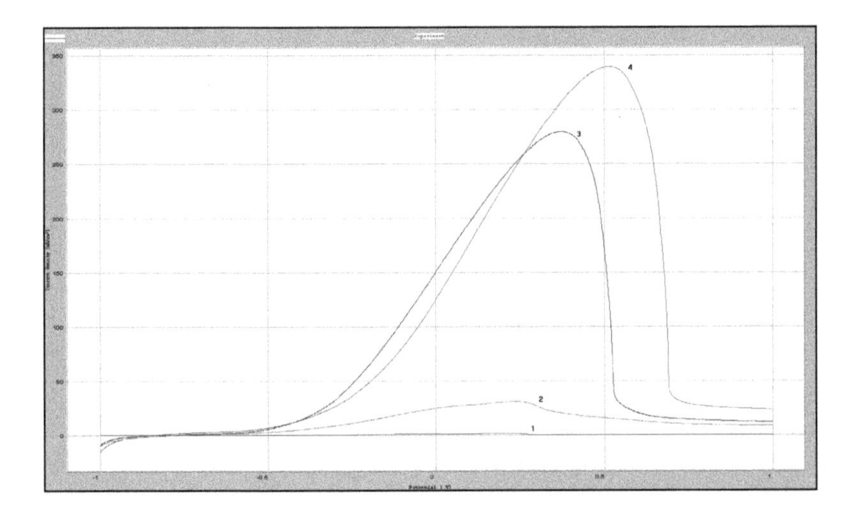

Fig. 17. Cyclic voltammograms of the corrosion process for various electrodes: 1- coated with paint; 2- system I (immersion time 5 minutes); 3- system II (immersion time 5 minute); 4- uncoated.

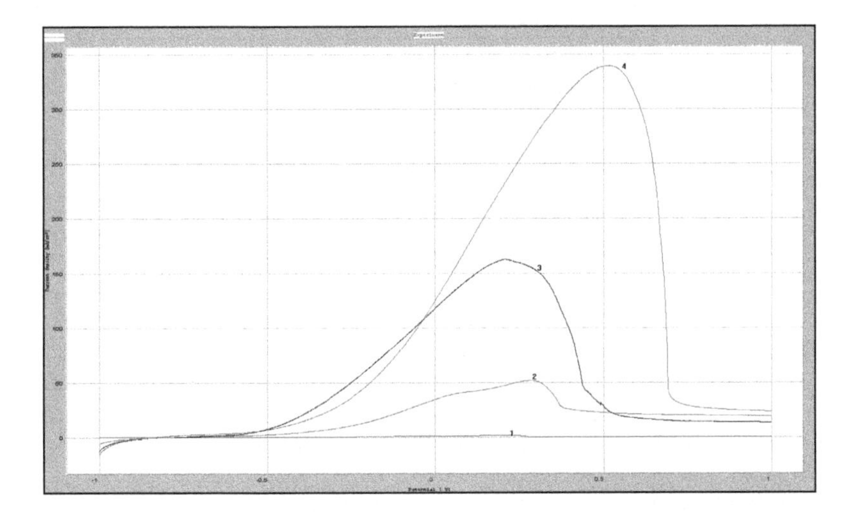

Fig. 18. Cyclic voltammograms of the corrosion process for various electrodes: 1- coated with paint; 2- system I (immersion time 60 minutes); 3- system II (immersion time 60 minute); 4- uncoated.

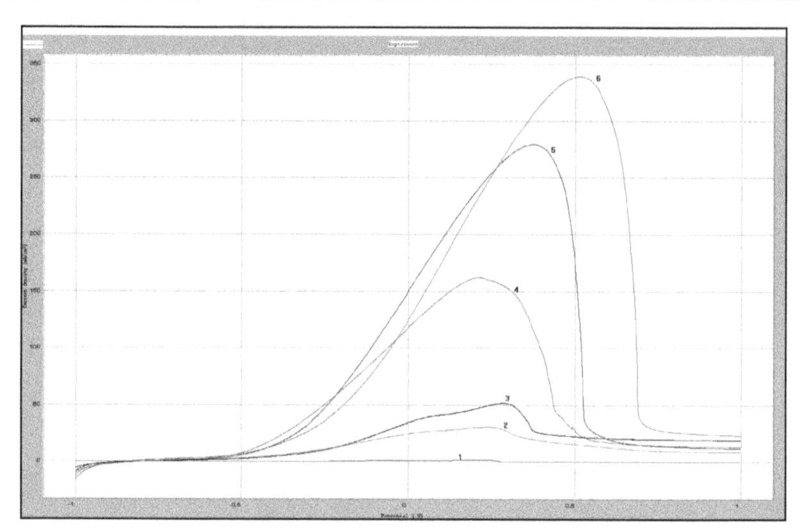

Fig. 19. Cyclic voltammograms of the corrosion process for various electrodes: 1- coated with paint; 2- system I (immersion time 5 minutes); 3- system I (immersion time 60 minute); 4- system II (immersion time 5 minute); 5- system II (immersion time 60 minute); 6- uncoated

Cyclic voltammograms and Tafel tests for 5 minutes and 60 minutes immersion time are presented in Table 3.

Parameters	Electrodes					
	Uncoated	System I		System II		Coated with paint
		Immersion time (5 minutes)	Immersion time (60 minutes)	Immersion time (5 minutes)	Immersion time (60 minutes)	
i_{peak} [mA/cm^2]	337.9	30.68	160.2	279.5	51.67	1.541
ε_{peak} [mV]	0.5178	0.2219	0.2172	0.3747	0.2824	0.21
i_{cor} [mA/cm^2]	0.7791	0.6062	0.5516	0.3983	0.4394	0.7689
v_{cor} [mm/year]	91.37	7.109	6.468	4.670	5.153	9.18
Rp	34.62	68.15	59.81	76.07	93.97	46.42
C	1.0000	1.0000	1.0000	1.0000	1.0000	1.0000

Table 3. Results obtained from cyclic voltammograms and Tafel tests.

The notations from the table:i_{peak} – peak current density; ε_{peak} – peak potential; i_{cor} - corrosion current density; v_{cor} - corrosion rate; C – correlation coefficient; Rp – polarization resistance

Polarization resistance can be related to the rate of general corrosion for metals at or near their corrosion potential, $E_{corr.}$

These can be obtained from a *Tafel* plot or estimated from the experimental data.

The polarization resistance or R_P is defined by the following equation:

$$R_p = \left(\frac{\Delta E}{\Delta i}\right)_{\Delta E \to 0}$$

where, ΔE variation of the applied potential around the corrosion potential and Δi is the resulting polarization current.

Polarization resistance, R_P, behaves like a resistor and can be calculated by taking the inverse of the slope of the current potential curve at open circuit or corrosion potential. High R_P of a metal implies high corrosion resistance and low R_P implies low corrosion resistance.

From electrochemical studies it was observed that the electrodes treated with system I in a immersion time of 5 minutes gives better results.

For salt spray chamber tests the electrodes were investigated visually; special attention was given to the appearance of the first corrosion signs and a similar or identical evolution of corrosion rates has resulted.

The electrodes were divided into three categories namely, 3 untreated electrodes, 3 electrodes painted with anti-corrosion paint and 12 electrodes treated according with system I and system II.

These 12 fall into four subcategories, namely:

- 3 treated according to system I, with immersion time of 5 minutes, and respectively other three electrodes with an immersion time of 60 minutes.
- 3 treated according to system II, with immersion time of 5 minutes, and respectively other three electrodes with an immersion time of 60 minutes.

Visual observations:

Untreated electrodes:

- After 24 hours there is a brown coloration on the entire surface of the electrode
- The brown coloration, after 48 hours, intensifies, the electrode's surface becomes more rough
- Specific symptoms appear after 120 hours, that is uniform corrosion throughout the surface of the electrode
- Corrosion progresses, symptoms are increasing after 192 hours. Rust formed is still adherent
- After 264 hours, the rust layer becomes more voluminous
- After 336 no major changes occur.

Painted electrodes:

- After 24 and 48 hours respectively, there are no reported signs of corrosion
- Only after 120 hours, there is loss of the initial gloss paint
- After 192 hours localized corrosion can be seen as brown spots on the surface of the electrodes

- The occurrence of pitting corrosion can be observed after 264 hours; pitting spots occur among previously localized brown spots. At one of the three electrodes, the paint swells.
- After 336 hours, in addition to the initial symptoms, few and very small points of pitting are observed

Electrodes treated as system I:

- no changes can be observed after 24 hours or 48
- After 120 hours the surface of electrodes become more matte
- After 192 hours the surface of electrodes become more rough
- After 264 hours signs of localized corrosion occur
- After 336 hours pitting corrosion occurs; the number of pitting spots is very small

Electrodes treated as system II:

- no changes can be observed after 24 hours or 48
- After 120 hours the surface of electrodes become more matte
- After 192 hours pitting corrosion occurs in several points; on the electrodes with immersion time of 60 minutes the number of pitting points is lower
- After 264 hours no major differences can be observed, the pitting points do not multiply
- After 336 hours uniform corrosion is observed on 2 of the 3 electrodes, with an immersion time of 60 minutes; on the electrodes with an immerse time of 5 minutes, the rust is adhering.

4. Conclusion

In conclusion, from the three types of organic inhibitors studied, it can be said that 5,10,15,20 tetrakis(1-methyl-4pyridyl)21H,23H-porphine,tetra-p-fosylate salt, having the immersion time of 5 minutes was almost similar with the paint used and gives an anticorrosive protection much better than the porphyrin previously tested (H2TPP porphyrin (5,10,15,20 tetrakis 4 phenyl-21H,23H). Similar results can be seen, from the visual observations of the salt spray chamber test.

5. References

Ahmad Z., (2006), Principles of Corrosion Engineering and Corrosion Control; *Institution of Chemical Engineers (Great Britain) - Elsevier/BH,* 656 pages ISBN 0750659246;

Bose A., He P., Liu C., Ellman B.D., Twieg R. J.,Huang S.D., (2002), *Journal of the American Chemical Society,* 124,4-5;

Monteiro PJ.M., Chong K.P., Larsen-Basse J., Komvopoulos K., (2001), Long Term Durability of Structural Materials, *Elsevier Science Ltd.;*

Rahimi A., (2004), Inorganic and Organometallic Polymers, *Iranian Polymer Journal* 13 (2), 149-164;

Standard practice for operating salt spray (fog) apparatus, B117-02

Oxidation Resistance of Nanocrystalline Alloys

Rajeev Kumar Gupta[1], Nick Birbilis[1] and Jianqiang Zhang[2]
[1]*Department of Materials Engineering, Monash University,*
[2]*School of Materials Science & Engineering, The University of New South Wales,*
Australia

1. Introduction

Nanocrystalline (nc) materials are single or multi-phase polycrystalline solids with a grain size of a few nanometers, typically less than 100 nm. Owing to the very fine grain size, the volume fraction of atoms located at grain boundaries or interfaces increases significantly in nanocrystalline materials [1]. A simple geometrical estimation, where the grains are assumed as spheres or cubes, yields the following values for the volume fraction of the interfaces: 50% for 5 nm grains, 30% for 10 nm grains and about 3% for 100 nm grains [2-5]. These values of interface volume fraction are several orders of magnitude higher than those of conventional microcrystalline materials. Consequently, nanocrystalline materials exhibit properties that are significantly different from and often improved over, their conventional microcrystalline (mc) counterparts. For example, nanocrystalline materials exhibit increased mechanical strength [6-10], enhanced diffusivity [11], improved corrosion resistance (some nanocrystalline materials) [12-20], optical, electrical and magnetic properties [21-24]. Due to their unique properties, nanocrystalline materials have attracted considerable research interests and the field of nanocrystalline materials has now become one of major identifiable activities in materials science and engineering.

Metals exposed to high temperature oxygen-containing environments form oxides. If an oxide scale can form and the oxide is dense and adherent, then this scale can function as a barrier isolating the metal from the external corrosive atmosphere. This oxide scale is called protective oxide scale. On the other hand, if a non-protective oxide scale is formed, oxygen can penetrate through the scale and the oxidation will extend further into the metal substrate, causing a rapid metal recession. Of all oxides, chromia and alumina are two kinds of oxides thermodynamically and kinetically feasible to meet the requirement for resisting high temperature oxidation.

Alloys containing aluminium or chromium can be selectively oxidised to form alumina or chromia. To form an external oxide scale, the concentration of aluminium or chromium should reach a critical value. Nanocrystalline alloys promote this selective oxidation process by reducing the critical value. For example, in the conventional microcrystalline Ni-20Cr-Al alloy system, > 6 wt% Al is required to form a protective Al_2O_3 scale at 1000°C [25]. If the Al content is lower than 6 wt%, complex oxide mixtures consisting of Cr_2O_3, $NiCr_2O_4$ and internal Al_2O_3 form, resulting in high reaction rates and poor oxidation resistance. With a nano-crystalline alloy structure, this value can be substantially reduced to 2 wt% Al, when the grain size is 60 nm [26]. This promotion effect is also evident for the K38G alloy containing 3.5-4 wt% Al and 16% Cr, which forms external Cr_2O_3 scale and internal Al_2O_3 precipitates in the cast form (large grains) but only Al_2O_3 when in the form of sputtered nano-crystalline structure [27-30].

The unique structure and high grain boundary fraction, enhances diffusion of impurities and alloying elements, and changes materials thermodynamic properties [31-33] which are expected to cause a considerable difference in the resistance of nanocrystalline materials to environmental degradation (oxidation) at high temperatures. For practical application of these nanocrystalline materials, an acceptable level of resistance to environmental degradation is required. However, the effect of the nanocrystalline structure on the high-temperature oxidation resistance has attracted only a limited research attention. Oxidation resistance of nanocrystalline Ni-Cr-Al [34-36], Fe-Cr [37-39], and Zr [40-43] based alloys have been mainly investigated and in most of the cases, oxidation resistance has been reported to improve due to nanocrystalline structure.

The properties of oxides (Al_2O_3, SiO_2 or Cr_2O_3) formed during oxidation [44,45,46] also depend upon the grain size of the alloy and nanocrystalline structure alters the properties of oxide. For example, more uniform oxide scale with finer grain size and higher Cr or Al content is formed on the nanocrystalline alloys [34-36]. The oxide scales formed from nanocrystalline materials exhibit enhanced plastic deformation (due to fine grain size of formed oxide), which can release the stresses accumulated in the scales, and therefore the scale spallation tendency is reduced. It was reported that cyclic and long-time oxidation resistance was significantly improved by applying nanocrystalline coatings on type 304 stainless steel [47,48], Ni-Cr-Al [27], Co-Cr-Al [49], and Ni-(Co)-Cr-Al [28-30].

In order to investigate the possible differences in oxidation resistance along with any underlying mechanisms, understanding the nanocrystalline structure of a material is essential. This chapter will therefore first describe the structure of nanocrystalline materials, their thermodynamic properties and the possible effects of changes in the material structure (caused by such fine grain size) that may influence the oxidation resistance of a material.

2. Structure and properties of nanocrystalline alloys

2.1 Dual phase model

The unique properties of nanocrystalline materials are associated with very fine grain size, whereby, depending upon the grain size, interfaces can include up to 50% of the atoms in the material [2-5]. Therefore determination of the structure and associated properties of individual features of a nanocrystalline structure becomes very important. Various models representing the structure of nanocrystalline materials, such as "gas like" model as suggested by Birringer et al. [5] and a "frozen gas like" model suggested later [1,50], are proposed in the literature. However, the structure of nc-materials, in general, may be described as a composition of two components: a crystalline component (CC), which is formed by small equiaxed single crystals each with random crystallographic orientations and the intercrystalline component (IC), which is formed by the interfaces between the crystallites (grain boundaries) and intersection points of these interfaces (triple junctions). The second component may be characterized by the reduced atomic density and inter-atomic spacing deviating from those in the perfect crystal lattice. The IC surrounds the nanometer-sized crystals and forms a network between them [1,21]. As grain size reduces, the IC increases and it may even exceed CC.

Various researchers support the view of the two phase model of nanocrystalline metals. For example, extended x-ray-adsorption fine structure (EXAFS) and Mössbauer spectroscopy of ball-milled iron indicated the presence of two phases as characterized by significantly different atomic arrangements. These different atomic arrangements can be attributed to the presence of interfacial region and crystalline region [51]. Similarly, EXFAS investigation of nanocrystalline Fe and Pd indicated a large reduction in the atomic coordination number, supporting the idea of a very disordered structure at the interfaces [52-56]. Positron-lifetime spectroscopy measurements showed a large density of vacancy-like defects in grain boundaries and relatively large free volume at the triple points arising from misorientation-induced atomic instability of these sites [57]. Elastic relaxation of the interfaces occurred with a very different parameter than conventional coarse grain size polycrystalline materials. Modelling of thermoplastic properties and structure demonstrates that two phase model is an appropriate mean to account for the vibrational density of states and excess energy density in terms of grain boundary [58]. Similar to experimental findings, computer simulation of nanocrystalline iron has shown that grain boundary component in nanocrystalline material is very high and is a strong function of grain size [59,60].

In discussing the structure of nanocrystalline materials further, the following terminology will be used in the text. Three types of grain contacts are possible in a polycrystalline material. They include, a) contact surfaces, b) contact lines and c) contact points. Surfaces of two grains which contact one another are called as contact interfaces. A contact line may represent a common line for three or more adjacent grains. A contact line of three grains is called triple junction. The boundary of grain is its surfaces. Grain boundaries which are seen in the metallographic slides are the section of interfaces by slide plane. A triple point is a section of triple junction by a plane. A detailed description of the terminology used here can be found elsewhere [61].

2.2 Volume fraction of crystalline and intercrystalline regions

Mutschele and Kirchheim [62] proposed following relation to evaluate the volume fraction (C_{ic}) of nanocrystalline materials associated with intercrystalline regions,

$$C_{ic} = 3\delta / d \tag{1}$$

where, δ is the average grain boundary thickness and d is the average diameter of the grains and grains are considered to be cubes. Later, Palumbo et. al. [2] have shown that equation 1 was not suitable for the calculation of volume associated with triple points and to make it more general (to account for the triple points associated with the intercrystalline component) they proposed following relationship for the calculation of the intercrystalline component (V_t^{ic}):

$$V_t^{ic} = 1 - \left[\frac{d-\delta}{d}\right]^3 \tag{2}$$

where, d is the maximum diameter of an inscribed sphere. This yields the following relation for grain boundary volume fraction (V_t^{gb}),

$$V_t^{gb} = \left[\frac{3\delta(d-\delta)^2}{d^3}\right] \tag{3}$$

The volume fraction associated with triple points (V_t^{tp}) is then given by,

$$V_t^{tp} = V_t^{ic} - V_t^{gb} \tag{4}$$

Using above equations and applying a boundary thickness (δ) of 1 nm [2, 62,63], the effect of grain size (*d*), in the range of 2 nm to 1000 nm, on the calculated volume fractions corresponding to intercrystalline regions, grain boundaries, and triple junctions, is shown in Figure 1. The intercrystalline component increases from a value of 0.3% (at a grain size of 1000 nm) to a maximum value of 87.5% at a 2 nm grain size (Figure 1). The volume fractions associated with intercrystalline regions and perfect crystal are equivalent (i.e., 50%) at a grain size of ~ 5 nm. In assessing the individual elements of the intercrystalline fraction, it is noted that the triple junction volume fraction displays greater grain size dependence than that of the grain boundaries. In the range 100 nm to 2 nm, the triple junction volume fraction increases by three orders of magnitude, while the grain boundary volume fraction increases by little over one order of magnitude. In the nanocrystalline range (i.e., d ~ 10 nm), the grain boundary fraction only increases from ~27% at 10 nm, to a maximum value of ~ 44% at 3 nm. Over the same range of grain sizes, the triple junction fraction increases from ~3% to a value of 50%.

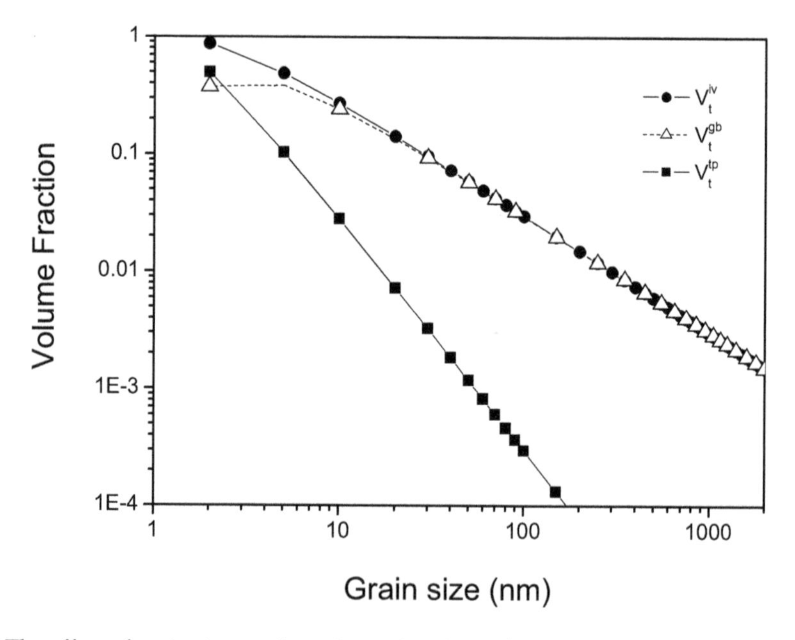

Fig. 1. The effect of grain size on the volume fractions of intercrystalline region, grain boundaries and triple points; calculated from equations 2-5 and assuming the grain boundary thickness to be 1 nm [2].

Based on Figure 1, properties which are influenced by the grain boundary and triple points are expected to be altered significantly when grain size is refined to, or below, 100 nm. The effect of triple points is more pronounced when the grain size is less than 10 nm; consequently material properties with grain size less than 10 nm would be significantly different than those with grain size > 10 nm.

2.3 Thermodynamic properties of nanocrystalline materials

Many researchers have described grain boundaries in nanocrystalline materials as more disordered than those in conventional microcrystalline materials [64-67]. For example, investigations on nanocrystalline Fe have demonstrated that grain boundaries in nanocrystalline Fe differ significantly from the grain boundaries in conventional polycrystalline Fe [64-69]. Thermodynamic properties (specific heat at constant pressure, heat of fusion and stored enthalpy) of ball-milled Fe and other nanocrystalline materials, investigated by Fecht suggested [64-69] that the grain boundaries' energy in nanocrystalline materials to be considerably greater than in the case of equilibrated grain boundaries in microcrystalline materials. In conventional polycrystalline materials, grain boundary energy, as determined by experiments, as well as static and dynamic simulations, is approximately 1 J/m^2, whereas, this value soars to 4 J/m^2 in nanocrystalline materials [70-75].

Significantly different thermodynamic properties of nanocrystalline materials are expected to increase the Gibbs free energy of the materials alloys which can be represented as per following relationship:

$$G= V_t^{gb}.G^{gb}+ V_t^{tp}.G^{tp}+ V_t^{cc}.G^{cc} \tag{5}$$

where, G^{gb}, G^{tp} and G^{cc} are the standard gibbs free energies of grain boundaries, triple points and grains.

Increases in the interfacial energy may lead to a significant increase in the free energy which can be described simply as (neglecting second order contributions due to specific heat differences):

$$\Delta G= \Delta H - T\Delta S \tag{6}$$

Enthalpy difference, ΔH, is shown to be quite higher in nanocrystalline materials than conventional microcrystalline materials. For example it has been shown that ΔH of nanocrystalline Fe increases with decrease in grain size [64-67,76]. Similar behaviour was reported for nanocrystalline copper as well. For example, nanocrystalline Cu, prepared as a powder by vapour deposition followed by compaction releases 300 J/mol at 430K when analysed immediately after compaction and 53 J/mol at 450K when analysed five days after preparation. Such values of enthalpy release have also been confirmed by a study on nanocrystalline Cu prepared by electrodeposition and cold rolling [77-79]. Comparison of these data show that nanocrystalline materials are far from equilibrium, not only because they contain a large amount of interfaces but also because these interfaces are not equilibrated. Therefore, these materials should have high value of free energy which may result in higher reaction rate at the nanocrystalline surfaces.

The total free energy also depends upon entropy term (equation 1), however, evaluation ΔS is not straightforward since there is a little reported in the literature on the entropy contribution from grain boundaries and interfaces for crystals of any size. Although it is expected that this entropy contribution is small and it can also be conceived that non-equilibrated grain boundaries have higher entropy. A value of 0.36 mJ/m^2 K has been estimated for as-prepared nanocrystalline Pt, in contrast to the value for conventional grain boundaries of 0.18 mJ/m^2 K [80]. In fact, the excess entropy per atom sitting in a grain boundary is a substantial part of the entropy of fusion, but the overall entropy per mole of substance sums up to a limited amount even for materials with very small grains. Using this knowledge of enthalpy and entropy, the free energy of nanocrystalline copper has been reported to be higher than that of coarse grain copper [80,81].

2.4 Diffusion in nanocrystalline materials

In general, atomic transport in nanocrystalline materials differs substantially from that in coarse-grained material, due to the crystallite interfaces providing paths of high diffusivity. In conventional microcrystalline materials, crystal volume self-diffusion or substitutional diffusion dominates, at least at temperatures higher than approximately half of the melting temperature. Interface diffusion, in combination with a high fraction of atoms in interfaces, gives rise to modified physical properties of nanocrystalline solids. Furthermore, diffusion processes may control the formation of nanocrystalline materials, for example, by means of crystallization of amorphous precursors, as well as the stability of nanocrystalline materials (relaxation, crystallite growth), their reactivity, corrosion behaviour, or interaction with gases. The relevance of diffusion-controlled processes demands a comprehensive understanding of atomic diffusion in nanocrystalline materials. Detailed discussion of the diffusion processes in the nanocrystalline material is out of the scope. For the readers interest it could be found elsewhere [82-85]. A recent study on nanocrystalline Fe has shown that the diffusion coefficient of Cr in Fe can be enhanced by several orders of magnitude by reducing the grain size to nanometer level [11].

Diffusion in a material can be expressed as the combined effect of diffusion through the grain boundaries and lattice diffusion and can be written as:

$$D_B = fD_{gb} + (1-f)D_b \rightarrow D_B = f\left(D_{gb} - D_b\right) + D_b \tag{7}$$

where, f is the grain boundary fraction, D_{gb} is the grain boundary diffusion coefficient and D_b is the bulk diffusion coefficient of B in the alloy. Assuming the cubic shape of grains, the grain boundary area fraction (f) can be calculated as per equation (1). Because D_{gb} is much larger than D_b, the effective diffusion coefficient of nanomaterials increases significantly by their high area proportion of grain boundaries.

3. Factors effecting the oxidation behavior of a nanocrystalline alloy

An effective protection of metallic materials against high temperature oxidation is based on the protective oxide scale which acts as diffusion barrier, isolating the material from the aggressive atmosphere. The principle is simple, however its application is complex; to act as a real barrier the oxide film needs to be dense and homogenous and has to cover entire surface of the alloy. Oxide scale should possess mechanical properties as close possible to the base material and most importantly it should be adherent to the substrate even in the presence of large thermal shocks. These parameters largely depend upon the alloy composition and microstructure and can be optimized choosing the right combination of the two. The development of the nanocrystalline structure has provided a large scope of modification of the microstructure and to investigate the effect of nanocrystalline structure on the properties of oxide scale formed and therefore the resultant oxidation resistance. Since nanocrystalline materials are thought to be very reactive due to presence of large fraction of defects, it was supposed that they may possess poor oxidation resistance. Here both the possibilities of improvement and deterioration of oxidation resistance due to a nanocrystalline structure are discussed:

3.1 Deterioration of oxidation resistance caused by nanocrystalline structure

The following are possibilities which may lead to a higher oxidation rate in a nanocrystalline structure:

1. It was described in the previous section that free energy of a nanocrystalline alloy is increased as the atoms residing at the grain boundaries are more reactive than the atoms at grains. This increased free energy of nanocrystalline materials would accelerate the reactions occurring upon them and therefore oxidation rate of nanocrystalline structure is expected to increase, leading to poor resistance to environmental degradation.
2. Increase in oxide nucleation sites and, therefore, formation of oxide scale with comparatively finer grain size through which diffusion of oxygen and metal would be faster because of enhanced diffusion through the grain boundaries. Such phenomenon occurring in Ni-Cr-Al alloy accelerates diffusion of Al through the oxide which facilitates the formation of Al oxide [34] and leads to substantial improvement in oxidation resistance. However, such diffusion in a pure metal may lead to a significant higher oxidation rate if a non-protective oxide scale forms. For example in cause of pure Ni, nanocrystalline structure has reported to increase the oxidation rate because of increased diffusion of Ni and oxygen through the grain boundaries of oxide formed on the metal [86].
3. In the case of alloys where the concentration of solute atoms is lower than a critical value for external oxide scale formation, internal oxidation occurs. The oxidation rate is enhanced because of increased diffusivity of oxygen through grain boundaries leading to severe internal oxidation near the surface. Rapid oxidation occurs for these alloys.

3.2 Improvement in oxidation resistance caused by the nanocrystalline structure

Improvement in oxidation resistance of some engineering alloys where protective oxide scales are formed at high temperatures is noticed in their nanocrystalline forms. Improved oxidation resistance of FeBSi [87], Ni-based alloys [88-93], Zr and its alloys [40-43], Cr-33Nb [94], Fe-Co based alloys [95,96] and Cu-Ni-Cr alloys [97] is reported in their nanocrystalline form (in comparison to their microcrystalline counterparts). The mechanistic role of a nanocrystalline structure leading to the improved oxidation resistance is discussed below:

3.2.1 Enhanced diffusivity and oxidation resistance

Certain alloys can develop a continuous layer of the protective oxide of more reactive alloying elements which forms basis for the development of oxidation resistanct alloys. Such alloys are Fe, Ni, Co based, with Al, Cr or Si as the reactive alloying additives. For example, Iron-chromium alloys (such as stainless steels) are the most commonly employed oxidation resistant materials. It has been established that when time-dependent inward flux of oxygen is less than the time-dependent outward flux of solute (Cr), a continuous layer of Cr-oxide is formed at or very near the surface. Formation of such oxide layer and therefore oxidation resistance of Fe-Cr alloys depend upon the supply of solute from the alloy to alloy/oxide scale. It was established in the literature that a fine grain (~17 μm or less) stainless steel easily developed a uniform layer of Cr_2O_3. For an alloy with grain sizes greater than ~40 μm, this protective layer of Cr_2O_3 was difficult to form due to insufficient grain boundary diffusion and inadequate chromium supply [98-100]. In nanocrystalline materials where grain size is very fine and diffusion coefficients are high, such Cr oxide formation should be facilitated by a large extent. Since enhanced diffusion of alloying elements in the nanocrystalline structure facilitates the formation of protective oxides, therefore alumina, chromia and/or silica forming alloys should have improved oxidation resistance in their nanocrystalline form.

3.2.2 Nucleation sites and lateral growth of passive film

In most alloys, nucleation and growth are the mechanism of oxide scale formation during oxidation [44,45,101-103]. It is widely reported that nucleation of oxide is favoured at the high energy sites, i.e., surface defects in the form of dislocations, grain boundaries, triple points, impurities etc. Since nanocrystalline materials are composed of the large fraction of surface defects therefore they offer a large fraction of closely spaced nucleation sites. During the lateral growth of oxide, these nucleation sites become very important as the presence of closely spaced nucleation sites reduces the lateral distance necessary for the lateral growth of a uniform oxide layer to cover the entire surface.

Lobb and Evans [104] have reported an improvement in the oxidation resistance of fine grained conventional microcrystalline materials as a result of increased grain boundary area fraction as the grain boundaries acts as the preferential nucleation sites for Cr_2O_3. They reported that the oxide film nucleated at the grain boundaries needs to grow laterally for the formation of a uniform oxide layer and reported that the finer the grain size, the better the uniformity of oxide should be. Later it was shown that above a critical grain size, formation of a uniform Cr layer was not possible which could be understood based on the combined role of diffusion and nucleation site densities in the alloys [44,45,98-100]. However, these studies were performed on the material where minimum grain size was a few microns. Nanocrystalline materials offer a huge amount of nucleation sites. Therefore, formation of a homogenous compact layer able to cover whole surface is facilitated, which is expected to result in a significant improvement in the oxidation resistance.

3.2.3 Structure of the oxide scale

Mechanical properties, microstructure, adhesion and growth of oxide scale has been reviewed recently [105,106] and it has been found that oxidation resistance of a metal largely depends upon their physical properties (e.g., crystal size, morphology and crystallographic orientation of oxides formed, lattice mismatch with the base metal, adhesion of oxide layer) of oxide [105,106]. For example, fine grained oxide scales often show a fast creep rate at high temperatures, releasing the stresses accumulated in the scales and, therefore, decrease in scale spallation tendency. This may have important implication in reducing the spallation of oxides at high temperature. Since nanocrystalline materials provide several orders of magnitude more nucleation sites than the microcrystalline materials therefore grain size of the oxide developed on nanocrystalline materials is expected to be finer. This finer grain size of oxide suppresses oxide scale spallation in nanocrystalline form. This effect has been demonstrated in high temperature oxidation tests of several nanocrystalline alloys [27-30,47].

It was also proposed that fine nano-sized oxide structure reduces conductivity which suppresses the transport of the oxidizing species, enhancing oxidation resistance. This proposal was successfully used to explain improved oxidation resistance of Zr and Zr based alloys [107]. However, it is not clear if this model also applies for other alloys.

It is important to note that the above factors can operate both indifferently and in combination [12]. The nature of the influence of nanostructure on the diffusion-assisted corrosion (viz., high temperature oxidation) depends on the role of the predominantly diffusing species in a given alloy. For example, oxidation resistances of an iron-aluminide and an Fe-B-Si alloy in the nanocrystalline state are reported to be superior to that in their microcrystalline state [87,108]. This behaviour is attributed to Al and Si, the well-known protective oxide film formers, being the predominantly diffusing species respectively in the two alloys, and the nanostructure facilitating their diffusion and expedited formation of protective films (of Al/Si oxide). Therefore, it is important to consider all the factors effecting the oxidation resistant, and the net effect of all the parameters would determine the change in the oxidation rate caused by nanocrystalline structure.

4. Critical concentration of solute required for the transformation of internal to external oxidation as a function of grain size

Oxidation of engineering alloys is very complex as the components of the alloys has different affinities for the oxygen, and reacting atoms do not diffuse at the same rates in the oxides or alloy substrates. Various types of oxides can be formed on and in the alloy. Atomic ratios of the elements in the oxide scale may differ significantly from those in the alloy. When oxygen and metal atoms diffuse and react at the surface of an alloy, an external oxide layer is formed on the surface and this is termed as the "external oxidation". For external oxidation, outward flow of metal atoms must exceed the inward flow of oxygen, whereas, when inward flow of oxygen exceeds the outward flow of the metal atoms, oxygen diffuses inside the metal and oxidation takes places within the alloy. This process is termed as "internal oxidation" which leads to catastrophic loss in the material property [44,45,109]. Figures 2 and 3 schematically show internal and external oxidation for alloy A-B under the conditions where only B oxidises and both A and B oxidise, respectively.

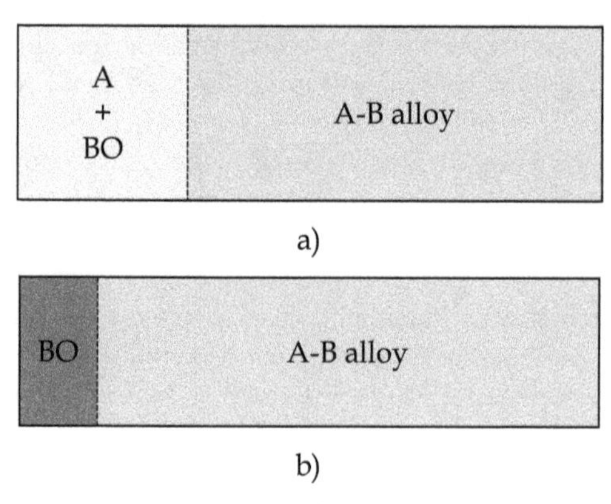

Fig. 2. Wagner's model of the transition from the internal to external oxidation of alloy A-B under the condition where only B can oxidize: a) B is less than the critical amount of B required for the transition and b) B is higher than the critical content of B required for the transition from the internal to external oxidation [44,45, 110-112].

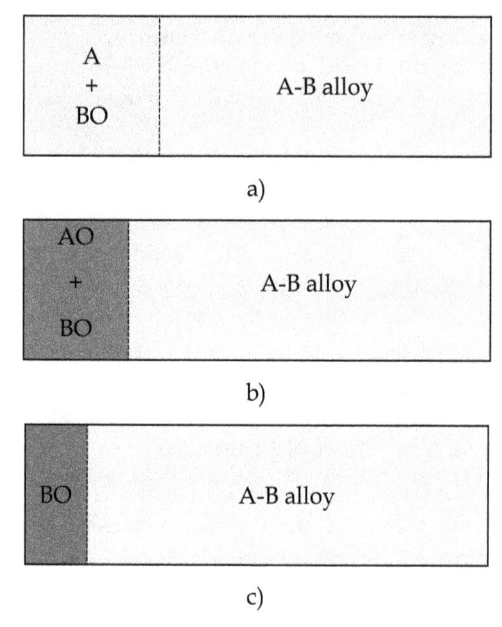

Fig. 3. Wagner's model of the transition from the internal to external oxidation of alloy A-B under the condition where both A and B can oxidize: transition from internal oxide (a) to external composite oxide scale (b) to external exclusive oxide scale (c) [44,45,110-112].

From a kinetic point of view, there is a minimum concentration of oxide former required to ensure that the alloy chromium (or any other oxide former) diffusion flux must be sufficient to outweigh inward oxygen diffusion, thus favouring external scale growth over internal oxide precipitation. The required concentration, $N_M(Crit)$, is found from Wagner's analysis [45,110-112] to be:

$$N_B(Crit1) = \left(\frac{g.\pi}{3} \cdot \frac{V_{AB}}{V_{BO_v}} \cdot \frac{N_O^{(s)} D_O}{D_B}\right)^{1/2} \tag{8}$$

Here $N_O^{(s)} D_O$ is the oxygen permeability in the alloy matrix, V_{AB}, V_{BO_v} are the molar volumes of alloy and oxide, D_B is the diffusion coefficient of chromium or aluminium in the alloy, and g the critical volume fraction of oxide required to form a continuous layer. According to this equation, the critical value for external chromia formation in Ni-Cr at 1000°C is $N_{Cr} = 0.29$ and for alumina formation in Ni-Al at 1200°C is $N_{Al} = 0.11$ [45], both in agreement with the experimental measurements. The requirement for these relatively large concentrations of Cr or Al to form a complete protective scale will, in many cases, change other alloy properties, which limits the applicability of this approach, particularly at lower temperatures.

No sooner has the continuous scale of chromia or alumina become established than the steady-state growth of this oxide starts. For chromia/alumina scale growth to be maintained, an alloy must supply chromium/aluminium by diffusion from its interior at a rate sufficient to balance the rate at which the metal is consumed in forming new oxide. Wagner's diffusional analysis [112] leads to the requirement:

$$N_B(Crit2) = \frac{V_{AB}}{V_{BO_v}} \left(\frac{\pi k_p}{2D_B}\right)^{1/2} \tag{9}$$

where V_{AB}, V_{BO_v} are the molar volumes of alloy and oxide, k_p is the parabolic rate constant for scale thickening:

$$x^2 = 2k_p t \tag{10}$$

with x the scale thickness formed in time, t, and D_B the alloy interdiffusion coefficient. The quantity $N_B(Crit2)$ is the minimum original alloy chromium concentration necessary to supply metal to the alloy-scale interface fast enough to support exclusive chromia scale growth.

To avoid internal oxidation and maintain external chromia/alumina scale, the concentration of aluminium/chromium should be higher than both N_B(Crit1) and N_B(Crit2). In nanocrystalline materials, D_B could be very high [11,12] according to Eqn 6, and therefore both N_B(Crit1) and N_B(Crit2) decrease significantly which will lead to substantial increase in the formation of a protective alumina/chromia external scale, therefore increased oxidation resistance. Assuming the alloy grains are cubic, the area proportion of grain boundary f can be calculated f=2δ/d. Also considering Dgb>>Db, Eqn 7 can be simplified as [113]:

$$N_B(Crit1) = A(D_b + \frac{2\delta}{d} D_{gb})^{-1/2} \tag{11}$$

where

$$A = (\frac{g\pi N_O^{(s)} D_O V_{AB}}{2V_{BO_v}})^{1/2},$$

thus reduction in grain size results in the reduction of N_B(Crit1).

Using Equations (1) and (6), the critical concentration N_B(Crit2) is calculated as:

$$N_B(crit2) = B.\left(D_b + \frac{2\delta}{d}D_{gb} - \frac{2\delta}{d}D_b\right)^{-\frac{1}{2}} = B\left(\frac{\sqrt{d}}{\sqrt{2\delta(D_{gb}-D_b)+d.D_b}}\right) \quad (12)$$

Where,

$$B=\frac{V_{AB}}{V_{BO_v}}\left(\frac{\pi k_p}{2}\right)^{1/2}.$$

At a given temperature and external oxygen partial pressure, N_B (crit1) and N_B (crit2) are a function of only grain size *(d)* of the alloy, diffusion coefficients (D_{gb} and D_b) and grain boundary width (δ) of the alloy.

The critical concentration of B required, where only bulk diffusion can control the oxidation (i.e., the area fraction of grain boundary is very small such that $D=D_b$), can be calculated as:

$$N_{OB}(crit2) = B.(D_b)^{-\frac{1}{2}} \text{ and } N_{OB}(crit1) = A.(D_b)^{-\frac{1}{2}} \quad (13)$$

From equations (10), (11) and (12), for an alloy of grain size, *d*, the ratio *(X)*, critical concentration of B required for external oxidation in material with a grain size *d* to that when the bulk diffusion coefficient dominates, can be given as:

$$X = \frac{N_B(crit1)}{N_{OB}(crit1)} = \frac{N_B(crit2)}{N_{OB}(crit2)} = \sqrt{\frac{d}{2\delta\left(\frac{D_{gb}}{D_b}-1\right)+d}} = \sqrt{\frac{1}{f\left(\frac{D_{gb}}{D_b}-1\right)+1}} \quad (14)$$

Since, the values of D_{gb} and D_b are available from the literature, the value of *X* can be calculated as a function of grain size, time and temperature. Above relationship may be very useful in comparing the change in the critical amount of B (Al or Cr) caused by nanocrystalline structure at a given temperature.

Given that the critical amount of B (N_B(crit1)) required for a material with grain size d_0 is *w*, according to equation (13), the ratio for the critical amount of B required for a grain size of *d* to that for the grain size d_0 can be expressed as X′:

$$X' = \frac{N_B}{w} = \sqrt{\frac{d(2\delta(D_{gb}-D_b)+d_0.D_b)}{d_0(2\delta(D_{gb}-D_b)+d.D_b)}} \quad (15)$$

For Fe-Cr alloy, X′ is calculated based on available data listed in the Table 1 (extrapolated from the literature [114,115] with d_0 assumed to be 5 micron, and plotted it as a function of grain size in Fig 4. Figure 4 clearly shows that Cr required for such transition decreases substantially when grain size is below 100 nm.

Temperature	Lattice (D_b) m^2/s	GB (D_{gb}) m^2/s
300 °C	1.2×10^{-26}	8.6×10^{-22}
840°C	1.5×10^{-15}	3.7×10^{-12}

Table 1. Diffusion coefficients of Cr in Fe-Cr alloy [114,115]

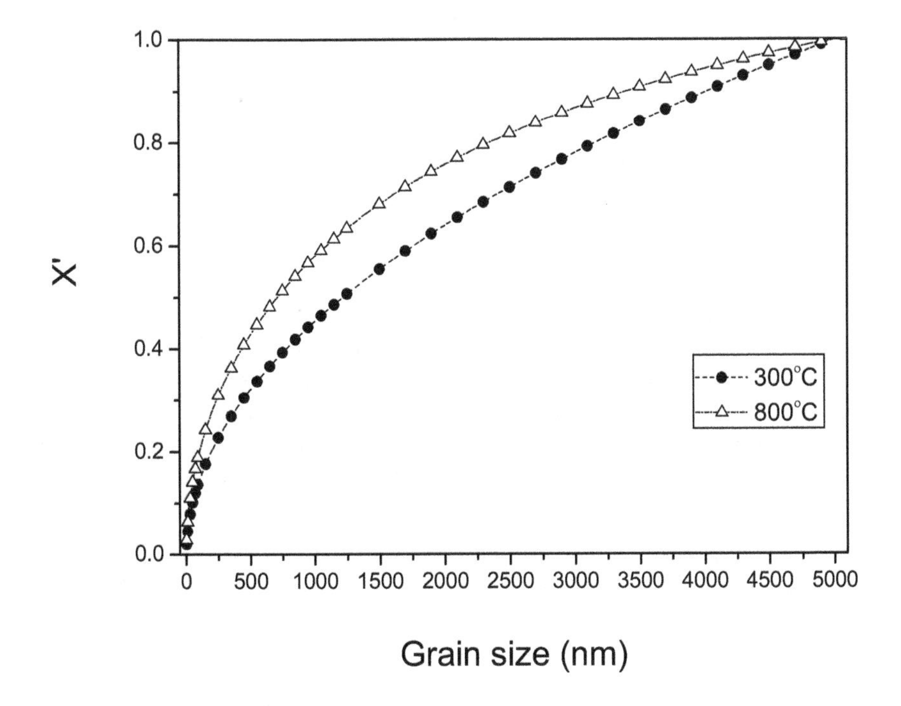

Fig. 4. Plot of the grain size with X′ (ratio of critical amount of Cr required for external passive oxide layer formation in a Fe-Cr alloy of grain size d to critical amount of Cr required for such transition in a Fe-Cr alloy with a grain size of 5 μm) [12,39].

5. Oxidation resistance of nanocrystalline alloys – Some examples

5.1 Fe-Cr alloys

5.1.1 Oxidation resistance of nanocrystalline and microcrystalline Fe-10Cr alloys

The oxidation behaviour of Fe-Cr alloys have been extensively investigated in the literature and reviewed by Wood [102]. Fe-Cr alloys are known for their high oxidation resistance due to formation of a Cr rich oxide layer which largely depends upon the selective oxidation of Cr. As already demonstrated before, a nanocrystalline structure is expected to cause a significant reduction in the Cr concentration required for chromia scale formation. For the validation of this hypothesis various nanocrystalline and microcrystalline Fe-10Cr and Fe-20Cr alloys were prepared by high energy ball milling followed by compaction and sintering which are described elsewhere [116]. The oxidation behaviour of such Fe-Cr alloys was investigated in a temperature range of 300 to 400°C by continuous weight gain experiments. The experimental details can be found elsewhere [12,38,39].

Weight gain curves representing the oxidation kinetics of nanocrystalline and microcrystalline Fe10Cr alloys, in the temperature range of 300-400°C are presented in

Figures 5-7 [12,38,39]. Oxidation kinetics at 300°C shows the microcrystalline alloy to be oxidizing at a considerably greater rate than the nanocrystalline alloy. After 3120 minutes of oxidation, weight gain of microcrystalline Fe-10Cr alloy was found to be nearly seven times greater than that of nanocrystalline alloy of same chemical composition.

Besides the considerably higher weight gain of the microcrystalline alloy, the evolution of oxidation kinetics was also different. Both nanocrystalline and microcrystalline Fe10Cr alloys follow parabolic kinetics for the first 240 minutes of oxidation (as evident in the weight-gain^2 versus time plot in Figures 5b and 6b). However during subsequent oxidation, nanocrystalline Fe10Cr alloy show considerable deviation from the parabolic behaviour whereas, microcrystalline alloy of same chemical composition continued to follow the parabolic kinetics (Figures 5c and 6c). The marked deviation of the nanocrystalline Fe-10Cr alloy from the parabolic behaviour is accounted for the insignificant increase in the weight-gain of this material after the first 240 minutes of oxidation (Figures 5a, 6a, 5b and 6c). This behaviour could be attributed to some critical change in the chemical characteristic of the oxide scale formed on both nanocrystalline and microcrystalline alloys before and after 240 minutes of oxidation which was described by Gupta et al using SIMS [12,38,39] analysis of oxide formed during various period of oxidation.

Oxidation kinetics of nanocrystalline and microcrystalline Fe-10Cr alloys at 350 and 400°C are presented in Figures 6 and 7. The trend of greater oxidation rate of the microcrystalline alloy, as seen at 300°C is also followed at the two higher temperatures. However, the influence of nanocrystalline structure in improving the oxidation resistance was extraordinarily enhanced at these higher temperatures as indicated by the comparative weight gains after 3120 minutes of oxidation: weight gain of microcrystalline Fe-10Cr alloy was found to be 18 times greater than that of the nanocrystalline Fe-10Cr alloy at 350°C, and nearly 17 times greater at 400°C.

A close observation of the data as presented in Figures 5-7 show that both nanocrystalline and microcrystalline Fe-10Cr alloys follow parabolic kinetics ,i.e., (weight grain per unit area)^2 = kt. The rate constants (k) in nanocrystalline alloy changes with time (Table 2). Oxidation kinetics of nanocrystalline Fe-10Cr can be divided in the two stages, each stage characterized by a unique k value (Table 2). Microcrystalline alloy, on the other hand follow a single parabolic rate constant. As presented in the Table 2, k value for microcrystalline (mc) Fe10Cr alloys are more than an order of magnitude greater than either of the k values for nanocrystalline (nc) Fe10Cr alloy at the three temperatures.

	nanocrystalline Fe10Cr	microcrystalline Fe10Cr
300°C	5.65×10^{-13} (1st stage) and 7.42×10^{-14} (2nd stage)	7.74×10^{-12}
350°C	1.04×10^{-12} (1st stage) and 1.7×10^{-13} (2nd stage)	1.46×10^{-10}
400°C	1.34×10^{-12} (1st stage) and 5.69×10^{-13}(2nd stage)	2.53×10^{-10}

Table 2. Parabolic oxidation rate constants (k) values in $g^2cm^{-4}s^{-1}$[12]

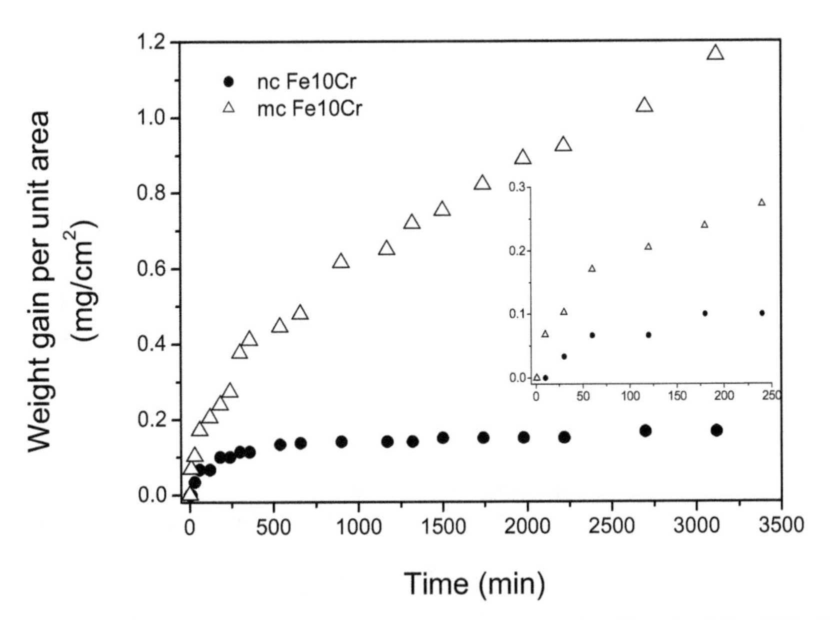

Fig. 5a. Oxidation kinetics of nanocrystalline (nc) and microcrystalline (mc) Fe-10Cr alloys at 300°C as represented by weight-gain vs time plots for 3120 minutes [12,37,39]. Inset shows the zoom of the region representing initial periods of oxidation (up to 240 minutes of oxidation).

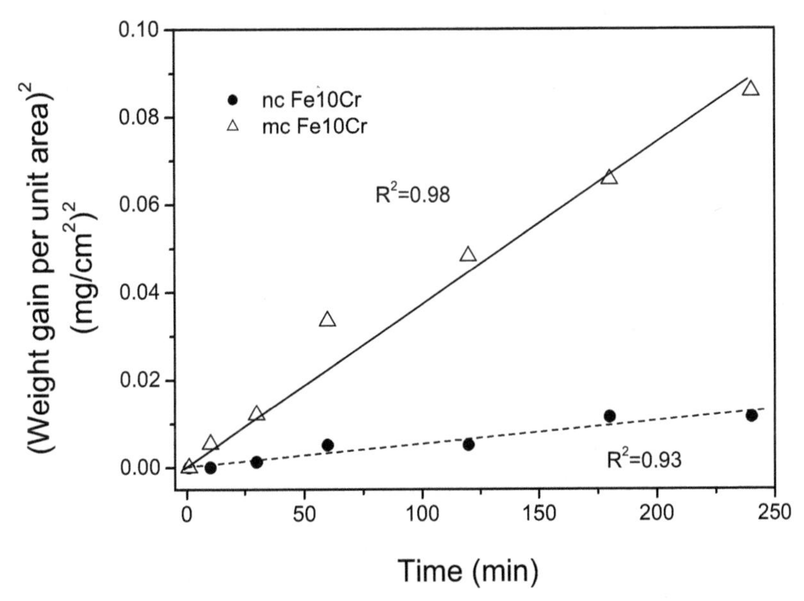

Fig. 5b. Oxidation kinetics of nanocrystalline (nc) and microcrystalline (mc) Fe-10Cr alloys oxidised at 300°C: weight-gain^2 with time (up to 240 minutes) suggesting parabolic kinetics for both mc and nc alloys [12,37,39]

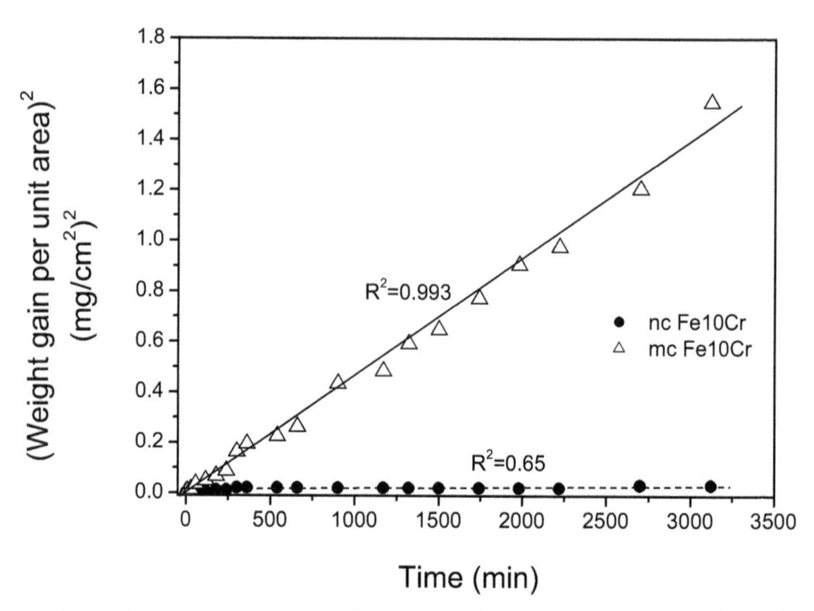

Fig. 5c. Oxidation kinetics of nanocrystalline (nc) and microcrystalline (mc) Fe-10Cr alloys, oxidised at 300°C: weight-gain^2 with time, suggesting parabolic kinetics for mc alloy but departure from parabolic kinetics for nc alloy [12,37,39].

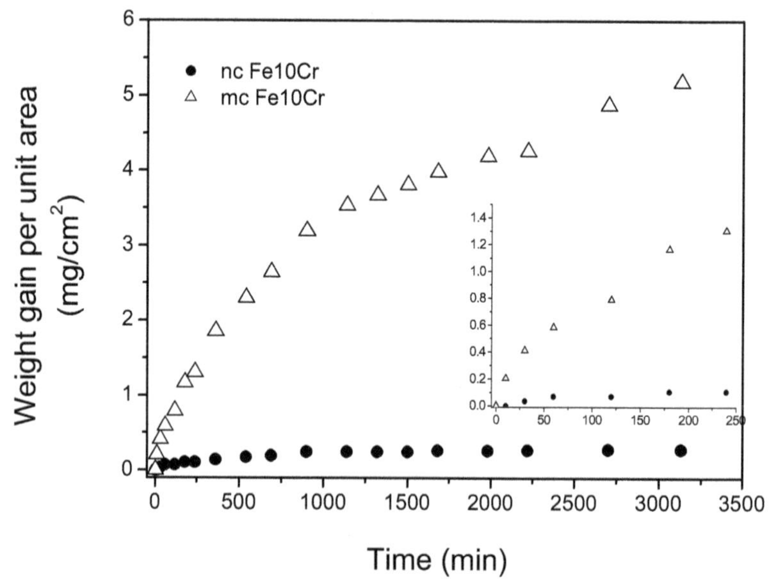

Fig. 6a. Oxidation kinetics (weight-gain vs time plot) of nanocrystalline (nc) and microcrystalline (mc) Fe-10Cr alloys, during oxidation at 350°C for 3120 min in air [12,39]. Inset shows the zoom of the region showing initial periods of oxidation (up to 240 minutes of oxidation).

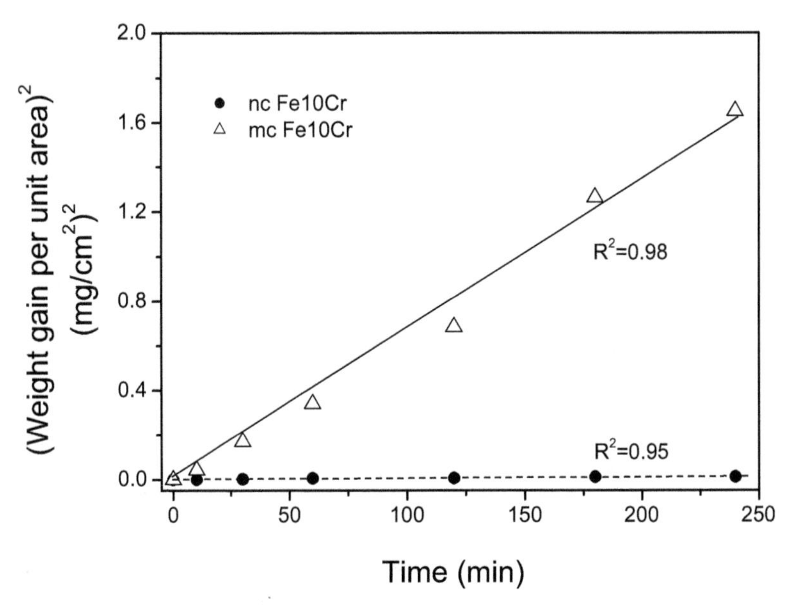

Fig. 6b. Oxidation kinetics of nanocrystalline (nc) and microcrystalline (mc) Fe-10Cr alloys oxidised at 350°C: weight-gain^2 with time (up to 240 minutes) suggesting parabolic kinetics for both mc and nc alloys [12,39]

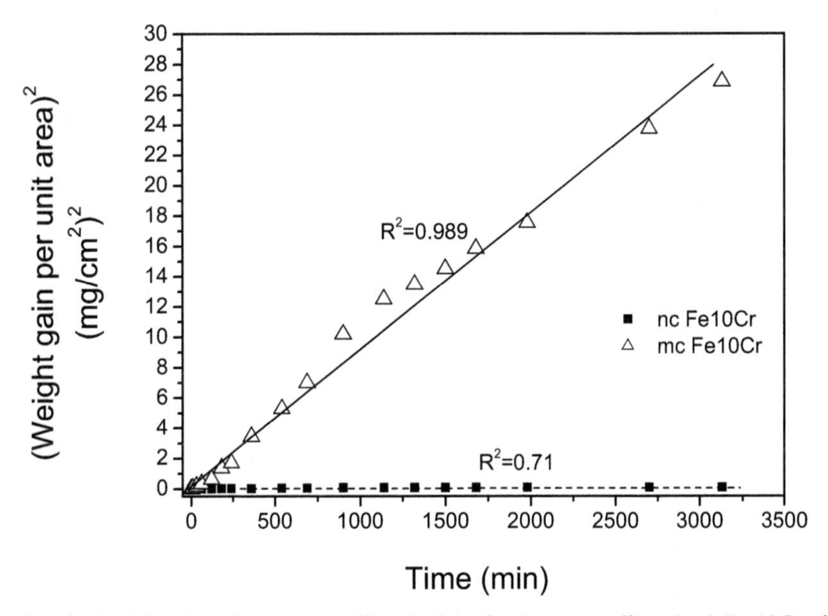

Fig. 6c. Oxidation kinetics of nanocrystalline (nc) and microcrystalline (mc) Fe-10Cr alloys, oxidised at 350°C: weight-gain^2 with time, suggesting parabolic kinetics for mc alloy but departure from parabolic kinetics for nc alloy [12,39].

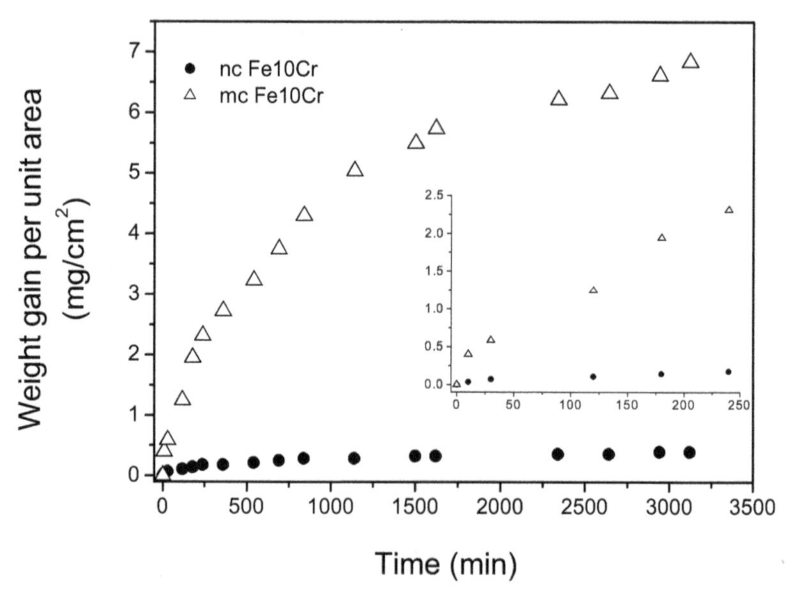

Fig. 7. Oxidation kinetics (weight-gain vs time plot) of nanocrystalline (nc) and microcrystalline (mc) Fe-10Cr alloys, during oxidation at 400°C for 3120 min in air [12,39]. Inset shows a zoom of the region of initial periods of oxidation (up to 240 minutes of oxidation).

5.1.2 SIMS depth profile of oxidized Fe-10Cr alloys

In order to understand the mechanism of improved oxidation resistance of nanocrystalline Fe-10Cr alloy, the composition, including Cr content of the thin oxide films developed on the nanocrystalline and microcrystalline alloys was characterized. The thin oxide films formed over nanocrystalline Fe-10Cr and microcrystalline Fe-10wt%Cr alloys at 300, 350 and 400°C in air were characterised by SIMS depth profiling [12,39].

Oxidation resistance of Fe-Cr alloys was associated the development of a protective layer of Cr_2O_3. Depth profiles for Cr, O and Fe for the nanocrystalline and microcrystalline Fe-10Cr alloys oxidized at the three temperatures for 30, 120 and 3120 min were obtained [12,32,33]. It was found that the oxide film developed on microcrystalline Fe-10Cr alloy is considerably thicker than that on nanocrystalline Fe-10Cr alloy at the three test temperatures [12,39].

The most relevant findings of the SIMS analyses as reported in the literature [12,38,39] are the depth profiles of chromium and their consistency with the trends of oxidation kinetics. The Cr depth profiles obtained after 52 hours of oxidation are presented in the Figures 8-10. At each of the oxidation temperatures, Cr content of the inner layer of nanocrystalline Fe-10Cr alloy was invariably found to be considerably higher than the highest Cr content in the inner layer of microcrystalline Fe-10Cr alloy. This provides an explanation for the greater oxidation resistance of the nanocrystalline Fe-10Cr alloy (as shown in Figures 8-10), since oxidation resistance of Fe-Cr alloys is governed primarily by the Cr content of the thin oxide scale.

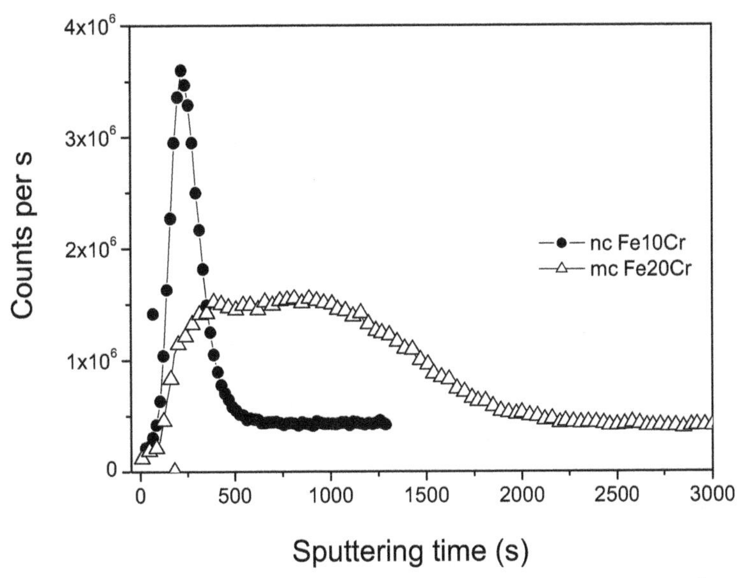

Fig. 8. SIMS depth profiles for Cr the in the oxide scale developed during oxidation of nanocrystalline (nc) and microcrystalline (mc) Fe-10Cr alloys at 300°C for in air for 3120 minutes, using a Cameca ims (5f) dynamic SIMS instrument. SIMS parameters were: Cs^+ ion primary beam (10 nA), depth profiling of craters of 250 μm × 250 μm area [12,39].

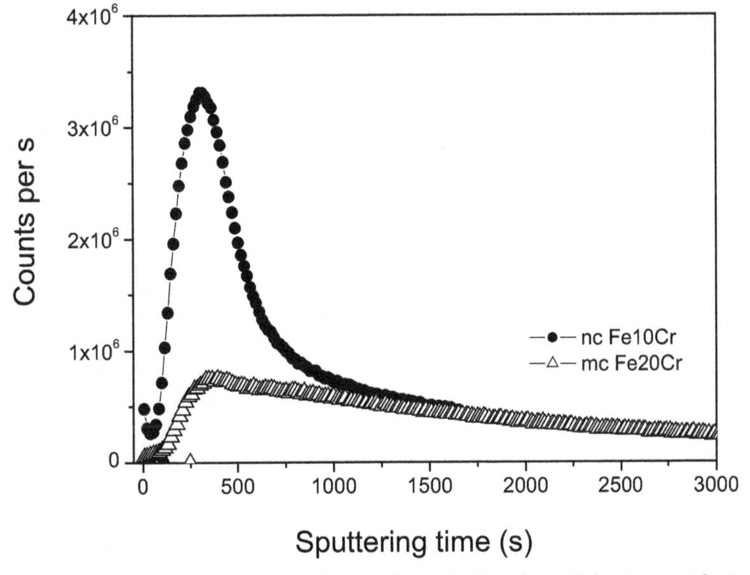

Fig. 9. SIMS depth profiles for Cr the in the oxide scale developed during oxidation of nanocrystalline (nc) and microcrystalline (mc) Fe-10Cr alloys at 350°C for in air for 3120 min, using a Cameca ims (5f) dynamic SIMS instrument. SIMS parameters were: Cs^+ ion primary beam (10 nA), depth profiling of craters of 250 μm × 250 μm area [12,39].

SIMS analysis as carried out in our previous work [12,39] provides a qualitative analysis of Cr enrichment of the surface. Based on such qualitative analysis of Cr content, a Cr_2O_3 oxide layer was proposed to develop in nanocrystalline alloy, whereas, it was proposed that a mixed Fe-Cr oxide layer forms in case of microcrystalline alloy. A Future study quantifying the Cr, Fe and O contents of oxide layer and their oxidation states using techniques such as X-ray photoelectron spectroscopy (XPS) must provide a better understanding of the effect of nanocrystalline structure on the chemical composition of oxide layer.

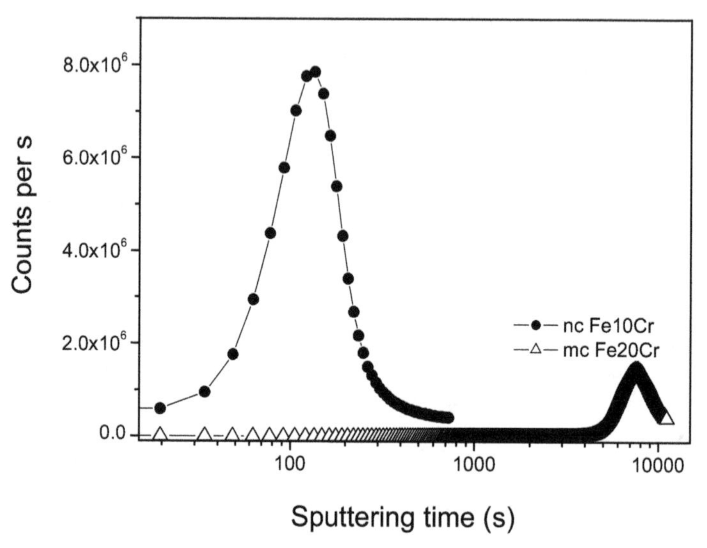

Fig. 10. SIMS depth profiles for Cr the in the oxide scale developed during oxidation of nanocrystalline (nc) and microcrystalline (mc) Fe-10Cr alloys at 400°C for in air for 3120 min, using a Cameca ims (5f) dynamic SIMS instrument. SIMS parameters were: Cs^+ ion primary beam (10 nA), depth profiling of craters of 250 μm × 250 μm area [12,39].

5.1.3 Oxidation resistance of nanocrystalline Fe10Cr versus oxidation of microcrystalline Fe20Cr alloy

For developing an understanding of how the considerably greater oxidation resistance of nanocrystalline Fe-10Cr alloy (in comparison with microcrystalline Fe-10Cr alloy) compares with the resistance of an alloy with much higher Cr content, samples of microcrystalline Fe-20Cr alloys were also oxidized at 350 °C for durations up to 3120 minutes [12,39]. However, what is most relevant to note is that the weight gain at the end of 3120 minutes of oxidation of microcrystalline Fe-20Cr alloy is similar to that of the nanocrystalline Fe-10Cr alloy at 350 °C for same period of time (shown in Figure 11), suggesting the degree of oxidation resistance conferred due to nanocrystalline structure at only 10% chromium to be similar to that of the alloy with 20% chromium but microcrystalline structure. This finding may have wide industrial applications in developing steel with low Cr but very high oxidation resistance as exhibited by Fe20Cr alloy.

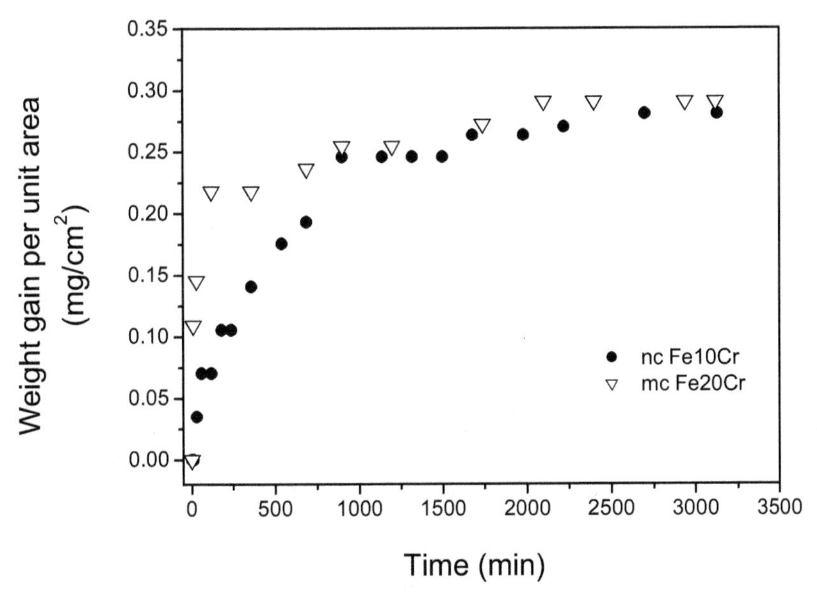

Fig. 11. Comparison of oxidation resistance of nanocrystalline Fe10Cr alloy with that of microcrystalline Fe20Cr alloy at 350°C in air [12,39].

5.2 Ni-Cr-Al based alloys

Oxidation resistance of Ni-Cr-Al based alloys largely depend upon the chemical and physical properties of alumina scale formed during high temperature oxidation. Most of the Ni-Cr-Al alloy contains enough Cr to form external Cr_2O_3 scale but for the application over 1000°C, the chromia scale does not provide any beneficial effect due to the volatilisation problems. Formation of a compact protective Al_2O_3 scale is most efficient in protecting material from the high temperature oxidation. In general, a conventional Ni-20Cr-Al system requires more than 6 wt% of Al to form protective oxide scale which largely depends upon the a) diffusion coefficient of the Al from the bulk to alloy/oxide interface and b) diffusion of Al in the formed oxide scale [45,118].

Since nanocrystalline materials possess significantly higher diffusion coefficient caused by higher fraction of grain boundaries therefore Al required for formation an exclusive Al oxide film can be reduced significantly and nanocrystalline materials should show improved oxidation resistance [88-93,119]. Wang et al [119] were among first researchers to investigate the oxidation resistance of nanocrystalline Ni-Al-Cr alloys and reported a significant improvement in the oxidation resistance of NiCrAl alloys due to nanocrystalline structure. Later, various authors have investigated the oxidation resistance of nanocrystalline NiCrAl alloys with various Al and Cr contents and produced by different methods but in all the cases nanocrystalline structures was reported to enhance the oxidation resistance. Most comprehensive work showing effect of the nanocrystalline structure on NiCrAl alloys was that of Gao et al [88] who reported excellent oxidation behaviour of a nanocrystalline coating of Ni20CrAl alloy over its microcrystalline counterpart. They reported a coating with grain size less than 70 nm may increase the oxidation rate by 4 times. Increment in oxidation resistance due to grain refinement was more pronounced when grain size was below 100 nm. It has been shown than Al content required to prevent external oxidation can be reduced from 6% to only 2% by reducing the grain size to ~ 60 nm [88-93]. These findings may have large industrial implications as it would provide an opportunity to achieve the desired oxidation resistance with lower Al content.

6. Challenges for application of nanocrystalline alloys at high temperature

For high temperature application, a major problem is thermal stability of the nanocrystalline structure. Thermodynamically, such fine structures could not survive for long times at high temperatures because of the large specific surface energy (driving force) for grain growth. An effective way to prevent this grain growth is by introducing stable second-phase particles in the microstructure, where they play a particle pinning effect (Zener effect) on grain growth [120]. Surprisingly, the coarsening kinetics for nanocrystalline alloys prepared by sputtering is rather slow even without apparent second-phase particles in the microstructure. Lou et al. [19] found that after 100 h oxidation in air at 1000°C, the grains of a sputtered K38G nanocrystalline coating were somewhat coarsened, but still rather fine-grained, growing from 20-100 nm to 200-1000 nm. It is not clear if this high stability against grain growth comes from grain boundary segregation of alloy elements (e.g. K38G alloy contains 3.6% Ti, 1.7% Mo, 2.6% W, 1.7% Ta in the Ni-Cr-Al base), which anchor or retard the grain boundary movement (similar to Zener effect) although no particle formation was apparent, or from the sputtered coating structure itself (columnar structure). Unfortunately, no systematic investigation of temperature and alloy composition effects on the thermal stability of these nanocrystalline alloys is available to date, future research should be carried out for high temperature application of nanocrystallline alloys.

Another problem limiting the application of nanocrystalline materials is preparation of nanocrystalline alloys. Currently, the bulk metallic nanomaterials can only be prepared at the laboratory scale, usually by compacting prepared nanocrystalline powders. However, consolidation of the nanopowders into bulk materials needs high temperature and pressure which may considerably coarsen the structure. Because of this difficulty, surface nanocoating has been considered a potential industry application. Nanocrystalline costing are often prepared by chemical vapour deposition (CVD), physical vapour deposition (PVD), electrochemical deposition, electro-spark deposition, and laser and electron beam surface treatment.

7. Future research

All studies reporting excellent oxidation resistance of nanocrystalline NiCrAl alloys were conducted at high temperatures. However, the grain growth behaviour of the alloys at these elevated temperatures was not determined and therefore many questions remain unanswered. Detailed studies investigating the effect of such fine grain size on the oxide scale formation along with grain growth of material will be helpful in understanding the underlying mechanism of the improved oxidation resistance.

Oxidation behaviour of Fe-Cr alloys as described in the section 5.1 was investigated at moderate temperatures [12,37-39]. Choice of moderate temperature was motivated by: 1) a very high difference in the grain boundary and lattice diffusion coefficient values at moderate temperatures and 2) higher grain growth at elevated temperatures. Grain growth of the nanocrystalline materials at high temperatures limits their use for high temperature applications. However, it was shown recently that addition of Zr to Fe-Cr based alloys prevents grain growth of these materials [120,121] and therefore such alloys with small amount of Zr (i.e., Fe-Cr-Zr alloys) will be ideal for investigation of oxidation resistance in the temperature range of 600-800°C.

Grain size of Fe-Cr alloys (used for the investigation of the effect of nanocrystalline structure on oxidation resistance) was limited to 54 (±4) nm which could be further decreased with the recent advancements in the sample preparation techniques such as one recently developed by Gupta et al [122] where an artefact free FeCrNi alloy with a grain size less than 10 nm was produced by in-situ consolidation technique. Further investigations on such alloys with grain size below 10 nm will demonstrate pronounced effect of triple points and grain boundaries and it may be possible to develop stainless steels with further improved oxidation resistance but less Cr content.

Improved oxidation resistance of nanocrystalline Fe-Cr or Ni-Cr-Al alloys have been attributed to the greater Cr and/or Al enrichment of the oxide scale (i.e., change in chemical composition of the oxide scale) due to faster diffusion of Cr and/or Al. However, physical properties of oxide scale, which are very important in determining the oxidation resistance of an alloy, have attracted only a little research attention. Investigation of the physical properties (grain size, morphology, crystallographic details etc.) of the oxide scale formed on the nanocrystalline alloys will further help in understanding the effect of nanocrystalline structure on the oxidation resistance of an alloy.

8. Concluding remarks

Nanocrystalline materials are being investigated due to their unique properties. More generally, development of materials resistant to environmental degradation is not the main focus of nanocrystalline metals research to date, but it seems possible that nanocrystalline metallic materials may lead to a substantial increment in oxidation resistance; caused by promoted oxide scale formation, improved adherence and reduced spallation tendency of the oxide scale. Nanocrystalline Fe-Cr and M-Cr-Al alloys have demonstrated improved oxidation resistance and present potential to be used for high temperature applications in future. More fundamental investigations are required to fully characterise the oxidation phenomenon and underlying principles for nanocrystalline materials.

9. References

[1] H. Gleiter, Progress in Materials Science, 33 (1989) 223.
[2] G. Palumbo, S. J. Thorpe, K. T. Aust, Scripta Metall. Mater. 24 (1990) 2347.
[3] C. C. Koch, K.M. Yousef, R. O. Scattergood, K. L. Murty, Advanced Eng. Mater. 7 (2005) 787
[4] R. W. Siegel, NanoStructured Materials 4 (1994) 121
[5] R. Birringer, H. Gleiter, H. P. Klein, P. Marquardt, Phys. Lett. A 102 (1984) 365.
[6] M. A. Meyers, A. Mishra, D. J. Benson, Progress in Materials Science 51 (2006) 427.

[7] T. Zhu, J. Li, Progress in Materials Science 55 (2010) 710.
[8] C. C. Koch, Encyclopedia of Materials: Science and Technology (2001) 5901.
[9] C. C. Koch, Scripta Materialia, 49 (2003) 657.
[10] S. Cheng, E. Ma, Y. M. Wang, L. J. Kecskes, K. M. Youssef, C. C. Koch, U. P. Trociewitz, K. Han, Acta Materialia 53 (2005) 1521.
[11] Z. B. Wang, N. R. Tao, W. P. Tong, J. Lu, K. Lu, Acta Materilia 51 (2003), 4319.
[12] R. K. Gupta, PhD thesis, Synthesis and corrosion behaviour of nanocrystalline Fe-Cr alloys, Monash University (2010).
[13] K. D. Ralston, N. Birbilis, Corrosion 66 (7) (2010) 075005 - 1
[14] G. Meng, Y. Li, F. Wang, Electrochimica Acta 51 (2006) 4277.
[15] C. T. Kwok, F. T. Cheng, H. C. Man, W. H. Ding, Materials Letters 60 (2006) 2419.
[16] Y. Li, F. Wang, Electrochimica Acta 51 (2006) 4426.
[17] X. Y. Wang, D. Y. Li, Electrochimica Acta 47 (2002) 3939.
[18] Sh. Hassani, K. Raeissi, M. Azzi, D. Li, M. A. Golozar, J. A. Szpunar, Corrosion Science 51 (2009) 2371.
[19] Kh. M. S. Youssef, C. C. Koch, P.S. Fedkiw, Corrosion Science 46 (2004) 51.
[20] L. Wang, Y. Lin, Z. Zeng, W. Liu, Q. Xue, L. Hu, J. Zhang, Electrochimica Acta 52 (2007) 4342.
[21] H. Gleiter, Acta Mater 48 (2000) 1.
[22] J. Y. Fan, X. L. Wu, K. Paul, K. Chu, Progress in Materials Science 51 (2006) 983.
[23] T. Sourmail, Progress in Materials Science 50 (2005) 816.
[24] M. E. McHenry, M. A. Willard, D. E. Laughlin, Progress in Materials Science 44 (1999) 291.
[25] G.R. Wallwork, A.Z. Hed, Oxidation of Metals, 3 (1971) 171.
[26] G. Chen, H. Lou, Surface and Coatings Technology 123 (2000) 92.
[27] Z. Liu, W. Gao, K.L. Dahm, F. Wang, Acta Materialia 46 (1998) 1691.
[28] L. Hanyi, W. Fuhui, X. Bangjie, Z. Lixin, Oxidation of Metals, 38 (1992) 299.
[29] H. Lou, S. Zhu, F. Wang, Oxidation of Metals, 43 (1995) 317.
[30] H. Lou, F. Wang, S. Zhu, B. Xia, L. Zhang, Surface and Coatings Technology, 63 (1994) 105.
[31] W. Xu, X. Song, N. Lu, C. Huang , Acta Materialia 58, (2010) 396.
[32] C. X. wang, C. W. Yang, Materials Science and Engineering R 49 (2005) 157.
[33] P. Keblinski, S.R. Phillpot, D. Wolf, H. Gleiter, Physics Letters A 226 (1997) 205.
[34] G. Chen, H. Lou, NanoStructured Materials, 11 (1999) 637
[35] G. Chen, H. Lou, Surface and Coatings Technology 123 (2000) 92.
[36] G. Chen, H. Lou, Corrosion Science 42 (2000) 1185.
[37] R. K. Singh Raman, R. K. Gupta, Corrosion science 51 (2009) 316.
[38] R. K. Gupta, R. K. Singh Raman, C. C. Koch, Journal of Materials Science 45 (2010) 4884.
[39] R. K. Singh Raman, R. K. Gupta ,C. C. Koch, Philosophical Magazine 90 (2010) 4884.
[40] U. Köster, L. Jastrow, Materials Science and Engineering A 449-451 (2007) 57.
[41] U. Köster, D. Zander, Triwikantoro, A. Rüdiger, L. Jastrow, Scripta Materialia 44 (2001) 1649.
[42] Triwikantoro, D. Toma, M. Meuris, U. Köster , Journal of Non-Crystalline Solids 250-252 (1999) 719.
[43] K. Mondal, U.K. Chatterjee, B.S. Murty, Journal of Non-Crystalline Solids, 334-335 (2004) 544.
[44] P. Kofstad, High Temperature Corrosion, Elsevier Applied Scicnece, New York (1988).
[45] D.J. Young, *High Temperature Oxidation and Corrosion of Metals,* Elsevier, Amsterdam (2008).

[46] M.K. Hossain, Corrosion Science 19 (1979) 1031.
[47] X. Peng, J. Yan, Y. Zhou, F. Wang, Acta Materialia 53 (2005) 5079.
[48] M.D. Merz, Metallurgical and Materials Transactions A, 10 (1979) 71.
[49] F. Wang, H. Lou, W. Wu, Vacuum, 43 (1992) 752.
[50] X. Zhu, R. Birringer, U. Herr, H. Gleiter, Phys. Rev. B 35 (1987) 9085.
[51] L. D. Bianco, A. Hernando, E. Bonetti, E. Navarro, Phys. Rev. B 56 (1997) 8894.
[52] T. Haubold, R. Birringer, B. Lengeler, H. Gleiter, Phys. Lett. A 135 (1989) 461.
[53] T. Haubold, W. Krauss, H. Gleiter, Philos. Mag. Lett 63 (1991) 245.
[54] A. DiCicco, M. Berrettoni, S. Stizza, E. Bonetti, G. Cocco Phys. Rev. B 12 (1994) 386.
[55] A.D. Cicco, M. Berrettoni, S. Stizza, E. Bonetti, Physica B 208-209 (1995) 547.
[56] R. K. Islamgaliev, R. Kuzel, E.D. Obraztsova, J. Burianek, F. Chmelik, R.Z. Valiev, Mater Sci and Engineering A 249 (1998) 152.
[57] H. E. Schaefer, R. Wurschum, R. Birringer, H. Gleiter, Phys. Rev. B 38 (1988) 9545.
[58] J. Wang, D. Wolf, S. R. Phillippot, H. Gleiter, Philos. Mag. A 73 (1996) 517
[59] D. Chen, Mater Sci and Engineering A 190 (1995) 193.
[60] D. Chen, Materials Letters 21 (1994) 405.
[61] R.E. Read-Hill, Physical Metallurgy Principles, Litton Educational Publishing, Inc. 1973
[62] T. Mutschele, R. Kirchheim, Scripta materialia 21 (1987) 1101.
[63] R. Kirchheim, T. Mutschele, W. Kieninger, H. Gleiter, R. Birringer, T. D. Koble, Mater Sci and Engineering A 99 (1988) 457.
[64] H. J. Fecht, NanoStructured Materials 1 (1992) 125.
[65] H. J. Fecht, Phys. Rev. Lett 65 (1990) 610.
[66] H. J. Fecht, E. Hellstern, Z. Fu, W.L Johnson, Adv. Powd. Met 2 (1989) 111.
[67] H. J. Fecht, Materials Science and Engineering A 179-180 (1994) 491.
[68] E. Hellstern E, H. J. Fecht, Z. Fu Z, W. L. Johnson, J. appl. Phys 65 (1989) 305.
[69] G. Wallner, E. Jorra, H. Franz, J. Peisl, R. Birringer, H. Gleiter, T. Haubold. W. Petry, MRS Symp. Proc 132 (1989) 149.
[70] J. Horvath, R. Birringer, H. Gleiter, Solid State Comm 62 (1987) 319.
[71] R. Birringer, P. Zimmer, Acta Materialia 57 (2009) 1703.
[72] A. Seeger, G. S. Acta metall. 1959;7:495.
[73] D. A. Smith, V. Vitek, R.C. Pond, Acta metal 24 (1977) 475.
[74] H. J.Fros, M. F. Ashby, F. Spaepen, Scripta matallurgica 14 (1980) 1051.
[75] D. Wolf, Philos. Mag. B 69 (1989) 667.
[76] Y. H. Zhao, H. W. Sheng, K. Lu, Acta Mater 49 (2001) 265.
[77] A. Lucci, G. Riontino, M. C. Tabasso, V. G. Tamanini, Acta Metall. 26 (1978) 615.
[78] B. Gu¨ nther, A. Kupmann, H. D. Kunze, 27 (1992) 833.
[79] L. Lu, M. L. Sui, K. Lu, Acta Mater. 49 (2001) 4127
[80] A. Tscho¨ pe, R. Birringer, Acta Metall. 41 (1993) 2791.
[81] A.T. Dinsdale, Calphad 15 (1991) 317.
[82] Y. K. Huang, A. A. Menovsky, F. R. Boer, NanoStructured Materials Materials 2 (1992) 587.
[83] R. Würschum, S. Herth, U. Brossmann, Advanced Engineering Materials, 5 (2003) 365
[84] I.V. Belova, G.E. Murch, J. of physics and chemistry of solids 64 (2003) 873
[85] H. Mehrer, Diffusion in Nanocrystalline Materials, springer Series in Solid-State Sciences (2007)
[86] G. Shujiang , F. Wang, S. Zhang, Surface and Coatings Technology 167 (2003) 212.
[87] H. Y. Tong, F. G. Shi , E. J. Lavernia, Scripta Metall Mater 32 (1995) 511.
[88] W. Gao, Z. Liu, Z. Li, Advanced Materials, 13 (2001) 1001.

[89] G. Chen, H. Lou, Scripta materialia 43 (2000) 119.
[90] G. F. Chen, H. Y. Lou, Materials Letters 45 (2000) 286.
[91] L. Liu, F. Wang, Materials Letters 62 (2008) 4081.
[92] F. Wang, X. Tian, Q. Li, L. Li, X. Peng, Thin and Solid Films 516 (2008) 5740.
[93] G. Cao, L. Geng, Z. Zheng, M. Naka, Intermetallics 15 (2007) 1672.
[94] H. Z. Zheng, S. Q. Lu, X. J. Dong, D. L. Quyang, Materials Science and Engineering A 496 (2008) 524.
[95] J. E. May, S. E. Kuri, P. A. P Nascente, Mater Sci and Engineering A 428 (2006) 290.
[96] J. E. May, G. Galerie, T. P. Busquim, S. E. Kuri, Materials and Corrosion 58 (2008) 87
[97] Y. Niu, Z. Q. Cao, F. G, Farne, G. Randi, C. L. Wang, Corrosion Science 45 (2003) 1125.
[98] S. Leistikow, I. Wolf and H.J. Grabke, Werkst. Korros., 38 (1987) 556.
[99] Y. Shida, N. Ohtsuka, J. Muriama, N. Fujino and H. Fujikawa, Proc. JIMS-3, High Temp. Corros., Trans. Jap. Inst. Met., 1983, 631.
[100] H. J. Grabke, E. M. Muller-Lorenz, S. Strauss, E. Pippel, J. Woltersdorf, Oxidation of Metals, 50 (1998) 314.
[101] M. Martin, N. Lakshmi, U. Koops, H.-I. Yoo, Z. Phys. Chem. 221 (2007) 1449.
[102] G. C. Wood, Oxidation of Metals, 2 (1970)
[103] V. R. Howes, Corrosion Science, 7 (1967) 735
[104] R. Lobb, H. Evans, Metal Science 26 (1981).
[105] M. Schutz, Oxidation of Metals, 44, (1995), 29
[106] H. Hindam, D. P. Whittle, Oxidaiton of Metals, 18 (1982) 245
[107] X. Y. Zhang, M.H. Shi, C. Li, N. F. Liu, Y.M. Wei, Mater Sci and Engineering A 448 (2007) 259.
[108] O. E. Kedim, S. Paris, C. Phigini, F. Bernard, E. Gaffet, Z. A. Munir, Mater Sci and Engineering A 369 (2004) 49.
[109] W. W. Smeltzer, D. P. Whittle, J. Electrochem. Soc. 125 (1978) 1116.
[110] C. J. Wagner, J. Electrochem. Soc. 99 (1952) 369.
[111] C. J. Wagner, J. Electrochem. Soc. 103 (1956) 571.
[112] C. J. Wagner, J. Electrochem. Soc. 63 (1959) 772.
[113] X. Peng, F. Wang, Oxidation – resistant nanocrystalline coatings, Development in high-temperature corrosion and protection of materials, W. Gao, Z. Li (Eds.), CRC Press, 2008.
[114] I. Kaur, W. Gust, L. Kozma, Handbook of grain and interphase boundary diffusion data: Stuttgart: Zigler Press, (1989), 523.
[115] A. W. Bowen, G .M. Leak, Metall. Trans. 1 (1970), 1695.
[116] R. Gupta, R. K. Singh Raman, C. C. Koch, Materials Science and Engineering A 494 (2008) 253.
[117] J. G. Goedjen, D. A. Shores, Oxidation of Metals 37 (1992) 125
[118] F. Wang, Oxidation of Metals 47 (1997) 247.
[119] F.J. Humphreys, M. Hatherly, Recrystallization and related annealing phenomena, Pergamon, 1996.
[120] R. K. Gupta, R. K. Singh Raman, C. C. Koch, TMS 2008 Annual Meeting 1 (2008) 151-157.
[121] K.S. Darling, R.N. Chan, P.Z. Wong, J. E. Semones, R.O. Scattergood, C.C. Koch, Scripta Materlialia 59 (2008) 530.
[122] R. K. Gupta, K. S. Darling, R. K. Singh Raman, K. R. Ravi, C. C. Koch, B. S. Murty, R. O. Scattergood, Journal of Materials Science (2011) DOI 10.1007/s10853-011-5986-6.

10

Corrosion Behavior of Stainless Steels Modified by Cerium Oxides Layers

Emilia Stoyanova and Dimitar Stoychev
Institute of Physical Chemistry, Bulgarian Academy of Sciences
Bulgaria

1. Introduction

The modifying of the surface, which involves altering only the surface layers of a material, is becoming increasingly important with the aim to enhance the corrosion resistance of many kinds of materials. The advantage of this approach lies in the fact that the natural physical and mechanical properties of the material are retained, while at the same time the corrosion resistance is increased. It is well known that electroplated zinc coating is employed as active galvanic protection for low and middle-content alloyed steels (Almeida et al.,1998; Hagans & Hass, 1994; Kudryavtsev, 1979; Lainer, 1984; Zaki, 1988). However, zinc is a highly reactive element, and therefore high corrosion rates of this coating are observed in cases of indoor and outdoor exposures. For this reason a post-treatment is needed to increase the lifetime of zinc coatings. This kind of treatment is applied in the current industrial practice to prolong the lifetime of zinc coatings and the steel substrates, respectively. This treatment consists of immersion in a chemical bath, which forms a conversion layer over the plated zinc. The so formed layer is a dielectric passive film with high corrosion resistance and it is also a better surface for paint adherence (Zaki, 1988). The main problem with the traditionally applied post-treatment procedures is the presence of Cr^{6+} salts that are considered to be carcinogenic substances, which are known to be very harmful to human health and environment (Schafer & Stock, 2005) and whose use is forbidden by European regulations (Hagans & Hass, 1994).

Molybdates, tungstates, permanganates and vanadates, including chromium-like components, were the first chemical elements to be tested as hexavalent chromium substitutes (Almeida et al., 1998a, 1998b; Korobov et al, 1998; Schafer & Stock, 2005; Wilcox & Gabe, 1987; Wilcox et al., 1988). Recently many alternative coatings have been developed, based on zirconium and titanium salts (Barbucci et al., 1998; Hinton, 1991), cobalt salts (Barbucci et al., 1998; Gonzalez et al., 2001) and organic conductive polymers (Gonzalez et al., 2001; Hosseini et al., 2006). The use of salts of rare-earth metals as the main component in the electrolytes, developed for the formation of cerium, lanthanum and other oxide protective films is also a very promising alternative to the chromate films and it is one of the advanced contemporary methods for corrosion protection of metals and alloys (Bethencourt et al., 1998; Crossland et al., 1998; Davenport et al., 1991; Fahrenholtz et al., 2002; Forsyth et al., 2002; Hinton, 1983, 1992; Hosseini et al., 2007; Liu & Li , 2000; Montemor et al., 2002; Montemor & Ferreira, 2008; Pardo et al., 2006; Wang et al., 1997). However, some aspects of the preparation and of the corrosion behavior of these coatings are not quite clear yet and their practical utilization is still uncertain. In order to find an attractive alternative to Cr^{6+} conversion coating, several treatment procedures that should manifest both efficient anti-corrosive behavior as well as an optimal benefit/cost ratio, and mainly insignificant environmental impact, have yet to be developed. It has been found out that cerium species can be successfully applied to protect zinc from corrosion (Aramaki, 2001, 2002; Arenas et al., 2003, 2004; Ferreira et al., 2004; Otero et al., 1996, 1998; Wang et al., 2004; Virtanen et al., 1997), aluminum and aluminum containing alloys (Aldykiewicz et al., 1995; Amelinckx et al., 2006; Arnott et al., 1989; Davenport et al., 1991; Mansfeld et al., 1989,1991, 1995; Pardo et al., 2006; Zheludkevich et al., 2006; Di Maggio et al., 1997; Lukanova et al.. 2008), stainless steels (Breslin et al., 1997; Lu & Ives, 1993, 1995), magnesium containing alloys (Arenas et al., 2002; Liu et al., 2001) even SiC/Al metal matrix composites. All these can be used to reduce the rate of general corrosion, pitting and crevice corrosion as well as stress corrosion (Breslin et al., 1997; Lu & Ives, 1993, 1995). The oxide films of rare-earth elements and refractory compounds can be formed mostly by means of chemical or electrochemical methods (Amelinckx et al., 2006a, 2006b; Avramova et al., 2005; Balasubramanian et al., 1999; Di Maggio et al., 1997; Marinova et al., 2006; Montemor et al., 2001, 2002; Schmidt et al., 1997; Stefanov et al., 2000a, 2000b, 2004a, 2004b; Stoychev et al., 2000, 2003, 2004; Tsanev et al., 2008; Tyuliev et al., 2002; Valov et al., 2002; Zheludkevich et al., 2005). It is supposed that cerium oxide/ hydroxide formation is the main reason for the corrosion protection property of cerium compounds. In spite of the growing number of investigations during the last years, focused on the mechanisms via which the oxides of rare-earth metals (mainly cerium oxides) lead to improvement of the corrosion stability of the systems "oxide(s)/protected metal", still a series of issues remain problematic. The first hypotheses in this respect have been put forward by Hinton (Hinton & Wilson, 1989; Hinton, 1992). He supposed in his early works that the cathodic reactions (reduction of oxygen and evolution of hydrogen) lead to alkalization of the near-to-the-electrode layer, which in its turn results in precipitation of the oxide of the rare-earth element, respectively in formation of protective film on the electrode surface.

The modern technologies for surface treatment, aimed at modifying the surface composition and structure of metals and alloys, including stainless steels, are becoming more and more important instruments for improving their stability to corrosion and for attributing the desired outside appearance and/or functional properties. Wang and coworkers (Wang et al., 2004) have studied the corrosion resistance of stainless steel SS304 after immersion treatment in electrolytes, containing Ce^{3+} ions, $KMnO_4$ and sulfuric acid. The obtained experimental results prove the considerable increase in the corrosion stability of steel in 3.5% NaCl solution. The corrosion potential of the steel, treated by immersion, has more positive values than that of the non-treated steel, while the potential of pitting formation is also shifted in the positive direction, which is the criterion for weakening the tendency of the studied steel to undergo pitting formation. It has also been observed that the values of the current of complete passivation are decreased by one order of magnitude. As far as the cathodic reaction is concerned, the cerium conversion coatings blocked the matrix steel, which caused the reduction of the oxygen and protons to take place at a higher over-potential and the cathodic reaction was inhibited. The analysis of the chemical state of cerium in the conversion film indicated that the prevailing amount of cerium is in trivalent state. Aldykiewicz and coworkers (Aldykiewicz et al, 1996) have investigated the influence of cerium oxide, deposited on aluminum alloys and they ascertained that the trivalent cerium is oxidized to tetravalent by the oxygen dissolved in the electrolyte, which leads to precipitation/formation of non-soluble CeO_2 on the cathodic sections of the electrode surface. Montemor and coworkers (Montemor et al., 2001, 2002) studied the effect of the composition of the electrolytes (based on $Ce(NO_3)_3$ and the regime of preparing conversion layers on galvanized (zinc coated) steel. The increase in the thickness of the cerium conversion films in the process of formation leads to their enrichment in Ce^{4+}. According to the same authors the conversion films, formed in $La(NO_3)_3$, are more efficient in view of anticorrosion protection, compared to those formed in electrolytes, containing $Ce(NO_3)_3$ and $Y(NO_3)_3$ (Montemor et al, 2002). The mechanism, involved in such a process of reducing the corrosion rate of the substrate, may be related to precipitation of cerium oxides and hydroxides in the vicinity of the anodic areas. These precipitates reduce the cathodic activity and hinder the transfer of electrons from the anodic to the cathodic spots.

The mechanism of zinc corrosion inhibition, when zinc is treated in solutions of $Ce(NO)_3$, has been studied by Aramaki (Aramaki, 2001a, 2001b, 2002a; 2002b). He established the formation of hydrated or hydroxylated Ce-rich layer. This process, in its turn, leads to the formation of Ce_2O_3 on the electrode/the protected surface, respectively to inhibition of the cathodic reactions of the corrosion process in solutions of NaCl. The work by Lu and Ives (Lu & Ives, 1995) has been extended further to study the effect of cerium salt solution treatment. Rotating disk assemblies were employed to monitor the cathodic electrode process and its inhibition by cerium salt treatment on austenitic stainless steels in a solution simulating sea water. The reduction of oxygen and hydrogen cations on both kinds of non-treated steels has been shown to be controlled by mass transfer processes in the solution. Cerium treatment effectively inhibits the cathodic reduction of oxygen, which is controlled primarily by charge transfer on the electrode. The over-potential for cathodic reduction of hydrogen cations is increased after the cerium treatment and the electrode reaction is controlled both by the mass transfer process in solution and by the charge transfer on the electrode. As a result of inhibition of the electrode processes, cerium improves the localized corrosion resistance, and in particular the crevice corrosion resistance, of stainless steels.

The electrochemical behavior of stainless steels - SS304 and 316L, following various cerium and cerium/molybdenum prereatment steps, was studied aiming at gaining more information on the process, by which cerium and molybdenum can modify the properties of passive film formed on stainless steels (Breslin et al, 1997). The coatings were analyzed by electrochemical impedance spectroscopy and X-ray photoelectron spectroscopy in order to identify the cerium species, which play the main role in the promotion of the passivation behavior. The pre-treatment step, denoted as $Ce(CH_3CO_2)_3$ and $CeCl_3$, involved immersion treatment of the electrodes in the $Ce(CH_3CO_2)_3$ solution for 1 h and then in the $CeCl_3$ solution for one additional hour at approximately 92°C. Regardless of the nature of the cerium salt, no changes in the rate of the cathodic reduction reaction could be observed. The increase in the corrosion potential (E_{corr}). is mainly due to the decrease in the passive current density, suggesting that treatment in cerium solutions does not affect the rate of the cathodic reaction, but rather reduces the rate of the passive film dissolution. However a significant lowering of the oxygen reduction current could be observed, following the electrodeposition of small amount of cerium onto the electrode surface. Thus, it appears that efficient formation of cerium hydroxide/oxide does not occur upon immersion of electrodes at elevated temperatures in cerium solutions. It was possible to observe a yellow colored film indicative of cerium in the 4^+ oxidation state on the surface of the stainless steel, following a 24-h immersion time interval in the $Ce(NO)_3$ solution at room temperature. Breslin and coworkers (Breslin et al, 1997) proved that the treatment of SS304 in cerium-salt solutions gave rise to an increase in the value of the pitting potential E_{pit}, with the greatest increase resulting from immersion in $CeCl_3$ at 90-95°C for 30 min, followed by immersion in $Ce(NO)_3$ solution at 90-95°C for additional 60 min time interval. The enhanced resistance to the onset of pitting, according to these authors, could be due to the dissolution of surface MnS inclusions during the immersion in the chloride-containing solution and possibly chromium enrichment of the passive film during treatment in the sodium nitrate solution, which is highly oxidizing. The presence of cerium in the solution seemed to have only a minor effect on E_{pit}. The survey of the various mechanisms, proposed in the current literature, indicates that the role of rare-earth elements as inhibitors of corrosion and as protective coatings is not completely elucidated. It is accepted that their presence leads to improvement of the corrosion stability of metals and alloys and therefore they are a promising alternative, in conformity with the requirements for protection of the environment prohibiting the conventional Cr^{6+} conversion treatment.

At the same time it is known that thin films of Ce_2O_3-CeO_2, have also an important functional designation for the manufacture of catalytic converters, where ceria is widely used in such kind of catalytic processes as a reducible oxide support material in emission control catalysis for the purification of exhaust gases from various combustion systems (Trovarelli, 1996). In the so called "three-way automotive catalysis", for example, the reducibility of ceria contributes to oxygen storage/release capability, which plays an important role in the oxidation of CO and hydrocarbons catalyzed on the surface of precious metal particles (Bunluesin et al, 1997). It is because of their specific interactions with oxygen that the cerium oxides are included in the support layers (Al_2O_3, ZrO_2, etc.) of the proper catalytically active components of the converters (noble metals like Pt, Ro, Pd and others) and they participate directly in the decontamination of exhaust gases (reduction of NO_x, oxidation of CO and hydrocarbons, etc.) originating from internal combustion engines (Mcnamara, 2000). In this connection it is important to point out that during the process of operation the main construction elements of the catalytic converters, which are made of stainless steel (Lox et al, 1995; Nonnenmann, 1989) (for example steel OC 404), are subjected simultaneously to over-heating and at the same time to the aggressive action of the nitrogen oxides, being liberated in the course of the processes of combustion, of sulfur oxides, of water vapor and incompletely oxidized hydrocarbons etc. (respectively resulting in formation of HNO_3, H_2SO_4 etc). In this respect and in the light of the data available in the literature about the protective action of the cerium oxides and hydroxides, it is essential to know what is the intimate mechanism of their anti-corrosion action and to what extent they could contribute, in particular, to the prolongation of the exploitation life-time of the catalytic converters, made of stainless steel.

Our studies on the protective effect of mixed Ce_2O_3-CeO_2 films electrochemically deposited on stainless steel OC 404 (SS) in model media of 0.1N HNO_3 and 0.1N H_2SO_4 (Nikolova et al., 2006a, 2006b, 2008; Stoyanova et al., 2006a, 2006b, 2010), have shown that these films in their nature are in fact cathodic coatings. The influence of the change in the concentration of

Ce in the oxide films has been studied in regard to the corrosion potential of the steel in the same corrosion medium. Thereupon it was found out that the increase in the surface concentration of Ce in the oxide films results in a gradual shift of the corrosion potential of steel in the positive direction - from the zone characteristic of anodic dissolution to the zone of deep passivity - defined by the anodic potentiodynamic curve. Moreover it has been proved that there occurs a cathodic reaction of reduction of the electrochemically active CeO_2 - one of the components of the electrodeposited mixed Ce_2O_3-CeO_2 film.

The present work discusses a hypothesis, aimed at elucidation of the question: how the change in the surface concentration of Ce in the mixed Ce_2O_3-CeO_2 oxide film electrodeposited on OC 404 steel influences the processes of anodic passivity of the studied steel, respectively the values of the potentials of complete passivation and the potentials of pitting formation, as well as the current density in passive state, determining the corrosion behavior of the steel under consideration. As far as the oxide of Ce^{3+} (i.e. Ce_2O_3) is chemically unstable and it is dissolved in sulfuric acid medium (Achmetov, 1988) the investigations carried out in ref.(Guergova et al, 2011) established an effective inhibitory action of the cerium ions (Ce^{3+}, Ce^{4+}), passing over from the system Ce_2O_3-CeO_2/SS into the corrosion medium. A possible inhibitory interaction has been supposed to occur on the surface of the steel.

2. Experimental

2.1 Specimen preparation and structure characterization

The stainless steel (SS) samples (SS type OC 404 containing 20% Cr, 5.0% Al, 0.02% C, the rest being Fe) were 10x10 mm plates of steel foil, 50 μm thick. The deposition of the films was carried out in a working electrolyte consisting of absolute ethanol saturated with 2.3 M LiCl and 0.3 M $CeCl_3x7H_2O$ salts. The cathodic deposition was performed in a galvanostatic regime at current density of 0.1mA.cm^{-2}. The deposition time interval was 60 min. Platinum coated titanium mesh was used as counter electrode (anode). It was situated symmetrically around the working electrode and its surface was chosen specially to ensure a low anode polarization, which hindered Cl^- oxidation. Because of the relatively low equivalent conductance of the working electrolyte (χ - 1.10^{-2} Ω^{-1} cm^{-1}), it becomes warmed up during the electrolysis. For this reason, the electrochemical measurements were carried out in a specially constructed electrochemical cell. The cell was kept at a constant temperature of 5-7°C by circulation of cooling water. The obtained CeO_2-Ce_2O_3 coatings had a thickness of 1μm (Avramova et al., 2005; Stefanov et al., 2004). The system CeO_2-Ce_2O_3/SS was investigated prior to and after thermal treatment (t.t.) at 450°C for 2 h in air. The model aggressive solution (0.1N H_2SO_4) was prepared by dilution of analytical grade 98% H_2SO_4 ("Merck") with distilled water. In order to evaluate the inhibitory effect of lanthanide salt, variable concentrations of $Ce(SO_4)_2.4H_2O$ from 0.1 to 1500 ppm were added to 0.1N H_2SO_4.

The morphology and structure of the samples was examined by scanning electron microscopy using a JEOL JSM 6390 electron microscope (Japan) equipped with ultrahigh resolution scanning system (ASID-3D) in regimes of secondary electron image (SEI) and back scattered electrons (BEC) image. The pressure was of the order of 10^{-4} Pa.

2.2 Chemical characterization

The chemical composition and state of the surfaces being formed was studied using X-ray photoelectron spectroscopy (XPS). The XPS studies were performed on an Escalab MkII system (England) with Al K_{α} radiation ($h\nu$ = 1486.6 eV) and total instrumental resolution of ~ 1 eV. The pressure in the chamber was 10^{-8} Pa. The binding energy (BE) was referred to the C1s line (of the adventitious carbon) at 285.0 eV. The element concentrations were evaluated from the integrated peak areas after Shirley-type of linear background subtraction using theoretical Scofield`s photoionization cross-sections.

2.3 Electrochemical (corrosion) characterization

The electrochemical behaviour of the samples (plates 10 x 10 x 0.05mm) was studied in a standard three-electrode thermostated cell (100 ml volume). The model corrosion medium was 0.1 N H_2SO_4 ("p.a." Merck) after deaeration with additionally purified argon at 25°C. A counter electrode, representing a platinum plate (10x10x0.6mm), and a mercury/mercurous sulfate reference electrode (MSE), ($E_{Hg/Hg2SO4}$= +0.642V versus SHE) were used. All potentials in the text are related to MSE. The anodic and cathodic polarization curves were obtained using a 273 EG&G potentiostat/galvanostat and computer-aided processing of the results according to an "Echem" programme, with a potential sweeping rate of 10 mV/s within a potential range from -1.500 to +1.500V. The recording of the potentiodynamic curves was carried out starting from the stationary corrosion potential (E_{st}), measured in the absence of external current (at open circuit) in the anode and cathode directions. We used a separate electrode for each recorded curve. The stationary corrosion potential of the samples under investigation was determined by direct measurement of the function "E_{st}–time" at open circuit (with respect to the same reference electrode) after immersing the samples in 0.1 N H_2SO_4 in the absence and in the presence of Ce^{4+}. The E_{st} was established after a sufficiently long time interval - from few minutes to several decades of minutes until the moment, when the corrosion potential change did not exceed 1-3mV for 5 min.

3. Theoretical background

It is known that one of the basic approaches to promote the corrosion resistance of alloys is to enhance their passivity. It has been established that upon introducing a new component with higher inclination to passive state into the metal or into the alloy, it transfers this property to the main metal or to the alloy. The formation of a system, which is more stable to corrosion (i.e more easily passivated system) could be achieved through promoting the effectiveness of the cathodic process. At first glance this is a self-contradictory statement, however it can be easily explained in the following way. It can be seen from the corrosion diagram, represented in Fig. 1, that if the anodic potentiodynamic curve of the alloy remains one and the same, the rate of corrosion can be changed considerably at the expense of the changing effectiveness of the occurring cathodic process. It is important to note that in the case of non-passivating systems (i.e. systems characterized by anodic behavior until point B of the anodic curve) the corrosion rate is always increasing with the increase in the cathodic efficiency (for example the transition from cathodic curve K_1 to K_2 in Fig.1). In the cases when the corrosion systems are passivating ones and the anodic polarization curve is not a monotonous dependence between the potential and the current and when there exists a region of potentials somewhere between the passivation potential (E_p) and the potential of transpassivity (E_t) (or the potential of pitting formation (E_{pit})), in which the increase of the effectiveness of the cathodic process is leading not to enhancement but rather to abatement of the corrosion rate (for example in the course of the transition from one cathodic process K_2 to another one K_3) one can observe a system more stable to corrosion (easily passivating system). Obviously in this case of minimal corrosion currents there will appear a cross-point between the anodic and the cathodic curves of the corrosion diagram within the zone of stable passive state. Under these conditions it is quite possible that a smaller corrosion current is corresponding to a more efficient cathodic process in comparison to the system displaying a lower cathodic efficiency (if we exclude the conditions where the potential of the system is reaching the value of the potential of transpassivation and the potential of pitting formation - K_4). Therefore one can conclude that during the occurring of an efficient cathodic process the system will pass over spontaneously to a stable passive state and it will be corroding at a much lower rate, corresponding to the current of complete passivation. The stationary corrosion potential of such a system will be more positive than the potential of complete passivation (E_{cp}) and at the same time more negative than the potential of break through the passive film and the potential of transpassivation. In this way the rate of corrosion can be reduced to a considerable extent by the correct use of the phenomenon "passivation".

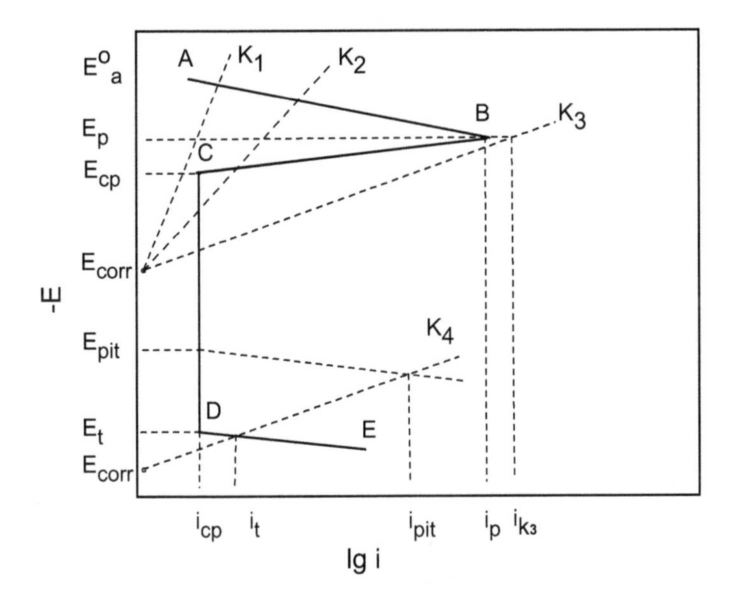

Fig. 1. Schematic polarization diagram explaining the action of the effective cathodic coatings on the steel corrosion: i_p, i_{cp}., i_{pit}, i_t- respectively currents of initial passivation, complete passivation, pitting formation and corrosion in transpassive state.

So, in order to create a system stable to corrosion and to decrease the rate of corrosion, it is necessary to find a way to promote the cathodic effectiveness (for example as it is in this specific case of investigations, carried out by us, to modify the steel surface with CeO_2-Ce_2O_3

oxides as cathodic coating). This theoretical approach has been used in this work, with a view to explain the obtained results, in view of stabilization and restoration of the passive state of OC 404 steel as a consequence of electrochemical formation of the surface CeO_2-Ce_2O_3 layers. We agree with the assumption that for a similar type of modified systems there exists only one passive state of the system (even without applying any external anodic current). Or, in other words, the result is a spontaneously self-passivating system and if in some way it is being led away from its passive state (for example in the case of cathodic polarization or by exerting a mechanical impact), after the termination of the external effect, again the system will pass over to its passive state.

As a matter of fact the increase in the effectiveness of the cathodic process is connected with the character of the cathodic process. The dilemma is whether the promotion of the cathodic efficiency is due only to the process of hydrogen depolarization, on the cerium oxide cathodic coating (in case of steel corrosion in acidic medium) or it is possible that there exists another cathodic process, owing to the oxidative properties of the electrochemically active CeO_2.

Figure 2 represents an example of a corrosion diagram, illustrating the changes in the behavior of the corrosion system upon increasing the surface concentration of the effective cathodic coating (for example in the case of modifying the steel surface with cerium oxides). It follows from the diagram that the shift in the corrosion potential of the system is associated with the increase in the concentration of the cathodic depolarizer (the cerium oxides), which will facilitate the transition from active state into passive state of the system, under the conditions of disturbed passivity. At concentration of the cerium oxides, corresponding to the cathodic curve C_1 (Fig. 2), the rate of dissolution of the steel will

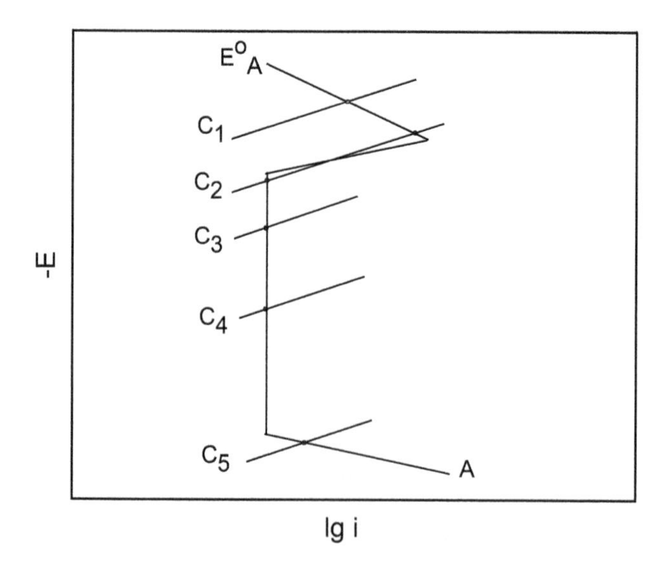

Fig. 2. Schematic polarization diagram illustraiting the influence of the effective cathodic coating on the steel corrosion, respective cathodic curves : c_1, c_2, c_3, c_4, c_5 in case of increasing the surface concentration of cerium oxides.

become commensurable with that of the pure steel. At concentration of the cerium oxides, corresponding to the cathodic curve C_2 it is possible to establish two corrosion potentials of the steel, located respectively in the passive and active regions of dissolution of the anodic potentiodynamic curve. At the higher concentrations of the cerium oxides, represented in the corrosion diagram by means of the respective cathodic potentiodynamic curves of the system cerium oxides/steel - C_3 and C_4, the steel is characterized by a stable passive state and under these conditions the rate of corrosion of the steel is no longer dependent on the surface concentration of the cerium oxides. The influence of the further increase in the concentration of the cerium oxides can be illustrated through the corrosion diagram by means of the cathodic curve C_5 – the rate of steel corrosion will grow up due to the fact that the stationary corrosion potential of the steel will be shifted to the region of transpassivity (Tomashov & Chernova, 1963; 1993). The experimental results, obtained by us, confirm these theoretical concepts.

4. Experimental results and discussion

4.1 Potentiodynamic polarization studies

Figure 3 shows a typical experimentally obtained corrosion diagram E-lg i, illustrating the kinetics of the cathodic and anodic processes on the studied steel in the absence of electrochemically deposited cerium oxides film (the curves 2) and after the deposition of thin oxide films with different surface concentrations of Ce (curves 3-5).

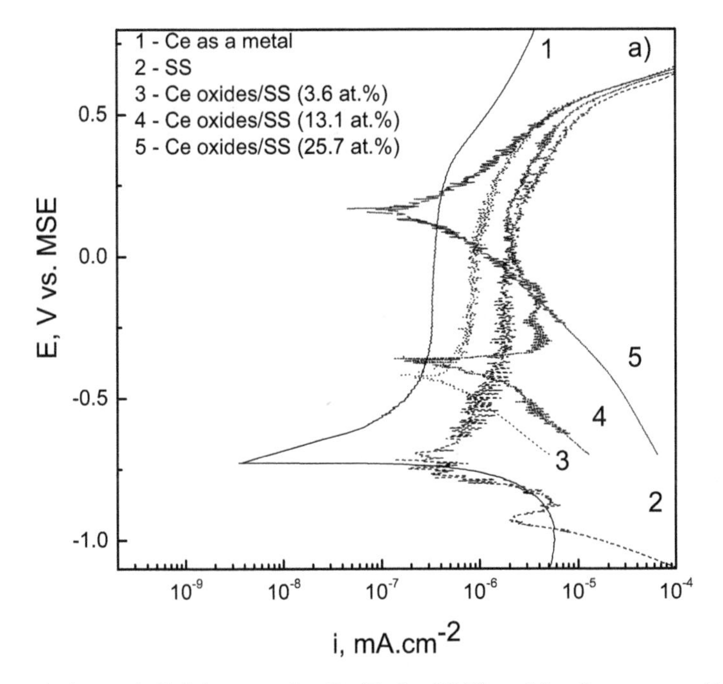

Fig. 3a. Potentiodynamic E–lgi curves for Ce (1), for SS (2) and for the systems CeO_2-Ce_2O_3/SS containing different concentrations of Ce (3-5), obtained in 0.1 N H_2SO_4.

The plotting of the model polarization curves enabled us to follow the changes in E_{corr}, estimated on the basis of the cross-point of the anodic polarization curve of the studied steel (SS) with the cathodic polarization curves of the studied systems, having different surface concentrations of Ce (ranging from 4% up to 30%). Such an approach of considering the partial polarization curves allows us to make the connection between the occurring anodic and cathodic corrosion processes, localizing the cathodic reaction on the rich in cerium zones on the electrode surface. For the sake of comparison the figure represents also the anodic and cathodic potentiodynamic curves of the metallic Ce (curves1). It follows from Fig. 3a that with the increase in the surface concentration of Ce (curves 3-5) the values of the corrosion potential E_{corr} are shifted strongly in the positive direction - from ~ -0.900V (for the non-coated with cerium oxides steel) up to ~ + 0.160 V. Obviously for the non-coated with CeO_2-Ce_2O_3 steel surface the cathodic depolarizing reaction, occurring at voltage ~ -0.900V, is connected with the evolution of hydrogen. As far as the respective cathodic branches of the potentiodynamic curves of the system Ce_2O_3-CeO_2/ SS are concerned, they are also shifted strongly in positive direction in the zone of passivity of the investigated steel.

The slope of the cathodic Taffel's curves grows up considerably from 0.250V up to 0.319 V with the increase in the surface concentration of cerium (Table 1). The change in the slope of these curves confirms the supposition about the occurring of another cathodic reaction, different from that of hydrogen evolution on the heterogeneous electrode surface.

	SS			
Samples	**E_{corr}, V**	**i_{corr}, mA.cm^{-2}**	**$i_{c.p}$, mA.cm^{-2}**	**b, V**
SS non-covered with cerium oxides	-0.890	2.24 x 10^{-7}	1.89 x 10^{-6}	0.107
SS covered with 3.6 at. % cerium oxides	- 0.432	2.85 x 10^{-7}	9.10 x 10^{-7}	0.250
SS covered with 13.1 at. % cerium oxides	- 0.371	2.93 x 10^{-7}	3.63 x 10^{-6}	0.276
SS covered with 25.7 at. % cerium oxides	+0.161	1.36 x 10^{-7}	-	0.319
	$SS_{t.t.}$			
Samples	**E_{corr}, V**	**i_{corr}, mA.cm^{-2}**	**$i_{c.p.}$, mA.cm^{-2}**	**b, V**
SS non-covered with cerium oxides	- 0.975	1.74 x 10^{-7}	3.21 x 10^{-6}	0.074
SS covered with 4.2 at. % cerium oxides	- 0.486	7.96 x 10^{-7}	2.96 x 10^{-6}	0.151
SS covered with 20.7 at. % cerium oxides	- 0.269	7.98 x 10^{-8}	1.33 x 10^{-6}	0.176
SS covered with 29.6 at. % cerium oxides	+ 0.090	8.19 x 10^{-7}	-	0.304

Table 1. Electrochemical characteristics of coated steel before and after thermal treatment compared to bare steel.

The strong shifting of E_{corr} of the steel surface, covered with cerium oxides in the positive direction depending on the surface concentration of cerium could also be associated with the occurrence of another cathodic process. The values of the corrosion potential in the presence of Ce are more positive than the Flade potential and more negative than the potential of transpassivity of the steel under consideration. Therefore we can conclude that even at surface concentration of Ce about 4% the corrosion potential of the steel is shifted in positive direction reaching potentials more positive than the potential of complete passivation. The improvement of the corrosion stability of the steel as a result of the action of the effective cathodic coating is expressed in the stabilization of the passive state of the steel. One can conclude from Figure 3a that the steel samples with electrochemically deposited cerium oxide film will corrode under the conditions of passivity. Thereupon with the increase in the surface concentration of Ce from 0 up to 3.6 at % a tendency is observed - a decrease in the currents of complete passivation ($i_{c.p.}$,see Table 1). At 13.1%concentration of the cerium oxides the current of complete passivation is of the same order with that of the non-covered steel, while at 25,7% concentration as a result of the strong shifting of E_{corr} in positive direction and its approaching the values within the zone of potentials of pitting formation and transpassivity the anodic potentiodynamic curve is not characterized by a well expressed region of passivity. The change in the currents of complete passivation depends to a considerable extent on the composition of the passive film on the steel. Therefore for the system Ce_2O_3-CeO_2/SS we can assume that it will pass over spontaneously into a stable passive state and that it will be dissolved at a much lower rate of corrosion, corresponding to the values of the anodic currents in the passive state (Fig. 3a).

If in one way or another the system is artificially taken out of its state of passivity (for example by means of cathodic polarization or by some mechanical impact), after discontinuing the external impact, it will restore again its passive state, i.e. what we obtain is a spontaneously self-passivating system.

A similar effect, expressed to an even greater extent, is also observed with the samples of thermally treated system Ce_2O_3-CeO_2/$SS_{t.t}$ (Fig. 3b), in which case it was established that as a result of disruption of the integrity of the oxide film the passive state of the steel is disturbed (Guergova et al., 2008) and conditions are created to increase the rates of the total and the local corrosion. The presence of electrochemically deposited cerium oxide film (in a way analogous to that for the samples of non-treated thermally system Ce_2O_3-CeO_2/SS) shifts strongly the corrosion potential of the system in positive direction (see curves 3-5). This effect determines the restoration of the passive state of the steel, disturbed as a result of its thermal treatment. Upon increasing the surface concentration of the cerium one can observe not only shifting of the corrosion potential of the samples in the positive direction, but also a tendency of decrease in the currents of complete passivation. An exception in this respect is observed at very high concentrations of the cerium oxides (≥ ~29%). Obviously in these cases the corrosion potential of the system CeO_2-Ce_2O_3/$SS_{t.t}$, which is still in the process of being established, starts approaching the value of the reversible redox potential of the couple Ce^{4+}/Ce^{3+}, whereupon the reaction of oxidation of Ce^{3} to Ce^{4+} is taking place. As a result of this the character of the anodic curve will be changed (Fig. 3b, curve 5) and the determination of the current of complete passivation of the steel based on this curve would be incorrect.

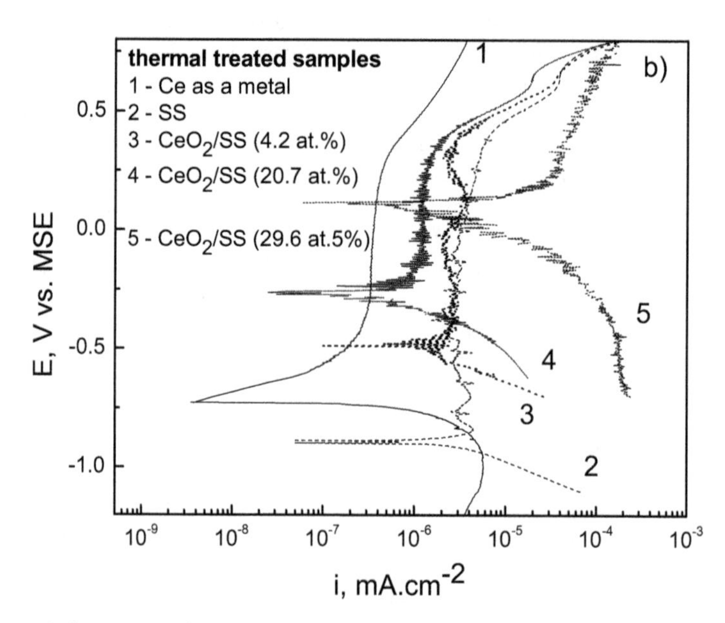

Fig. 3b. Potentiodynamic E–lgi curves for Ce(1), for $SS_{t.t.}$ (2) and for the systems CeO_2-$Ce_2O_3/SS_{t.t.}$ containing different concentrations of Ce (3-5), obtained in 0.1 N H_2SO_4.

4.2 Chronopotentiometric investigation

Fig. 4a and 4b illustrate the altering of the stationary corrosion potentials in the case of open circuit (open circuit potentials) with the SS and $SS_{t.t.}$ samples and with the systems Ce_2O_3-CeO_2/SS and Ce_2O_3-$CeO_2/SS_{t.t.}$ The juxtaposition of the values of the stationary corrosion potentials with the anodic potentiodynamic curves of SS and $SS_{t.t.}$ shows that in the presence of cerium oxide film on the surface of the steel one can observe a strongly manifested tendency to self-assivation. In the cases of non-thermally treated steel its high corrosion resistance and its ability to passivate itself is connected also with the high content of Cr, while the role of the cerium oxides is reduced to promoting the passivation ability and stabilization of its passive state in weakly acidic medium (Stoyanova et al, 2006). In the case of thermally treated steel, however, due to the cracking of the surface passive film, as a result of the thermal treatment its stationary corrosion potential reaches values ($E_{st.}$= - 0.975V), characteristic of the corrosion in the active state (Fig. 4b). The disrupted passive state is also a prerequisite for the development of local corrosion in the active anodic sections - pitting and/or inter-crystalline, which is characteristic for this type of steel. It is also seen in Fig. 4b that the electrochemically formed cerium oxide films on the surface of the steel samples lead to strong shifting of the stationary corrosion potential of the steel in the positive direction - to potentials more positive than the potential of complete passivation and more negative than the potential of transpassivity. The established experimental facts unanimously indicate that the electrochemically deposited oxide films on the surface of the steel lead to restoration of its passive state, due to promoted ability of the system to passivate itself under the conditions of the real corrosion process.

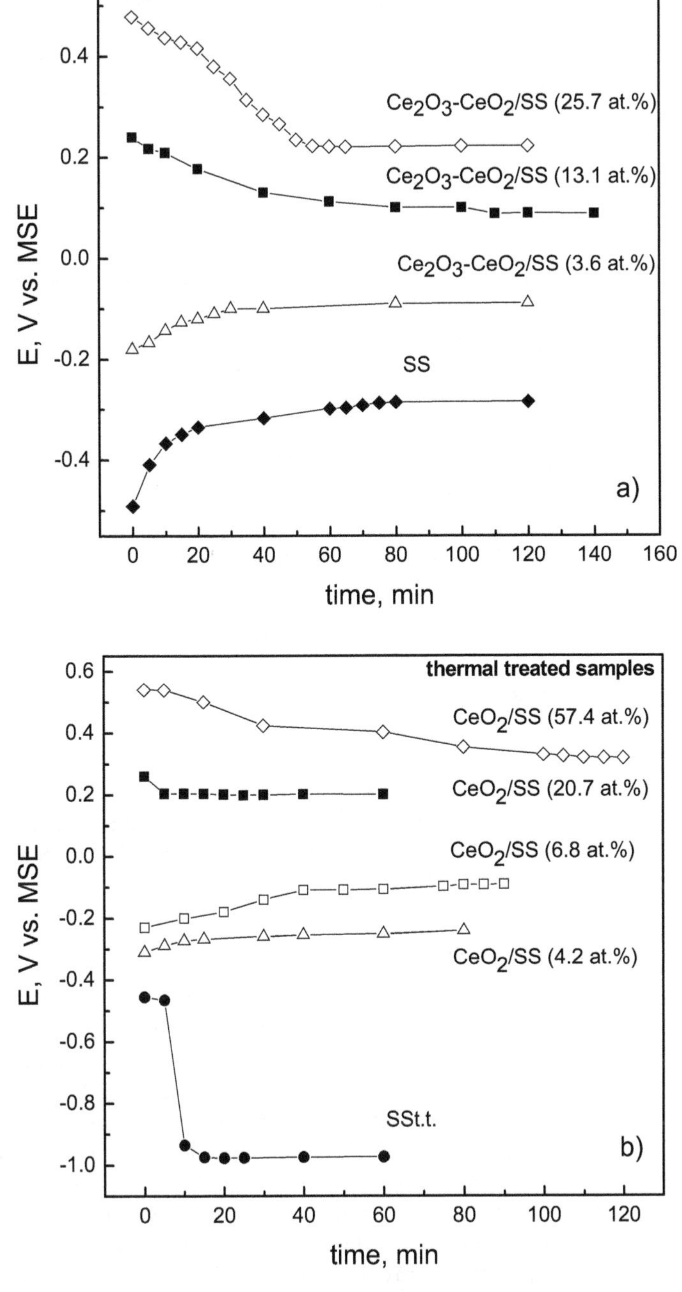

Fig. 4. Open circuit potential vs. time curves for SS (a) and $SS_{t.t.}$(b), as well as for the systems CeO_2-Ce_2O_3/SS (a) and CeO_2-Ce_2O_3/$SS_{t.t.}$(b), containing different concentrations of Ce, obtained in 0.1 N H_2SO_4.

The strong shifting of E_{corr} of the steel sample, modified with cerium oxide film, in the positive direction, depending on the surface concentration of cerium, can be attributed to the occurrence of another cathodic process in addition to the reaction of hydrogen evolution. We can assume that the effective cathodic sections of CeO_2-Ce_2O_3 will participate in the occurring cathodic depolarizing reaction according to the equations (1 and 2) given below:

$$CeO_2 + 2H^+ = Ce(OH)_2^{2+} \quad (1)$$

$$2Ce(OH)_2^{2+} + 2e^- = Ce_2O_3 + H_2O + 2H^+ \quad (2)$$

The above indicated reactions occur as a result of the extraordinary oxidation-reduction capability of the couple CeO_2-Ce_2O_3. The occurring of these reactions means that in the course of the corrosion process the surface film will be changing, enriching itself in Ce_2O_3. On the other side, the reactions (3) and (4) will also take place on the anodic sections of the steel surface and the latter one will lead to passivation:

$$Me \rightarrow Me^{n+} + ne \quad (3)$$

$$Me + H_2O \rightarrow Me_2O_n + ne \quad (4)$$

Taking into account the fact that the oxides of Ce^{3+} of the type Ce_2O_3 are soluble in acids, the reaction reported in (Achmetov, 1988) will also occur:

$$Ce_2O_3 + 6H^+ = 2Ce^{3+} + 3H_2O \quad (5)$$

As well as the respective conjugated reaction of oxidation:

$$Ce^{3+} + 2H_2O = Ce(OH)_2^{2+} + 2H^+ + e^- \quad (6)$$

Obviously, the cathodic reaction of reduction of CeO_2, which occurs at the corrosion potentials, established for the systems CeO_2-$Ce_2O_3/SS_{t.t.}$, is the main reason for restoring and preserving the passive state of the thermally treated steel samples (in accordance with equation 4) during their corrosion in solutions of sulfuric acid.

4.3 The inhibiting effect of cerium ions

In connection with the above statements a next step has been made in the investigations, namely studying the influence of the Ce^{3+} and Ce^{4+} ions as components of the corrosion medium (0.1N H_2SO_4) on the anodic behavior of stainless steel. These investigations were provoked by the observed occurrence of cathodic depolarization reaction of Ce^{4+} (CeO_2) reduction, as a result of which the surface concentration of cerium is decreasing and theoretically it should approach zero value (Stoyanova et al., 2010). For this purpose an inverse experiment was carried out at different concentrations of Ce^{4+} ions in the corrosion medium we monitored the changes in the stationary corrosion potential of the thermally treated steel by the chronopotentiometric method. The aim of this experiment was to prove the occurrence of a reversible reaction of reduction of Ce^{4+}: $Ce^{4+} - e \leftrightarrow Ce^{3+}$, (instead of the reaction of hydrogen depolarization), which in its turn creates also the option to form a film (chemically insoluble) of cerium hydroxides/oxides on the active sections of the steel surface.

Figure 5 illustrates the analogous E-τ dependences at open circuit, obtained upon immersion of SS_{tt} in 0.1 N H_2SO_4 solution, to which various concentrations of Ce^{4+} ions have been added. It was important to find out what is the influence of cerium ions on the corrosion behavior of the samples of thermally treated steel, when the character of the corrosion process is changed as a consequence of the thermal treatment of the steel. It should be reminded at this point that the stationary corrosion potentials (E_{st}) of SS and $SS_{t.t.}$in 0.1N H_2SO_4 solution have values for the non-thermally treated steel E_{st}= ~ -0.300 V, while for the thermally treated steel this value is E_{st}= ~ -0.980 V (Fig. 4)). The registered negative values of E_{st} in the absence of cerium ions for the $SS_{t.t.}$ samples, in our opinion, are connected with the strong cracking of the natural passive film on the surface of SS (Fig. 6). The most probable reason for this loss of the "stainless character" of the steel surface are the revealed sections, determining a several times higher concentration of iron-containing agglomerates in the surface layer (Table 2).

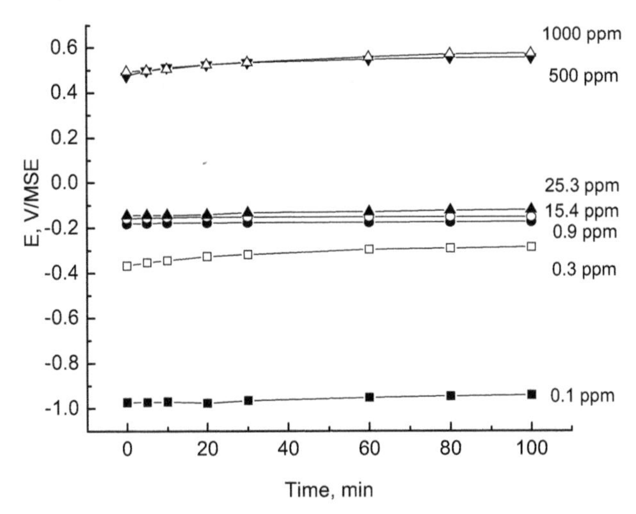

Fig. 5. Evolution of the open-circuit potential for $SS_{t.t.}$ at different concentrations of Ce^{4+} in 0.1 N H_2SO_4 solution.

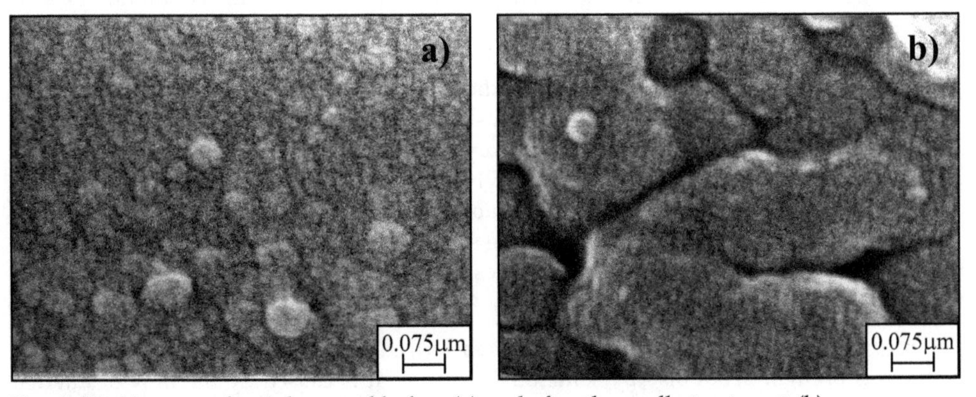

Fig. 6. SEM images of stainless steel before (a) and after thermally treatment (b).

Samples	O, at. %	Fe, at. %	Cr, at. %	Al, at. %	Ce, at. %	Cr/Fe, %	Al/Fe, %	Al/Cr, %	E_{st}, V
SS	58.3	3.1	3.4	35.2	-	1.09	11.35	10.35	_
SS 50h in 0.1N H_2SO_4	66.8	2.7	3.9	26.6	-	1.44	9.85	6.82	-0.209
$SS_{t.t.}$	64.9	7.2	7.0	20.9	-	0.97	2.90	2.99	_
$SS_{t.t.}$ 50h in 0.1N H_2SO_4	65.4	9.1	12.2	13.3	-	1.34	1.46	1.09	-0.300

Table 2. Distribution of the elements (in at. %) on the surface SS and $SS_{t.t.}$ before and after 50h immersion in 0.1N H_2SO_4.

For this reason this cycle of investigations has been carried out using samples of thermally treated steel, since the thermal treatment leads to change in the character of the corrosion process of steel. It should be taken into account that such kind of excessive heat treatment happen to take place both in the formation of catalytic converters, as well as in the course of their operation. In the latter case in the presence of cerium ions in the corrosion medium (Fig. 5), one observes a strong shifting of E_{st} in positive direction (from -0.942 V to -0.286 V), even at relatively low concentrations of Ce^{4+} (0.3 ppm) in the corrosion medium. The further increase in the concentration of Ce^{4+} ions (from 0.3 to 0.9 ppm) results in insignificant changes in E_{st}. Thereupon for SS_{tt} this shift jumps from -0.942 V (at Ce^{4+} ions concentration 0.1 ppm) up to -0.175 V (at Ce^{4+} ions concentration ~ 0.9 ppm). In the consecutive 20 – 30 fold increase in the concentration of Ce^{4+} ions (15-25 ppm) a preservation of the E_{st} value is observed, whereupon it manifests values ~ -0.150 - -0.120 V. The consecutive 20-50 fold increase in the concentration of Ce^{4+} ions (500-1000 ppm) leads to strong shifting of E_{st} in positive direction reaching values of about +0.510 - +0.570V.

These results prove that in the case of samples of thermally treated steel non-coated with Ce_2O_3-CeO_2 one observes analogous changes in the stationary corrosion potential of the steel electrode, which have already been registered for the system Ce_2O_3-CeO_2/$SS_{t.t.}$. The juxtaposition of the above-mentioned changes in E_{st} at open circuit (conditions of self-dissolution) with the characteristic zones (corrosion potential, Flade potential, zone of passivity, transpassivity region), defined by the cathodic and anodic potentiodynamic E-lgi polarization curves (conditions of external cathodic and anodic polarization) for SS_{tt} (Fig. 7.) in 0.1 N H_2SO_4 solution not-containing Ce^{4+}, shows the following. The addition of cerium ions causes shifting and establishing stationary corrosion potential (Fig. 7) in the zone of passivity of the steel. Evidently, this effect will lead also to improvement of the passivation ability, respectively to improvement of the stability to corrosion, of the steel in sulfuric acid medium, which is of great importance for the specific case of thermally treated steel, when the inhibitory action of the Ce^{4+} ions eliminates the negative influence of the cracking of the natural passive film on the steel.

The recovery of the passive state of SS_{tt}, characterized by disrupted passive film, in our opinion, is brought about also by some other reasons. It is caused by the flow of internal cathodic current (instead of external anodic current), which is determined by the occurring of a reduction reaction $Ce^{4+} \leftrightarrow Ce^{3+}$ in the redox system Ce^{4+}/ Ce^{3+}. Therefore the Ce^{4+} ions, as component of the corrosion medium, are acting as inhibitor, exerting an oxidative effect. It follows from (Fig. 7), that the increase in the concentration of the inhibitor in the corrosion environment leads to a substantial decrease in the corrosion current - from 1.10^{-6} (at inhibitor concentration of 0.1 ppm) - to $1.\ 10^{-8}$ $A.cm^{-2}$ (at inhibitor concentration ~0.9 ppm). What is making impression is the fact that the further increase in the concentration of the Ce^{4+} ions in the corrosion medium from ~0.9 ppm (which could be accepted as "critical") up to 1000 ppm influences to a smaller extent the rate of corrosion. It is necessary also to point out that with the increase in the concentration of cerium ions the corrosion potential is shifted in positive direction, whereupon its values remain more positive than the potential of complete passivation and more negative than the potential of depassivation of the steel within the entire interval of studied concentrations – an effect analogous to the one already established for the systems Ce_2O_3-$CeO_2/SS_{t.t.}$in 0.1 N H_2SO_4 solution.

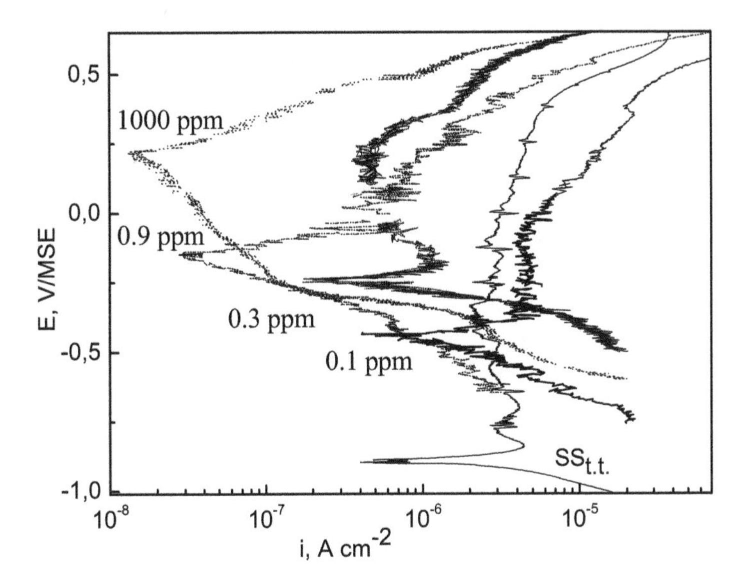

Fig. 7. Potentiodynamic E-lgi curves of $SS_{t.t.}$ at different concentrations of Ce^{4+} in 0.1 N H_2SO_4 solution.

Table 3 lists the electrochemical kinetic parameters: corrosion current density (i_{corr}), corrosion potential ($E_{corr.}$) and degree of inhibition efficiency (Z, %), characterizing the corrosion process in the presence and in the absence of cerium ions, determined on the basis of the results represented in Fig. 7. The degree of inhibition efficiency has been calculated on the basis of the equation:

$$Z\% = (i^{o}_{corr.} - i_{corr.}) / i^{o}_{corr.} x\ 100 \qquad (7)$$

where i^{o}_{corr} and i_{corr} are the values of the corrosion current density in the absence and in the presence of cerium ions.

Samples	E, V	i_{corr}, A m^{-2}	Z, %
	SS after thermal treatment		
$SS_{t.t.}$	-0.900	1.5×10^{-6}	-
with 0.1 ppm	-0.435	5.8×10^{-7}	37.8
with 0.3 ppm	-0.238	2.2×10^{-7}	86.8
with 0.9 ppm	-0.156	3.2×10^{-8}	98.2
with 1000 ppm	0.212	1.6×10^{-8}	99.4

Table 3. Electrochemical parameters characterizing corrosion behaviour of $SS_{t.t.}$.

It is seen from the table that upon increasing the concentration of Ce^{4+} ions in the corrosion medium the degree of protection reaches values up to 99% for the samples of thermally treated steel. The obtained data about the promotion in the efficiency of the inhibiting action with the increase in the concentration of Ce^{4+} in the corrosion medium for the thermally treated steel (Table 3) supposes an interconnection between the inhibitor concentration and the degree of surface coverage, Q following the equation (8):

$$Q = (i^{o}_{corr} - i_{corr}) / i^{o}_{corr} \quad (8)$$

where i^{o}_{corr} and i_{corr} are respectively the corrosion current density, obtained by extrapolation of anodic and cathodic potentiodynamic curves in the absence and in the presence of various concentrations of the inhibitor in the corrosion medium. On the basis of the obtained data about the fraction of surface coverage of steel electrode as a function of the concentration of the inhibitor, one can accept that the adsorption process obeys Langmuir's isotherm. According to this isotherm the interconnection between the fraction of surface coverage and the concentration of the inhibitor is the following:

$$Q = KC.(1+KC), \text{ and respectively:} \quad (9)$$

$$C/Q = 1/K + C \quad (10)$$

where K is the adsorption constant and C is the concentration of the inhibitor. The dependence C/Q as a function of C for the thermally treated steel is represented in Fig. 8. It is seen that the experimental data describe a linear dependence, whereupon the coefficient of the linear regression and the slope of the straight line of this dependence approach a value of 1, which proves the validity of Langmuir's isotherm in our case.

The constant K in the equation (9) is connected with the standard free energy of adsorption (Δ G) in accordance with the equation:

$$K = (1/55.5) \exp(-\Delta DG^{o}_{ads} / RT) \quad (11)$$

The value of K, determined graphically based on the plot of the dependence C/Q as a function of C, is $44,6 \times 10^{6} M^{-1}$, while the value of $(-\Delta G^{o}_{ads})$ amounts to 10.35 κ $call.mol^{-1}$.The relatively low value of ΔG^{o}_{ads}, is indicative of electrostatic forces of interaction between the ions of the inhibitor and the steel surface. Or in other words the interaction of the inhibitor with the surface of the thermally treated steel has physical nature. On the basis of the obtained electrochemical corrosion data from the potentiodynamic curves in the presence and in the absence of inhibitor and judging from the measurements of the stationary corrosion potential of the steel, depending on the concentration of the inhibitor at open circuit we could suppose that under the conditions of internal anodic polarization in the presence of inhibitor the nature of the passive layers remains the same as in the case of external anodic polarization. As far as we can judge the specific action of the inhibitor is manifested in the formation of an adsorption layer, which is transformed into bulk phase, on the active anodic sections of the surface of the metal. In view of the XPS analyses (Guergova, 2011) after 500-hour interval of staying of the thermally treated steel in the aggressive medium in the presence of Ce^{4+} ions (25 ppm), on the surface of the studied film in the region of the Ce3d XPS band one can observe the appearance of a certain amount of cerium (1.5 at. %) in the form of Ce_2O_3. The cerium is most probably incorporated into the surface film as a result of the stay of the steel sample in the inhibited corrosion medium, which leads to its modifying as a consequence of the formation of mixed oxides of the type of cerium aluminates and chromates (Burroughs et al., 1976; Hoang et al., 1993). In support of such hypotheses comes the absence of visible corrosion damages on the surface of $SS_{t.t.}$ exposed for 500 h in 0.1N H_2SO_4 solution in the presence of Ce^{4+} ions (Fig. 9). Of course, from purely electrochemical point of view, the ability of the inhibitor to define strongly positive oxidation-reduction potential of the steel is connected with the proceeding of reduction of Ce^{4+} into Ce^{3+}. In order to investigate the kinetics of reduction of the Ce^{4+} ions to Ce^{3+}, in the region of potentials, characteristic of the passive state of the steel under consideration, we plotted the anodic and the cathodic potentiodynamic curves, characterizing the behavior of the oxidative-reductive couple Ce^{4+}/Ce^{3+} at various concentrations of Ce^{4+} in 0.1 N H_2SO_4 solution, on an inert support of platinum (Fig. 10). Such an approach (Tomashov & Chernova, 1965), to our mind, enables the complete elucidation of the mechanism of inhibitory action of the cerium ions. It allows direct juxtaposition of the changes in the values of the corrosion potentials (respectively the corrosion currents) of the steel in their presence with the values of the reversible redox potentials (respectively the exchange currents) of the couple Ce^{4+}/ Ce^{3+} at comparable concentration levels.

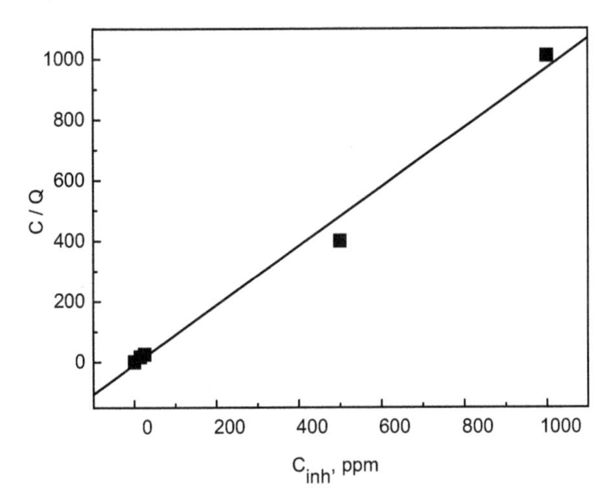

Fig. 8. Langmuir adsorption plots for $SS_{t.t.}$ in 0.1 N H_2SO_4 solution at different concentrations of Ce^{4+} ions.

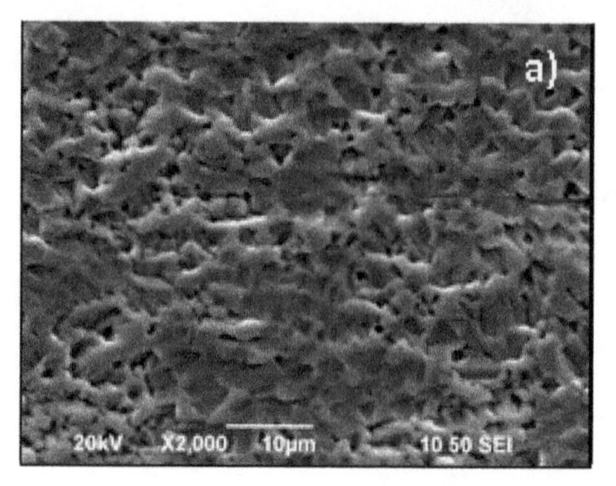

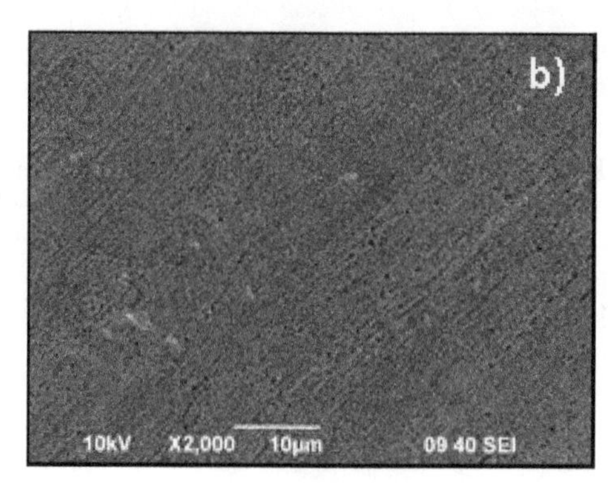

Fig. 9. SEM images on thermally treated stainless steel after 500 h immersion in 0.1 N H_2SO_4 without Ce^{4+}(a) and in the same media with 25 ppm Ce^{4+} (b).

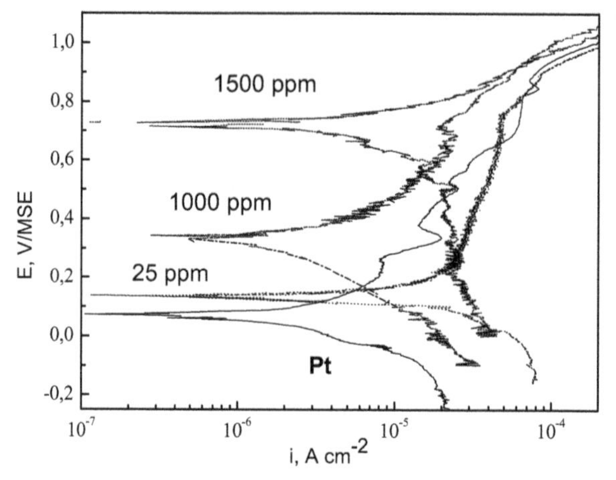

Fig. 10. Potentiodynamic E-lgi curves of Pt at different concentrations of Ce^{4+} in 0.1 N H_2SO_4 solution.

The comparison of the obtained results gives evidence that with the increase in the concentration of the cerium ions in 0.1 N H_2SO_4 solution the equilibrium oxidation-reduction potential of the system Ce^{4+}/ Ce^{3+} is shifted in positive direction (Fig.10 and Table 4), in correspondence with the equation of Nernst, whereupon at all the studied concentrations it is located in the zone of potentials, characteristic of the passive state of steel (Fig. 7). Thereupon the corrosion potentials of the steel are more negative than the equilibrium oxidation-reduction potentials of the system Ce^{4+}/Ce^{3+}. At the same time, the juxtaposition of the corrosion currents for the steel in the presence of cerium ions with the exchange currents for the system Ce^{4+}/Ce^{3+}on Pt, at comparable concentration levels of the cerium ions, shows that they have quite close values.

The so obtained data give us the reason to classify the studied oxidation-reduction couple as an inhibitor having an oxidative effect, which does not influence directly the kinetics of the anodic process. Its action is expressed in its participation in the depolarizing reaction of the corrosion process (i.e. oxidative depolarization), respectively in the establishment of oxidative-reductive potential of the medium more positive than the potential of complete passivation of the steel. This effect, in its turn, defines the value of the stationary corrosion potential of the steel to be more positive than the potential of complete passivation.

Samples	E_o, V	i_o, A cm^{-2}
Pt metal	0.070	2.17 x 10^{-7}
Pt with 0.3 ppm	0.107	8.99 x 10^{-6}
Pt with 25 ppm	0.135	6.35 x 10^{-7}
Pt with 1000 ppm	0.331	6.08 x 10^{-7}
Pt with 1500 ppm	0.722	7.67 x 10^{-7}

Table 4. Reversible redox potentials E_o, and equilibrium currents io, of the system Ce^{4+}/Ce^{3+} on Pt at different concentrations of the Ce^{4+} in the corrosion medium.

In order to prove the integral nature of cerium oxides films as efficient cathodic coating, involved directly in the corrosion process and the role of cerium ions as inhibitors possessing oxidative effect, participating also directly in the corrosion process and leading to the formation of phase layer of cerium oxides on the active cathodic sections of the steel surface, we compared the dependences E-lgi for the systems Ce_2O_3-$CeO_2/SS_{t.t.}$ as well as for the system $Ce^{4+}/Ce^{3+}/Pt$. – Fig. 11. It is seen in Fig. 11 that the corrosion currents of the systems Ce_2O_3-$CeO_2/SS_{t.t.}$are close in value to the exchange current of the oxidation-reduction current of the couple Ce^{4+}/Ce^{3+} on Pt. The differences between the corrosion potential of the systems Ce_2O_3-$CeO_2/SS_{t.t}$ and of the equilibrium oxidation-reduction potential of the couple Ce^{4+}/Ce^{3+} can be explained by the discrepancies in the surface and bulk phase concentrations of the components.

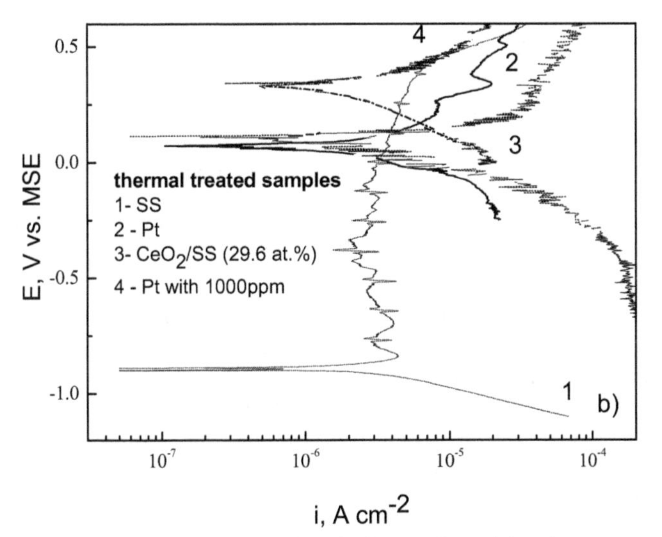

Fig. 11. Potentiodynamic E–lgi curves for $SS_{t.t}$ (1)., for Pt (2) and for the systems CeO_2-Ce_2O_3/Pt (3); Ce^{4+}/Ce^{3+}/Pt (4); obtained in 0.1 N H_2SO_4 solution.

4.4 XPS and SEM results

In confirmation of these results and the conclusions come also the data of the XPS analyses of the samples, having electrochemically deposited cerium oxide films, characterizing the changes in the chemical state and in the composition of the surface film, depending on the time interval of the immersion stay of the samples in $0.1NH_2SO_4$ solution (Stoyanova et al., 2010). Table 5 represents the results for the sample with surface concentration of electrochemically deposited cerium oxide layer 45.1 at.% .It is seen that after 1000 hours of exposure to the corrosion medium the surface concentration of cerium is decreased from 45.1aт.% down to 0.2 at.%. This result is convincing evidence for the occurring of depolarizing reaction involving the participation of the rich in CeO_2 sections of the surface, acting as effective cathodes, in accordance with the equations (1-4). It becomes evident that the presence of cerium oxide film determines the establishment of more positive stationary corrosion potential of the system, due to the proceeding of the reactions 1- 6, the surface passive film will become modified, whereupon its composition, respectively the ratio Cr/Fe, will become different.

Time of exposure, h	E_{st}, V	C, at. %	O, at. %	Al, at. %	Fe, at. %	Cr, at. %	Ce, at. %
as deposited	0.151	15.6	37.6	1.0	0.1	0.6	45.1
18	0.169	50.1	35.7	1.6	0.5	0.4	11.7
200	0.249	64.4	31.5	0	0	0	4.1
250	0.236	63.7	31.5	0	0	0.2	4.6
400	0.239	68.5	27.4	0	0	0	4.1
1000	0.060	31.4	42.9	19.9	0.7	4.9	0.2

Table 5. Concentration of the elements (in at. %) on the surface layers of the system CeO_2-Ce_2O_3/SS after thermal treatment and after corrosion test in 0.1 N H_2SO_4.

In this cycle of experimental runs, using the XPS method, the changes were monitored, occurring in the chemical composition of the passive film of the system CeO_x /SS, during prolonged exposure of the samples in 0.1 N H_2SO_4 solutions (Table 5). The analyses were carried out after the 18th, 200th, 250th, 400th, and 1000th hour - time intervals of exposure. Within the interval 200-400 hours E_{st} remains practically the same, while after 1000 hours of exposure it is shifted strongly in the negative direction, reaching a value of about ~ +0.060 V. To obtain further information about the influence of ceria on the corrosion behaviour of as-deposited sample we analyzed in depth the Ce3d and O1s XPS spectra. As it has already been discussed in our previous papers (Nikolova et al., 2006; Stoyanova et al, 2006), the Ce3d spectrum is a complex one, due to the fact that the peak is spin-orbital split into a doublet, each doublet showing extra structure due to the effect of the final state. There are 8 peaks assignments in the spectra labelled according to Burroughs (Burroughs et al., 1976), where the peaks V, V^{II}, V^{III} and U, U^{II}, U^{III} refer to the 3d5/2 and 3d3/2 respectively and they are characteristic of Ce(IV) 3d final states. The peaks labelled as V^{I} and U^{I} refer to 3d5/2 and 3d3/2 they are characteristic of Ce(III)3d final state (Fig.12).The literature data make it obvious that the chemical state of Ce could be evaluated based on the percentage of the area of the u‴ peak, located at 916.8 eV with respect to the total Ce3d area. So if the percentage of the u‴ peak related to the total Ce3d area varies from 0 to 14%, then the Ce^{4+} percentage related to the total amount of Ce varies from 0 to 100%. In our case the u‴% amounts change as a result of dipping the as-deposited sample into 0.1 N H_2SO_4 solution, so we observed also change in the percentage of Ce^{4+} and Ce^{3+} on the surface (Table 6). The obtained O1s X-ray photoelectron spectra, recorded after different time intervals of exposure, are quite complicated. These spectra had to undergo de-convolution procedure to analyze the contribution of the separate components in them.

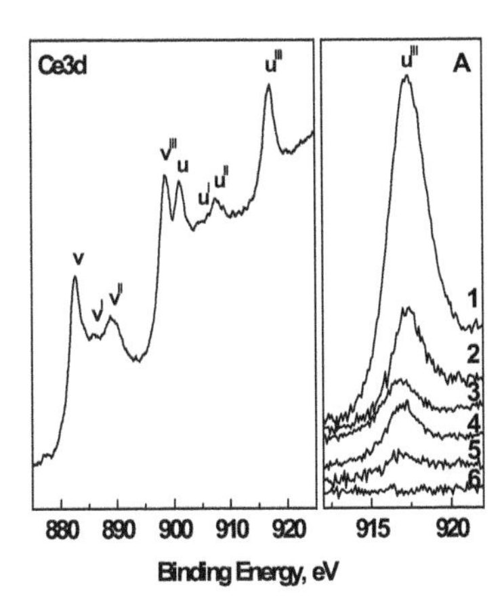

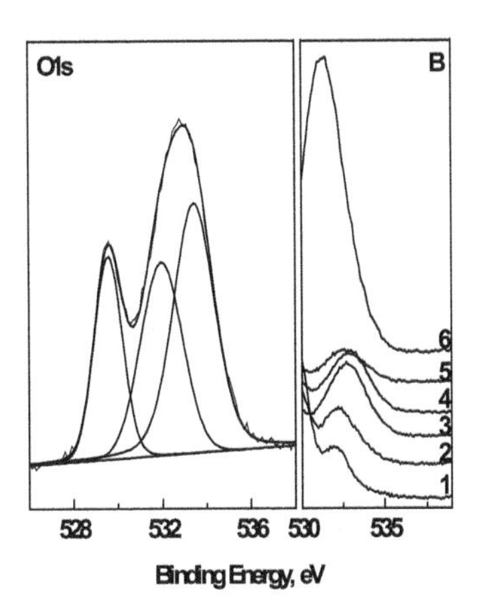

Fig. 12. Ce3d and O1s XPS spectra of CeO_2-Ce_2O_3/SS sample after 250 h exposition to 0.1 N H_2SO_4, Panel A and B correspond to high BE and high BE portion of Ce3d and O1s spectral regions taken at different step of corrosion test. (1) as-deposited; (2) after 18 h ; (3) after 200 h; (4) after 250 h; (5) after 400 h; (6) after 1000 h exposition to 0.1 N H_2SO_4 solution.

The chemical composition of the surface film of the CeO_2-Ce_2O_3/$SS_{t.t.}$ systems under consideration is shown in Table 5. It is important to note that on the surface of the 'as-deposited" samples, covered with cerium oxide film, we detected also the presence of iron, chromium and aluminum in addition to the cerium. The latter is mainly in the valence state Ce^{4+} i.e. in the form of CeO_2. After 18 hours of exposure to 0.1 N H_2SO_4 solution the chemical composition on the surface of the system has changed. The surface concentration of cerium drops down from 45.1 to 11.7 at. %, whereupon Ce^{3+} appears in the form of $Ce(OH)_3$ and Ce_2O_3. After continuing the exposure further (200-400 hours) one observes evolution of the spectra with respect to cerium and oxygen (Fig.12). The main peak in the spectrum of oxygen, having binding energy of about ~529.5 eV, is attributed to the presence of O^{2-} ions, which exist basically as Ce-O bonds in the crystal lattice of the cerium oxide being formed. The second peak, located at 531.7 eV, is associated with the existence of OH^- groups on the surface, while the presence of a peak at 533 eV shows that there is adsorbed water on the surface of the studied passive films (533 eV) (Hoang et al., 1993; Paparazzo, 1990). It can be seen in the spectra that after exposure of the samples in corrosion medium for 200-400 hours the respectively detected high-energy peak in the spectrum of oxygen for these samples is growing up initially, while afterwards it decreases its intensity. This effect, in our opinion, is owing to consecutive enrichment and then impoverishment of the surface layer of the film in OH^- groups i.e. adsorbed water molecules (Fig.12, Table 6). The obtained results gives us the reason to draw the conclusion that the surface passive film under these conditions at this stage consists of CeO_2, Ce_2O_3, $Ce(OH)_4$ as well as $CeO(OH)_2$ and $Ce(OH)_3$, whose existence has been ascertained also by some other authors (Huang et al., 2008).

Time of exposure, h	O1s, eV	Percentage of oxygen contribution to the total	Bonds	Ce3d, eV	Percentage of Ce^{4+} to the total Ce
as deposited	529.2 531.4	59 41	Ce-O Ce-OH	883.0	100
18	529.2 531.5 533.2	43 42.5 14.5	Ce-O Ce-OH Others	882.9	85.35
200	529.4 531.9 533.3	21.4 33.6 45	Ce-O Ce-OH Others	882.4	76
250	529.5 531.9 533.4	23 32 45	Ce-O Ce-OH Others	882.5	76
400	529.5 532.1 533.4	23 45 32	Ce-O Ce-OH Others	882.5	76
1000	531.2	100	OH	882.0	-

Table 6. Calculated contribution of oxygen and percentage of Ce^{4+}, depending of the exposure time in 0.1 N H_2SO_4. Types of the chemical bonds and values of binding energy.

After 1000 hours of exposure the quantity of cerium is drastically decreased, as a consequence of the occurring reactions 1-4 and only some insignificant amounts of cerium have been registered in the valence state Ce^{3+}, i.e. in the form of Ce_2O_3. Only a single peak has been detected in the spectrum of oxygen, having a binding energy of 531,2 eV. Chromium, aluminum and iron have also been detected (Table 5). On the basis of the values of their binding energies (Table 5), including also the location of the O1s peak (Fig.12), we can also conclude that they exist in the form of oxides and hydroxides: Cr_2O_3, $Cr(OH)_3$, Fe_2O_3, FeOOH, Al_2O_3 and $Al(OH)_3$. The high concentration of carbon registered in the surface film is owing to the considerable amount of carbonates adsorbed during the thermal treatment in a high-temperature oven.

In support of the conclusions, drawn on the basis of the above results, evidence is also given by the direct SEM observations carried out. It follows from the electron microscopic studies of the samples, exposed to the corrosion medium, that the disruption of the passive state of steel at the initial stages (until the 50th hour) leads to appearance of local corrosion and its propagation and spreading further to give total corrosion (Fig.13 a,b). For the sake of comparison Fig. 13 shows the same surface in the presence of cerium oxide coating after 50 hours of exposure to the corrosion medium. In this case no corrosion damages can be observed on the surface not only after the 50th hour, but even after 1000 hours of exposure (Fig.13) to the corrosion medium, as a result of its modifying, already discussed above.

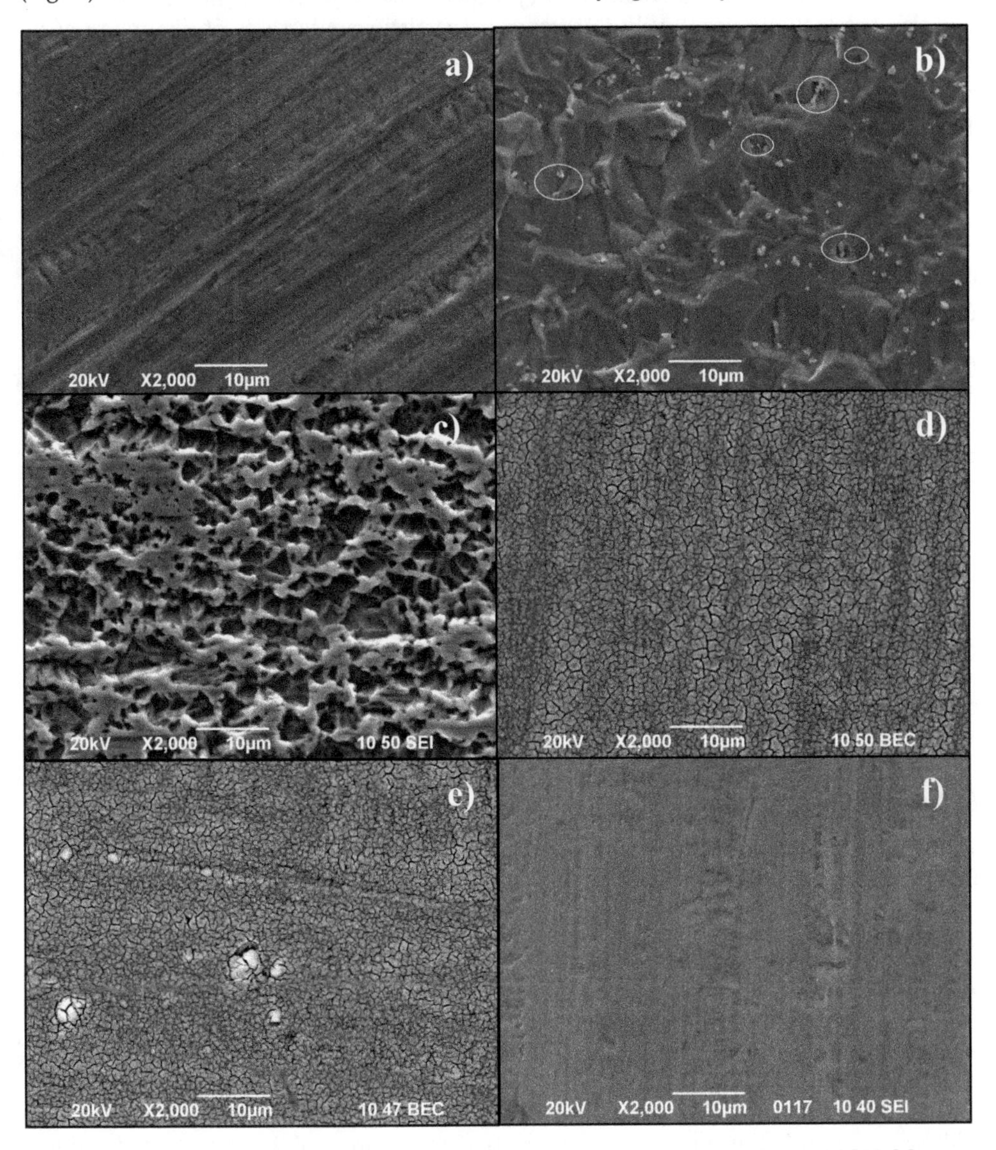

Fig. 13. SEM images for the samples: a) $SS_{t.t.}$; b) $SS_{t.t.}$ after 50 h exposition in 0.1 N H_2SO_4 (O - areas of local corrosion); c) $SS_{t.t.}$ after 1000 h exposition; d) CeO_2-Ce_2O_3/$SS_{t.t.}$; e) CeO_2-Ce_2O_3/$SS_{t.t.}$ after 50 h exposition in 0.1 N H_2SO_4; f) CeO_2-Ce_2O_3/$SS_{t.t.}$ after 1000 h exposition.

5. Conclusions

The above discussed results elucidate the mechanism of action of cerium oxide coatings as effective cathodes and of cerium ions (when they are present as a component of the corrosion medium) - as inhibitor having oxidative action, leading to improvement of the corrosion stability of stainless steels. They explain the improved ability of the steel to undergo passivation, respectively to recover its passive state in cases of disruption of its surface passive film - especially in this specific case, studied by us, i.e. disruption as a result of thermal treatment.

This effect is associated with a strong shifting of the stationary corrosion potential of the steel in positive direction, moving over from potentials, characteristic of corrosion in active state to potentials, located inside the zone of passivity. In this respect, another basic purpose of the investigations was the elucidation of the mechanism of action of the cerium oxides film and in particular collecting experimental evidence for the supposition about the occurring of an efficient depolarization reaction of CeO_2 reduction, resulting in a state of passivity, instead of hydrogen depolarization reaction. For this purpose we considered also the decrease in the surface concentration of Ce in the passive layer under the conditions of the actual corrosion process (self-dissolution) of the stainless steel, by means of XPS, EDS and ICP-AES analyses. A decrease in the surface concentration of CeO_2 (Ce^{4+}) has been observed, which is known to be chemically inert in acidic media. The obtained results prove the occurrence of an effective cathodic process of Ce^{4+} reduction into Ce^{3+} in the surface oxide film. These results elucidate in details the corrosion behavior of the system Ce_2O_3-CeO_2/steel in 0.1N H_2SO_4. They explain the improved ability of self-passivation, respectively the restoration of the passive state of the thermally treated steel in the presence of cerium oxides as components of the modified passive surface film.

It has also been shown that the couple Ce^{4+}/Ce^{3+}, as component (inhibitor) of the corrosion medium, displays analogous action. It is expressed in the occurrence of a reaction of reduction of the oxidative component of the medium - Ce^{4+}, leading to consumption of the electrons removed in the course of oxidation, respectively passivation, of the active anodic sections.

6. Acknowledgements

The authors gratefully acknowledge financial support by the Bulgarian National Science Fund under Contract DO 02-242/TK 01-185.

7. References

Achmetov, N.S., (1988), *General and Inorganic Chemistry* (in Russian, Second edition), Publishing House "Vyshaya Shkola", Moscow

Aldykiewicz Jr A.J., Davenport A.J., Isaacs H.S., (1995).Investigation of cerium as a cathodic inhibitor for aluminum-copper alloys. *Journal of the Electrochemical Society*, Vol. 142, No.10, pp. 3342-3350. ISSN: 0013-4651

Aldykiewicz Jr A.J., Davenport A.J. Isaacs H. S. (1996). Studies of the formation of cerium-rich protective films using X-ray absorption near-edge spectroscopy and rotating disk electrode methods. *Journal of the Electrochemical Society* Vol.143 No1, pp. 147-154, ISSN: 0013-4686

Almeida, E., Diamantino, T.C, Figueiredo M.O., Sa C. (1998). Oxidizing alternative species to chromium VI in zinc - galvanized steel surface treatment. Part I-A morphological and chemical study. *Surface and Coating Technology*, Vol. 106, pp. 8-17. ISSN: 0257-8972

Almeida, E., Fedrizzi, L., Diamantino, T.C. (1998). Oxidizing alternative species to chromium VI in zinc - galvanized steel surface treatment Part 2-An electrochemical study. *Surface and Coating Technology*, Vol. 105, pp. 97-101, ISSN: 0257-8972

Amelinckx, L., Kamrunnahar, M., Chou P, Macdonald D., (2006). Figure of merit for the quality of ZrO_2 coatings on stainless steel and nickel-based alloy surfaces. *Corrosion Science*, Vol. 48, (April, 2006) pp. 3646-3667, ISSN: 0010-938X

Aramaki, K. (2001). The inhibition effects of cation inhibitors on corrosion of zinc in aerated 0.5M NaCl. *Corrosion Science*, Vol. 43, (June, 2001), pp.1573-1588, ISSN: 0010-938X

Aramaki K. (2001). Treatment of zinc surface with cerium(III) nitrate to prevent zinc corrosion in aerated 0.5 M NaCl, *Corrosion Science*, Vol. 43, No.11 pp. 1201-1215, ISSN: 0010-938X

Aramaki, K. (2002). Preparation of chromate-free, self-healing polymer films containing sodium silicate on zinc pretreated in a cerium (III) nitrate solution for preventing zinc corrosion at scratches in 0.1 M NaCl. *Corrosion Science*, Vol. 44 (August 2001), pp. 1375-1389, ISSN: 0010-938X

Aramaki, K. (2002). Cerium (III) chloride and sodium octylthiopropionate as an effective inhibitor mixture for zinc in 0.1M NaCl. *Corrosion Science*, Vol.. 44, (June 2001), pp. 1361-1374, ISSN: 0010-938X

Aramaki, K.,(2002). Self-healing protective films prepared on zinc by treatment with cerium (III) nitrate and sodium phosphate. *Corrosion Science*, Vol. 44, No.11, pp.2621-2634. ISSN: 0010-938X

Arenas, M.A., Conde A., de Damborenea, J., (2002). Cerium: a suitable green corrosion inhibitor for tinplate. *Corrosion Science*, Vol.44, No.3, pp. 511-520, ISSN: 0010-938X

Arenas, M.A., de Damborenea, J.,(2003). Growth mechanism of cerium layers on galvanised steel. *Electrochimica Acta*, Vol. 48, No. 24, (October, 2003) pp. 3693-3698, ISSN: 0013-4686

Arenas, M.A, Garcia, I., de Damborenea, J., (2004). X-ray photoelectron spectroscopy study of the corrosion behaviour of galvanised steel implanted with rare earths. *Corrosion Science*, Vol. 46, No. 4, pp. 1033-1049, ISSN: 0010-938X

Arnott, D.R., Hinton, B.R.W & Ryan N.E, (1989). Cationic-film-forming inhibitors for the protection of the AA 7075 aluminum alloy against corrosion in aqueous chloride solution. *Corrosion*, Vol. 45, No. 1, pp. 12-18

Avramova, I., Stefanov, P., Nicolova, D., Stoychev, D., Marinova, T., (2005). Characterization of nanocomposite CeO_2-Al_2O_3 coatings electrodeposited on stainless steel. *Composites Science and Technology*, Vol. 65, pp. 1663-1667 ISSN: 0266 - 3538

Avramova, I., Stoychev, D., Marinova, T., (2006). Characterization of thin CeO_2-ZrO_2-Y_2O_3 films electrochemical deposited on stainless steel. *Applied Surface Science*, Vol. 235, pp. 1365-1370, ISSN: 0169-4332

Balasubramanian, M., Melendres, C.A., Mansour, A. N., (1999). An X-ray absorption study of the local structure of cerium in electrochemically deposited thin films. *Thin Solid Films*, Vol. 347, No. 1-2, (June 1999), pp. 178-183, ISSN: 0040-6090

Barbucci A., Delucchi M, Cerisola G. (1998), Study of chromate free pretreatments and primers for the protection of galvanized steels, *Progress In Organic Coatings*, Vol. 33, No. 2, pp. 131-138, ISSN: 0300-9440

Bethencourt, M., Botana, F.J., Calvino, J.J, Marcos, M., Rodríguez-Chacón, M. A., (1998). Lanthanide compounds as environmentally-friendly corrosion inhibitors of aluminium alloys : a review. *Corrosion Science* . Vol. 40, pp. 1803-1819, ISSN: 0010-938X

Breslin, C.B, Chen, C., Mansfeld, F., (1997). The electrochemical behavior of stainless steels following surface modification in cerium containing solutions. *Corrosion Science*, Vol. 39, No. 6, pp. 1061-1073 ISSN: 0010-938X

Bunluesin, T., Gorte, R.G., Graham, G.W., (1997). CO oxidation for the characterization of reducibility in oxygen storage components of three-way automotive catalysts. *Applied Catalysis B: Environmental,* Vol. 14, No. 1-2, (December 1997), pp. 105-115 ISSN: 0926 - 3373

Burroughs, P. Hamnett, A., Orchard A.F., Thornton G. J., (1976). Satellite structure in the X-ray photoelectron spectra of some binary and mixed oxides of lanthanum and cerium. *Journal of the Chemical Society, Dalton Transactions,* Vol. 17, pp. 1686-1698, ISSN: 1472 - 7773

Crossland, A.C., Thompson, G.E., Skeldon, P., Smith G. C. (1998) Anodic oxidation of Al-Ce alloys and inhibitive behaviour of cerium species. *Corrosion Science*, Vol.40, pp.871-885, ISSN: 0010-938X.

Davenport, A.J., Isaacs, H.S., Kendig, M.W. (1991). XANES investigation of the role of cerium compounds as corrosion inhibitors for aluminium. *Corrosion Science,* Vol. 32, No.5-6, pp. 653-663, ISSN: 0010-938X.

Di Maggio, R., Rossi, S., Fedrizzi, L., Scardi P. (1997). ZrO_2-CeO_2 films as protective coatings against dry and wet corrosion of metallic alloys. *Surface and Coatings Technology*, Vol. 89,No. 3, pp.292-298, ISSN:0257-8972.

Fahrenholtz, W.G., O'Keefe, M.J., Zhou, H., Grant, J.T., (2002). Characterization of cerium - based conversion coatings for corrosion protection of aluminum alloys. *Surface and Coatings Technology*, Vol. 155, No.2-3, pp.208-213, ISSN: 0257-8972

Ferreira, M.G., Duarte, R.J., Montemor, M. F., Simoes A.M., (2004). Silanes and rare earth salts as chromate replacers for pretreatments on galvanized steel. *Electrochimica Acta*, Vol. 49, No. 17-18, pp. 2927-2935, ISSN: 0013-4686

Forsyth, M., Forsyth, C.M., Wilson, K., Behrsing, T., Deacon, G.B., (2002). ART characterization of synergistic corrosion inhibition of mild steel surfaces by cerium salicylate. *Corrossion Science* Vol. 44, No. 11, (November 2002) pp.2651-2656, ISSN: 0010-938X

Gonzalez, S., Gil, M.A., Hemandez, J.O., Fox, V., Souto, R. M. (2001). Resistance to corrosion of galvanized steel covered with an epoxy - polyamide primer coating, *Progress In Organic Coatings,* Vol. 41, No. 1-3, pp. 167-170, ISSN: 0300-9440

Guergova, D., Stoyanova, E., Stoychev, D., Atanasova, G., Avramova, I., Stefanov, P. (2008). Influence of calcination of SS OC 4004 with alumina or ceria layers on their passive state in different acid media. *Bulgarian Chemical Communications,* Vol. 40, No. 3, pp227-232

Guergova, D., Stoyanova, E., Stoychev, D., Avramova, I., Stefanov, P., (2011). Investigation of the inhibiting effect of cerium ions on the corrosion behaviour of OC404 stainless steel in sulfuric acid medium. *Open Corrosion Journal* in press, ISSN: 1876-5033

Hagans, P.L, Hass, C.M., (1994) *ASM Handbook Volume 05: Surface Engineering*, Vol.5, pp. 405-411, ISBN: 978-087170-384-2

Hinton, B.R.W., (1991). Corrosion prevention and chromates. *Metal Finishing* Vol. 89, pp. 55-61, ISSN: 0026-0576

Hinton, B.R.W., Wilson, L. (1989). The corrosion inhibition of zinc with cerous chloride,*Corrosion Science,* Vol. 29, No. 8, 1989, pp. 967-975, 977-985, ISSN: 0010-938X

Hinton, B.R.W., (1992). Corrosion inhibition with rare earth metal salts, *Journal of Alloys and Compounds,* Vol. 180, No. 1-2, (25 March 1992), pp. 15-25, ISSN: 0925-8388

Hoang, M., Hughes, A.E., Turney, T.W., (1993). An XPS study of Ru-promotion for Co/CeO_2 Fischer-Tropsch catalyst. *Applied Surface Science.* Vol. 72, No. 1, pp. 55-65, ISSN: 0169 - 4332

Hosseini M.G., Sabouri, M., Shahrabi, T. (2006). Comparison between polyaniline - phosphate and polypyrrole- phosphate composite coatings for mild steel corrosion protection. *Material and Corrosion,* Vol. 57, pp. 447 - 453, ISSN: 0947-5117

Hosseini, M.G., Ashassi-Sorkhabi, H., Ghiasvand, H.A.Y., (2007). Corrosion protection of electro-galvanized steel by green conversion coatings. *Journal of Rare Earths,* Vol. 25, pp. 537-243, ISSN: 1002-0721

Huang Xingqiao, Ning Li, Huiyong Wang, Hanxiao Sun, Shanshan Sun, Jian Zheng, (2008).Electrodeposited cerium film as chromate replacement for tin plate, *Thin Solid Films,* Vol.516, No. 6, (January 2008) pp. 1037-1043, ISSN: 0040 - 6090

Korobov, V.I., Loshkarev, Y.M., Kozhura, O.V. (1998). Cathodic treatment of galvanic zinc coatings in solutions of molybdates, *Russian Journal of Electrochemistry,* Vol. 33, pp. 55-62, ISSN: 1023 -1935

Kudryavtsev, N.T. (1979). *Electrolytic Metal Coatings,* In Russian, Publ. House "Khimiya", Moscow

Lainer, V.I. (1984) *Protective Metal Coatings,* In Russian, Publ. House "Metalurgia", Moscow

Liu, R., Li, D.Y., (2000). Effects of yttrium and cerium additives in lubricants on corrosive wear of stainless steel 304 and Al alloy 6061. *Journal of Materials Science.* Vol. 35, No. 3, pp. 633-641 ISSN: 0022-2461

Liu, H., Yang, J., Liang, H.-H., Zhuang, J.-H., Zhou W.-F., (2001). Effect of cerium on the anodic corrosion of Pb-Ca-Sn alloy in sulfuric acid solution. *Journal of Power Sources,* Vol. 93, No. 1-2, pp.230-233, ISSN: 0378-7753

Lox, E.S., Engler, B.N., (1995) in: A. Frennet, J.M. Bastin (Eds.), Catalysis and Automotive Pollution Control III, Elsevier, Amsterdam, 1995, p.1559 (Chapter: Enviromental Catalyis – Mobile sources)

Lu, Y.C., Ives, M.B., (1993). The improvement of the localized corrosion resistance of stainless steel by cerium. *Corrosion Science,* Vol. 34, No. 11, pp. 1773-1781, ISSN: 0010-938X

Lu, Y.C., Ives, M.B.(1995). Chemical treatment with cerium to improve the crevice corrosion resistance of austenitic stainless steels. *Corrosion Science,* Vol. 37, No. 1 pp.145-155, ISSN: 0010-938X.

Lukanova, R., Stoyanova, E., Damyanov, M., Stoychev, D.(2008). Formation of protective films on Al in electrolytes containing no Cr^{6+}ions. *Bulgarian Chemical Communications*, Vol.40, No.3, pp.340-347.

Mansfeld, F., Lin, S. and Shin, H., (1989). Corrosion protection of Al alloys and Al-based metal matrix composites by chemical passivation, *Corrosion*, Vol. 45 No8, pp. 615-630.

Mansfeld, F., Wang. V., Shih, H., (1991).Development of 'stainless aluminum', *Journal of the Electrochemical Society*, Vol. 138, No.12, pp. L74-L75. ISSN: 0013-4651.

Mansfeld. F., Wang. Y, (1995). Development of "stainless" aluminium alloys by surface modification. *Materials Science and Enginerring A*, Vol.198, No.12 pp.51-61, ISSN: 0921-5093.

Marinova, T., Tsanev, A., Stoychev, D. (2006). Characterisation of Mixed Yttria and Zirconia Thin Films. *Materials Science and Engineering B*, Vol.130, No. 1-3, pp. 1-4. ISSN: 0921-51-07.

Mcnamara, J.M. (2000). Health effects of vehicle emissions a review from the second international conference, *Platinum Metals Review*, Vol.44, No.2, pp. 71-73.

Montemor, M.F., Simoes, A.M., Ferreira, M.G.S.(2001). Composition and behaviour of cerium films on galvanized steel, *Progress in Organic Coatings*, Vol.43, No.4, 274-281, ISSN: 0300-9440.

Montemor M.F, Simoes, A.M.., Ferreira, M.G.S., (2002). Composition and corrosion behaviour of galvanized steel treated with rare - earth salts: the effect of the cation. *Progress in Organic Coatings*, Vol.44, No2, pp.111-120. ISSN: 0300-9440.

Montemor, M.F., Ferreira, M.G.S., (2008). Analytical characterization of silane films modified with cerium activated nanoparticles and its relation with the corrosion protection of galvanized stell substrates, *Progress in Organic Coatings*, Vol.63, No.3 (October 2008) pp.330-337. ISSN: 0300-9440.

Nonnenmann, M.,(1989). New high-performance gas flow equalizing metal supports for exhaust gas catalysts, *Automobilitech. Z.*, Vol.No. 4, pp.185-192.

Nikolova, D. Stoyanova, E., Stoychev, D,. Stefanov, P., Marinova, Ts (2006). Anodic behaviour of stainless steel covered with an electrochemically deposited Ce_2O_3-CeO_2 film. *Surface & Coatings Technology*,Vol. 201, pp. 1559 - 1567, ISSN: 0257-8972.

Nikolova, D., Stoyanova, E., Stoychev, D., Avramova, I., Stefanov, P. (2006). Stability of the passive state of stainless steel OC 4004 in sulphuric acid solutions improved by additionally electrodeposited oxide layers, Book of papers of the International International Workshop *Nanostructured Materials in Electroplating*, Sandanski, Bulgaria, March, 2006, pp.127-131 Eds.D. Stoychev, E.Valova, I. Krastev, N. Atanassov, March 2006, Sandanski, Bulgaria

Nickolova, D., Stoyanova, E., Stoychev, D. P., Stefanov, P., Avramova I., (2008). Protective effect of alumina and ceria oxide layers electrodeposited on stainless steel in sulfuric acid media. *Surface & Coatings Technology*, Vol. 202, pp. 1876-1888, ISSN: 0257-8972

Otero, E., Pardo, A., Saenz, E., Utrilla, M.V, Hierro, P., (1996). A Study of the influence of nitric acid concentration on the corrosion resistance of sintered austenitic stainless steel. *Corrosion Science*, Vol. 38, No. 9, pp. 1485-1493, ISSN: 0010-938X

Otero, E., Pardo, A., Utrilla, M.V., Saenz, E., Alvarez, J.F. (1998). Corrosion behavior of 304I and 316I stainless steels prepared by powder metallurgy in the presence of sulfuric and phosphoric acid. *Corrosion Science*, Vol. 40, No. 8, pp. 1421-1434, ISSN: 0010-938X

Paparazzo, E., (1990). Surface XPS studies of damage induced by X-ray irradiation on CeO_2 surfaces. *Surface Science* Vol. 234, No.1-2, pp. L253-L25, ISSN: 0039 – 6028.

Pardo, A., Merino, M.C., Arrabal, R., Viejo, F., Carboneras, M., Munoz, J.A., (2006). A surface characterization of cerium layers on galvanised steel. *Corrosion Science*, Vol. 48, pp. 3035-3048, ISSN: 0010 - 938X

Pardo, A., Merino, M., Arrabal, C. R., Merino, S., Viejo, F., Carboneras, M. (2006). Effect of Ce surface treatments on corrosion resistance of A3xx.x/SiCp composites in salt fog. *Surface and Coatings Technology*, Vol. 200, No. 9, pp. 2938-2947, ISSN: 0257-8972

Schafer, H., Stock, H.R. (2005). Improving the corrosion protection of aluminum alloy using reactive magnetron sputtering, *Corrosion Science*, Vol. 47, (June 2004) pp.953-964, ISSN: 0010-938X

Schmidt, H., Langenfeld, S., Naß, R., (1997). A new corrosion protection coating system for pressure-cast aluminium automotive parts. *Materials and Design* , Vol. 18, No. 4-6, pp. 309-313, ISSN: 0261 -3069

Stefanov, P., Stoychev, D, Valov, I., Kakanakova-Georgieva, A., Marinova, T. (2000) Electrochemical deposition of thin zirconia films on stainless steels 316 L. *Materials Chemistry and Physics*, Vol. 65, pp.222-225, ISSN: 0254 - 0584

Stefanov, P., Stoychev, D, Stoycheva, M., Ikonomov, J., Marinova, T., (2000). XPS and SEM characterisation of zirconia thin films prepared by electrochemical deposition. *Surface and Interface Analysis* , Vol. 30, pp. 628-631, ISSN: 1096 - 9918

Stefanov, P., Stoychev, D., Atanasova, G., Marinova, T. (2004). Electrochemical deposition of CeO_2 on ZrO_2 and Al_2O_3 thin films formed on stainless steel. *Surface and Coatings Technology*, Vol. 180-181, pp. 446 – 449, ISSN: 0257-8972

Stefanov, P., Stoychev, D., Aleksandrova, A., Nicolova, D., Atanasova, G., Marinova, T., (2004). Compositional and structural characterization of alumina coatings deposited electrochemically on stainless steel. *Applied Surface Sciences*, Vol. 235, pp. 80-85 ISSN: 0169 - 4332

Stoyanova, E., Nikolova, D., Stoychev, D., Electrochemical behaviour of stainless steel OC4004 with modified passive film in nitric and sulphuric acids, Book of Papers of the International Workshop *Nanostructured Materials in Electroplating* pp. 122-126, Eds. D. Stoychev, E.Valova, I. Krastev, N. Atanassov, March 2006, Sandanski, Bulgaria

Stoyanova, E., Nikolova, D., Stoychev, D., Stefanov, P., Marinova, Ts., (2006). Effect of Al and Ce oxide layers electrodeposited on OC 4004 stainless steel on its corrosion characteristics in acid media. *Corrosion Science*, Vol.48,pp.4037-4052,ISSN:0010-938X

Stoyanova, E., Guergova, D., Stoychev, D., Avramova, I., Stefanov, P., (2010). Passivity of OC404 steel modified electrochemically with CeO_2 - Ce_2O_3 oxide layers in sulfuric acid media. *Electrochimica Acta*, Vol. 55, No. 5, pp. 1725 – 1732, ISSN: 0013 - 4686

Stoychev, D., Ikonomov, I., Robinson, K., Stoycheva, M., Marinova, T., (2000). Surface modification of porous zirconia layers by electrochemical deposition of small amounts of Cu, Co and Co+Cu. *Surface and Interface Analysis*, Vol. 30, pp. 69-73, ISSN: 1096 - 9918

Stoychev, D., Valov, I., Stefanov, P., Atanasova, G., Stoycheva, M., Marinova, T., (2003). Electrochemical growth of thin La_2O_3 films on oxide and metal surfaces, *Materials Science and Engineering* C, Vol. 23, No. 1-2, pp. 123-128 ISSN: 0928 - 4931

Stoychev, D., Stefanov, P., Nikolova, D., Aleksandrova, A., Atanasova, G., Marinova, T., (2004). Preparation of Al_2O_3 thin films on stainless steel by electrochemical deposition. *Surface and Coatings Technology*, Vol.180-181, pp.441-445, ISSN:0257-8972

Tomashov, N.D., Chernova, G.P., (1965). *Passivity and Protection of Metals from Corrosion*, (in Russian) Publ. House "Nauka", Moscow

Tomashov, N.D., Chernova, G.P., (1993). *Theory of Corrosion and Corrosion-stable Materials*, (in Russion) Publ. House "Metallurgia", Moscow

Trovarelli A., (1996). Catalytic properties of ceria and CeO_2-containing materials, *Catalysis Reviews - Science and Engineering* , Vol. 38, No. 4, pp. 439-520, ISSN: 0161 - 4940

Tsanev, A., Iliev, P., Petrov, K., Stefanov, P., Stoychev, D., (2008). Electrocatalytical activity of electrodeposited Zr-Ce-Y/Ni and Co/Zr-Ce-Y/Ni oxide systems at evolution of hydrogen and oxygen. Bulgarian Chemical Communications, Vol. 40, No. 3, pp. 348-354

Tyuliev, G., Panayotov, D., Avramova, I., Stoychev, D., Marinova T., (2002). Thin-film coating Cu-Co oxide catalyst on lanthana/zirconia films electrodeposited on stainless steel. *Materials Science and Engineering C*, Vol.23 No1-2, pp.117-121, ISSN: 0928 - 4931

Valov, I., Stoychev, D., Marinova, T., (2002). Study of the kinetics of processes during electrochemical deposition of zirconia from nonaqueous electrolytes. *Electrochimica Acta*, Vol. 47, No. 28, pp. 4419-4431, ISSN: 0013-4686

Virtanen, S., Ives, M., Sproule, G., Schmuki, P., Graham, M. (1997). A surface analytical and electrochemical study on the role of cerium in the chemical surface treatment of stainless steels. *Corrosion Science*, Vol. 39, No. 10-11, pp. 1897-1913, ISSN: 0010-938X

Wang, K.L., Zhang, Q.B., Sun, M.L., Zhu, Y.M. (1997). Effect of laser surface cladding of ceria on the corrosion of nikel –based alloys. *Surface and Coatings Technology*, Vol. 96, No. 2-3, pp.267-271, ISSN: 0257-8972

Wang, Ch., Jiang, F., Wang, F. (2004). The characterization and corrosion resistance of cerium chemical conversion coatings for 304 stainless steel. *Corrosion Science*, Vol. 46, No. 1, pp. 75-89, ISSN: 0010-938X

Wilcox, G.D., Gabe D.R. (1987). Passivation studies using group VIA anions. V. Cathodic treatment of zinc, *British Corrosion Journal.*, Vol. 22, pp. 254-256, ISSN: 0007-0599

Wilcox, G.D., Gabe, D.R., Warwick, M.E. (1988). The development of passivation coatings by cathodic reduction in sodium molybdate solutions. *Corrosion Science*, Vol. 28, No. 6, pp. 577-585, ISSN: 0010-938X

Zaki N. (1988). Chromate conversion coating for zinc. *Metal Finishing*, Vol. 86, pp.75-83 ISSN: 0026-0576

Zheludkevich, M.L., Serra, R., Montemor, M. F., Fereira, M.G. (2005). Oxide nanoparticle reservoirs for storage and prolonged release of the corrosion inhibitors. *Electrochemistry Communications*, Vol. 7, No. 8, pp. 836-840, ISSN: 1388 - 2481

Zheludkevich, M.L., Serra, R., Montemor, M.F., Ferreira, M.G., (2006) Corrosion protective properties on nanostructured sol-gel hybrid coatings to AA2024-T3. *Surface and Coatings Technology*, Vol. 200, No. 9, pp. 3084-3094, ISSN: 0257-8972

Permissions

The contributors of this book come from diverse backgrounds, making this book a truly international effort. This book will bring forth new frontiers with its revolutionizing research information and detailed analysis of the nascent developments around the world.

We would like to thank Hong Shih, Ph.D., for lending his expertise to make the book truly unique. He has played a crucial role in the development of this book. Without his invaluable contribution this book wouldn't have been possible. He has made vital efforts to compile up to date information on the varied aspects of this subject to make this book a valuable addition to the collection of many professionals and students.

This book was conceptualized with the vision of imparting up-to-date information and advanced data in this field. To ensure the same, a matchless editorial board was set up. Every individual on the board went through rigorous rounds of assessment to prove their worth. After which they invested a large part of their time researching and compiling the most relevant data for our readers. Conferences and sessions were held from time to time between the editorial board and the contributing authors to present the data in the most comprehensible form. The editorial team has worked tirelessly to provide valuable and valid information to help people across the globe.

Every chapter published in this book has been scrutinized by our experts. Their significance has been extensively debated. The topics covered herein carry significant findings which will fuel the growth of the discipline. They may even be implemented as practical applications or may be referred to as a beginning point for another development.

The editorial board has been involved in producing this book since its inception. They have spent rigorous hours researching and exploring the diverse topics which have resulted in the successful publishing of this book. They have passed on their knowledge of decades through this book. To expedite this challenging task, the publisher supported the team at every step. A small team of assistant editors was also appointed to further simplify the editing procedure and attain best results for the readers.

Our editorial team has been hand-picked from every corner of the world. Their multi-ethnicity adds dynamic inputs to the discussions which result in innovative outcomes. These outcomes are then further discussed with the researchers and contributors who give their valuable feedback and opinion regarding the same. The feedback is then collaborated with the researches and they are edited in a comprehensive manner to aid the understanding of the subject.

Apart from the editorial board, the designing team has also invested a significant amount of their time in understanding the subject and creating the most relevant covers. They scrutinized every image to scout for the most suitable representation of the subject and create an appropriate cover for the book.

The publishing team has been involved in this book since its early stages. They were actively engaged in every process, be it collecting the data, connecting with the contributors or procuring relevant information. The team has been an ardent support to the editorial, designing and production team. Their endless efforts to recruit the best for this project, has resulted in the accomplishment of this book. They are a veteran in the field of academics and their pool of knowledge is as vast as their experience in printing. Their expertise and guidance has proved useful at every step. Their uncompromising quality standards have made this book an exceptional effort. Their encouragement from time to time has been an inspiration for everyone.

The publisher and the editorial board hope that this book will prove to be a valuable piece of knowledge for researchers, students, practitioners and scholars across the globe.

List of Contributors

Hong Shih
Etch Products Group, Lam Research Corporation, Fremont, California, USA

Roman Ritzenhoff and André Hahn
Energietechnik-Essen GmbH, Germany

Alicia Esther Ares
Researcher of CIC, CONICET, Argentina
Materials, Modeling and Metrology Program, Faculty of Sciences, National University of Misiones, Posadas, Argentina
Materials Laboratory, Faculty of Sciences, National University of Misiones, Posadas, Argentina

Claudia Marcela Mendez
Materials Laboratory, Faculty of Sciences, National University of Misiones, Posadas, Argentina

Liliana Mabel Gassa
INIFTA, National University of La Plata, Faculty of Exact Sciences, La Plata, Argentina
Researcher of CIC, CONICET, Argentina

L.C. Tsao
Department of Materials Engineering, National Pingtung University of Science & Technology, Neipu, Pingtung, Taiwan

Pierre Ponthiaux and François Wenger
Ecole Centrale Paris, Dept. LGPM, Châtenay-Malabry, France

Jean-Pierre Celis
Katholieke Universiteit Leuven, Dept. MTM, Leuven, Belgium

Swe-Kai Chen
Center for Nanotechnology, Materials Science, and Microsystems (CNMM), National Tsing Hua University, Hsinchu, Taiwan

Carlos Alberto Giudice
UTN (Universidad Tecnológica Nacional), CIDEPINT (Centro de Investigación y Desarrollo en Tecnología de Pinturas), La Plata, Argentina

Adina-Elena Segneanu, Ionel Balcu, Nandina Vlatanescu, Zoltan Urmosi and Corina Amalia Macarie
National Institute of Research and Development for Electrochemistry and Condensed Matter, INCEMC-Timisoara, Romania

Rajeev Kumar Gupta and Nick Birbilis
Department of Materials Engineering, Monash University, Australia

Jianqiang Zhang
School of Materials Science & Engineering, The University of New South Wales, Australia

Emilia Stoyanova and Dimitar Stoychev
Institute of Physical Chemistry, Bulgarian Academy of Sciences, Bulgaria

Printed in the USA
CPSIA information can be obtained
at www.ICGtesting.com
JSHW011814301024
72690JS00002B/76

9 781632 381026